AI and Remote Sensing for Earth Sciences

Mehmet Ali Çelik
https://orcid.org/0000-0002-7729-6650
İgdir University, Turkey

IGI Global
Scientific Publishing
Publishing Tomorrow's Research Today

Vice President of Editorial	Melissa Wagner
Director of Acquisitions	Mikaela Felty
Director of Book Development	Jocelynn Hessler
Production Manager	Mike Brehm
Cover Design	Jose Rosado

Published in the United States of America by
IGI Global Scientific Publishing
701 East Chocolate Avenue
Hershey, PA, 17033, USA
Tel: 717-533-8845 | Fax: 717-533-7115
Website: https://www.igi-global.com E-mail: cust@igi-global.com

Library of Congress Cataloging-in-Publication Data

LCCN: 2026005272 (CIP Data Pending)
ISBN13: 9798337387451
Isbn13Softcover: 9798337387468
EISBN13: 9798337387475
British Cataloguing in Publication Data
A Cataloguing in Publication record for this book is available from the British Library.

Table of Contents

Detailed Table of Contents

Moisture of soils, evapotranspiration, and hydrology of the terrestrial water cycle have a powerful impact on agricultural productivity, water management, and ecosystem sustainability. Recent innovations in artificial intelligence (AI) enhance the processes of modeling by using complex nonlinear interactions of climate, soil, vegetation, and topography. The methods of machine learning and deep learning (random forest, support vector machines, artificial neural networks, and recurrent neural networks) combine in situ data, remote sensing, and reanalysis products to improve the estimates of soil moisture and evapotranspiration. The hydrological models are AI-assisted systems that enhance the simulation of runoff, groundwater recharging, and even prediction of drought and floods, especially in data-poor areas. The use of AI has a high potential for accurate, flexible, and climate-resistant water resource management despite the difficulties associated with data quality, interpretability, and uncertainty.

Climate-driven landform evolution is governed by complex deformation processes resulting from extreme climatic events and long-term environmental transitions. Conventional remote sensing approaches that rely on static imagery or single-sensor observations often fail to distinguish permanent geomorphic change from short-term spectral or seasonal variability. This chapter presents a multimodal temporal reasoning framework that integrates optical, SAR, LiDAR, and hyperspectral data within a unified sequence-learning architecture conditioned by climatic drivers. Modality-aware spatial encoding and attention-based temporal modeling enable coherent reconstruction of landform dynamics across riverine, coastal, glacial, and deltaic environments. The proposed approach enhances robustness to missing data, improves temporal consistency, and advances monitoring from retrospective change detection toward predictive geomorphology, supporting early-warning systems and climate-resilient land-use planning.

Mangroves, sea grasses, and salt marshes that form blue carbon ecosystems are important in the mitigation of climate change because they have the unsurpassed capacity of sequestraining carbon. Coastal wetlands deposit much more carbon than terrestrial forest, and therefore it is important in the management of carbon across the world. The nature of these ecosystems is however complicated and dynamic making it difficult to accurately monitor carbon stocks within these ecosystems. The conventional field techniques frequently cannot be sufficiently used in massive and high-resolution studies. This article introduces an innovative method of combining multisource remote sensing devices and tools especially unmanned aerial vehicles (UAVs) with hyperspectral sensors to improve the monitoring of blue carbon. This paper provides enhanced spatial and spectral resolutions through the use of UAVs and hyperspectral imaging to monitor the blue carbon ecosystems with high accuracy at small scales.

As cities evolve into intelligent cyber-physical ecosystems, infrastructure resilience and disaster preparedness are critical. Earthquake forecasting remains a challenging task, requiring advanced machine learning and non-seismic data integration. This study proposes a hybrid SSA-LSTM model that leverages satellite-derived Outgoing Longwave Radiation (OLR) as a seismic precursor. Singular Spectrum Analysis (SSA) is employed to decompose noisy OLR signals and extract dominant temporal features, which are then learned by a Long Short-Term Memory (LSTM) network. The proposed model is evaluated against a conventional LSTM using identical historical seismic and OLR datasets. Experimental results demonstrate that SSA-LSTM consistently outperforms the standard LSTM, achieving lower forecasting errors and improved robustness. Statistical analysis confirms the effectiveness of SSA in enhancing long-term dependency learning. The findings highlight the potential of combining signal decomposition and deep learning for reliable earthquake forecasting.

Remote sensing technology combined with artificial intelligence has revolutionized atmospheric pollution monitoring and air quality assessment. This chapter explores the integration of satellite-based remote sensing with machine learning and deep learning techniques for estimating ground-level pollutant concentrations including PM2.5, PM10, NO2, O3, and other atmospheric pollutants. The chapter examines various AI methodologies including Random Forest, XGBoost, Convolutional Neural Networks (CNN), Long Short-Term Memory (LSTM) networks, and Graph Neural Networks applied to data from multiple satellite platforms such as MODIS, Sentinel-5P TROPOMI, Himawari-8, and VIIRS. Through comprehensive analysis of

This work focuses on land use mapping, change detection, and analyzing LST using remote sensing data in Bangalore city. This study tested and evaluated the use of high-resolution LISS IV imagery and Landsat 8 data with TIRS data. Here, the SVM algorithm has been utilised for monitoring land use changes, and mono-window algorithms have been used for the calculation of LST. It has been used to classify different types of land cover features, such as urban areas, vegetation, agricultural land, water bodies, and open land. Between 2014 and 2023, the results showed that the urban area had increased by 70%, and the overall classification accuracy ranged between 61-81%. However, in LST, emissivity was also estimated with the help of the NDVI proportion of vegetation method, which was developed for use with the OLI Bands 4 & 5. From the result, it can be deduced that the LST of the urban region was higher than that of vegetative regions. It has been observed that the mono-window algorithm-based LST calculations and SVM-based image classification results are consistent.

Climate variability, extended droughts, and human-imposed land-use changes have also contributed to the increased severity of wildfires, underscoring the need for intelligent, large-scale risk forecasting systems that are highly spatially aware and timely. This new model builds region-wise graphs using the dynamic construction

Preface

INTRODUCTION

Artificial intelligence (AI) and remote sensing are no longer parallel domains operating in isolation; they have evolved into a deeply integrated scientific ecosystem that is transforming how Earth systems are observed, interpreted, and governed. Over the past decade, advances in multispectral and hyperspectral satellite platforms, synthetic aperture radar (SAR), thermal imaging systems, UAV technologies, and open geospatial infrastructures have produced an unprecedented volume of Earth observation data. Simultaneously, machine learning, deep learning, cloud computing, and high-performance analytical frameworks have matured into scalable tools capable of extracting meaningful patterns from complex environmental signals.

This convergence marks a paradigm shift. Environmental processes are inherently nonlinear, multi-scalar, and spatiotemporally heterogeneous. Traditional statistical approaches—while foundational—often struggle to capture emergent behaviors in coupled climate–land–water systems. AI-enabled modeling offers adaptive pattern recognition, high-dimensional feature extraction, probabilistic forecasting, and uncertainty quantification mechanisms that extend the analytical capacity of remote sensing workflows. Consequently, AI and remote sensing integration is no longer a methodological preference; it is becoming an operational necessity for addressing climate change, hydroclimatic risk, biodiversity loss, land degradation, disaster resilience, agricultural sustainability, and urban environmental stress.

THE SUBJECT MATTER IN TODAY'S WORLD

The relevance of this volume is situated within an era characterized by accelerating environmental pressures and heightened societal demand for reliable, near–

real-time environmental intelligence. Climate extremes are intensifying; droughts are increasingly energy-driven; wildfire regimes are expanding; urban heat islands are amplifying public health risks; and food security is under mounting strain due to hydroclimatic volatility. At the same time, global commitments such as the Sustainable Development Goals (SDGs), climate adaptation frameworks, and national resilience strategies require robust, data-driven decision support systems.

AI-enhanced Earth observation systems offer precisely this capacity. They enable predictive analytics for drought monitoring, probabilistic wildfire risk mapping, crop yield estimation under climatic uncertainty, coastal vulnerability assessment, and atmospheric pollution modeling. Yet the deployment of AI in environmental science also introduces methodological and ethical responsibilities: transparency, reproducibility, model interpretability, domain generalization, data bias awareness, and uncertainty communication are critical.

This book addresses these opportunities and challenges by presenting rigorously developed research and applied frameworks that demonstrate how AI can enhance Earth observation pipelines—from preprocessing and feature engineering to predictive modeling, benchmarking, validation, and policy-relevant decision support.

TARGET AUDIENCE

This volume is written for a multidisciplinary readership:

- **Researchers and Graduate Students** in climatology, hydrology, geomorphology, ecology, agriculture, environmental engineering, geoinformatics, and computer science who seek advanced methodological applications and reproducible workflows.
- **Practitioners and Policy Analysts** in environmental agencies, agricultural services, water management authorities, forestry departments, disaster risk reduction institutions, and urban planning offices who require operational AI-driven monitoring systems.
- **Data Scientists and AI Specialists** entering the geospatial domain who need contextual understanding of Earth system constraints, physical process coupling, and spatiotemporal data complexities.
- **Interdisciplinary Scholars** interested in bridging Earth system science with explainable and physics-informed AI frameworks.

By fostering dialogue between Earth scientists and AI researchers, this volume aims to reduce fragmentation between methodological innovation and environmental application.

OVERVIEW OF CHAPTER CONTRIBUTIONS

Chapter 1: Soil Moisture, Evapotranspiration, and Hydrological Process Modeling Using AI

Focusing on hydroclimatic processes, this chapter explores AI-based modeling of soil moisture dynamics and evapotranspiration estimation. It highlights the importance of coupling physical hydrological understanding with machine learning techniques to improve drought diagnostics and water balance assessments under climate variability.

Chapter 2: Multimodal Temporal AI Reasoning for Climate-Driven Landform Change Detection

Addressing geomorphological dynamics, this chapter integrates multimodal remote sensing data with AI-based temporal reasoning models to detect landform evolution under climatic forcing, emphasizing pattern recognition across heterogeneous datasets.

Chapter 3: Multisource Remote Sensing and UAV for Hyperspectral Blue Carbon Monitoring

This chapter examines AI-assisted coastal and marine ecosystem monitoring. By integrating UAV and satellite data, it develops blue carbon mapping approaches relevant to climate mitigation and ecosystem resilience.

Chapter 4: Integrating AI and Cloud-Enabled Deep Learning for Earthquake Forecasting

This chapter explores geophysical applications, focusing on cloud-based AI modeling for earthquake precursor analysis, demonstrating the scalability of AI-driven hazard forecasting.

Chapter 5: Remote Sensing–Based Air Quality and Atmospheric Pollution Modeling Using AI

This contribution applies machine learning to atmospheric monitoring, emphasizing pollutant dispersion modeling, urban air quality mapping, and public health implications.

Chapter 6: Artificial Intelligence and Remote Sensing for Coastal Ecosystem Service Assessment and Climate-Resilient Management: AI-Eyed Guardians of Resilient Coasts

Focusing on resilience planning, this chapter integrates ecosystem service valuation with AI-based spatial modeling to inform coastal adaptation strategies

Chapter 7: AI-Enhanced Graphene Gas Sensors for Smart Energy Systems, Data-Driven Sensitivity Enhancement in Multigas THz Platforms.: AI-Enhanced Graphene Gas Sensors for Smart and Renewable Energy Applications

Expanding beyond satellite sensing, this chapter demonstrates how AI improves sensitivity and performance in graphene-based THz gas sensors, linking nanotechnology with environmental monitoring.

Chapter 8: Artificial Intelligence in Agriculture and Food Security: Remote Sensing and Yield Estimation

Providing an applied agronomic perspective, this chapter discusses yield modeling under climatic stress and data constraints.

Chapter 9: Urbanization and Thermal Footprints: Mapping Bangalore's Growth and UHI Evolution

Using remote sensing and AI, this chapter examines urban heat island intensification and thermal landscape evolution in rapidly urbanizing environments.

Chapter 10: Wildfire Risk Prediction Using Graph Convolutional Networks and Google Earth Engine

The final chapter applies graph-based AI modeling within cloud geospatial platforms to predict wildfire risk, emphasizing operational deployment and hazard mitigation.

CONCLUSION AND CONTRIBUTION TO THE FIELD

AI and Remote Sensing for Earth Sciences contributes to the field in three fundamental ways:

1. **Methodological Integration** – It demonstrates how AI techniques can be systematically embedded into remote sensing workflows while maintaining scientific rigor and physical interpretability.
2. **Operational Relevance** – It links methodological innovation to applied environmental decision-making across agriculture, climate risk, urban systems, hazards, and ecosystem management.
3. **Responsible AI Deployment** – It emphasizes transparency, benchmarking, uncertainty quantification, and reproducibility as foundational principles in environmental AI research.

As environmental pressures intensify and the demand for reliable Earth system intelligence grows, AI-enhanced remote sensing will remain central to sustainable governance. It is our intention that this volume not only reflects the current state of interdisciplinary research but also sets methodological standards for future work. By encouraging reproducible practices, transparent evaluation, and cross-disciplinary collaboration, this book aims to help shape the next generation of AI-driven Earth observation science.

Mehmet Ali Çelik
İgdir University, Turkey

Chapter 1
Soil Moisture, Evapotranspiration, and Hydrological Process Modeling Using AI

Nilotpal Das
http://orcid.org/0009-0005-0083-4093
Uttaranchal University, India

Atin Kumar
http://orcid.org/0000-0002-1653-2146
Uttaranchal University, India

Sharad Sachan
Lovely Professional University, India

Arnab Maji
Visva-Bharati University, India

ABSTRACT

Moisture of soils, evapotranspiration, and hydrology of the terrestrial water cycle have a powerful impact on agricultural productivity, water management, and ecosystem sustainability. Recent innovations in artificial intelligence (AI) enhance the processes of modeling by using complex nonlinear interactions of climate, soil, vegetation, and topography. The methods of machine learning and deep learning (random forest, support vector machines, artificial neural networks, and recurrent neural networks) combine in situ data, remote sensing, and reanalysis products to improve the estimates of soil moisture and evapotranspiration. The hydrological

DOI: 10.4018/979-8-3373-8745-1.ch001

models are AI-assisted systems that enhance the simulation of runoff, groundwater recharging, and even prediction of drought and floods, especially in data-poor areas. The use of AI has a high potential for accurate, flexible, and climate-resistant water resource management despite the difficulties associated with data quality, interpretability, and uncertainty.

1. INTRODUCTION

The key components of the terrestrial water cycle are soil moisture, evapotranspiration, and hydrological processes, which determine the storage, transformation, and redistribution of precipitation within the soil-plant-atmosphere nexus and the ultimate control of water availability to the ecosystem, agriculture, and society (Amsalu et al., 2026). The soil moisture controls infiltration, surface runoff, groundwater recharge, and uptake of water by plants and hence mediates the land-atmosphere feedbacks, drought development, and the production of floods. The evapotranspiration of water connects the surface and groundwater to the atmosphere, affecting the balance of water, climate regulation, and agricultural water use (Çelik, 2026). Combined with the runoff generation process and groundwater flow, they define the spatial and temporal variations of water storage and fluxes that support the successful water resources planning and irrigation scheduling, drought and flood risks assessment, and ecohydrological functioning (Liu et al., 2024; Çelik et al., 2026). Nevertheless, traditional hydrological models, most often founded on conceptual or physically based lumped or semi-distributed models, have significant problems with modelling these processes in the real world. They frequently need a great deal of parameterization, severe simplifications of soil vegetation atmosphere interactions, and stationarity to be assumed, which is hard to maintain under climate and land-use change. They may fail to perform well in regions of scarce data, where sparse observations cannot be used to provide strong calibration or validation, nor will they be able to help reveal highly nonlinear, scale-dependent, and spatially heterogeneous relationships between meteorological drivers, soil properties, vegetation dynamics, and anthropogenic effects. Consequently, conventional models might make biased or too imprecise predictions of soil moisture, evapotranspiration, and associated fluxes, hence limiting their application in operational decision making and in solving the fine-scale hydrological fluctuations. To address these constraints, artificial intelligence (AI), including machine learning and deep learning, has become one of the most influential supplements and substitutes to traditional hydrological models within a short period of time. AI methods are useful at learning nonlinear input-output relationships that are also complex, directly from large multi-source data, such as in situ measurements, remote sensing products, reanalysis, and Internet of Things (IoT) sensor networks, and do not necessitate explicit closed-form process equations. Applications in hydrology and water resources AI has been used effectively to forecast soil moisture with satellite images, evapotranspiration with only a few meteorological inputs, streamflow and ground water, water quality, and hybrid frameworks that incorporate components based on data in a physically based framework to effectively capture the characteristics of the water cycle (Biazar et al., 2025; Liu et al., 2024). The developments allow more precise, high-resolution, and adaptive modelling of the important hydrological variables, particularly in areas where the traditional models are ill-posed. It is on this background that the current chapter revolves around soil moisture, the evapotranspiration, and the hydrological process modelling with four primary objectives and a scope. (i), It clarifies the hydrological and practical significance of soil moisture and evapotranspiration in the water cycle, its importance in the closure of the water balance, of the water, and its interaction with the climate and water resources management. (ii) It critically analyses how conventional hydrological models perform poorly in modelling such processes, especially in conditions of nonstationary climate, non-homogeneous landscapes, and in cases of data scarcity. (iii) It evaluates and integrates the state of the art of AI-based strategies in estimating soil moisture and evapotranspiration and modelling related hydrological processes such as standalone data-driven systems, hybrid physics/AI systems, and AI/remote sensing systems, with their strengths, weaknesses, and data needs (Biazar et al., 2025; Liu et al., 2024). (iv) It identifies the emerging directions in research, such as explainable AI, hybrid and physics-informed architectures, and interaction with high-resolution Earth observations, and how they can be developed into robust, interpretable, and operationally useful AI-based hydrological models to improve the sustainable management of soil and water resources.

2. FUNDAMENTALS OF SOIL–WATER–ATMOSPHERE INTERACTIONS

2.1 Soil Moisture Dynamics and Controlling Factors

Soil moisture is the quantity of water in the pores of soil and is an essential state variable that is a medium of relationship between precipitation, infiltration, evapotranspiration, runoff, and water use by plants in the soil-water-atmosphere continuum. It is extremely interchangeable in space and time and reacts to rainfall inputs, drainage, lateral flow, and demand of the atmosphere and is greatly modulated by soil texture and structure, bulk density, organic matter, topography, vegetation, land use, and climate forcing (Rasheed et al., 2022; Huang et al., 2015) (Figure 1).

The surface soil, individually on precipitation and drying, changes rapidly and is mainly determined by antecedent rainfall, potential evapotranspiration, and surface features, but deeper layers are characterized by slower changes, which are influenced by the thickness of soils, topographic wetness, slope, and long-term water balance (Huang et al., 2015; Suo et al., 2018). At the catchment to regional scales, more forces are precipitation and evaporative demand, which are accompanied by several other forces produced by vegetation cover and human activities, including irrigation and land management, that change infiltration, runoff generation, and root-zone water uptake (Ran et al., 2025; Hu et al., 2025). A combination of these controls creates significant vertical stratification (high-speed changing, active, sub-active, and rather stable layers of soil) and non-uniform spatial distribution, which puts soil moisture as a crucial, yet demanding variable to be considered in hydrological modelling, drought-related evaluation, and ecohydrological analysis (Huang et al., 2015; Suo et al., 2018).

Figure 1. Soil moisture dynamics and controlling factors

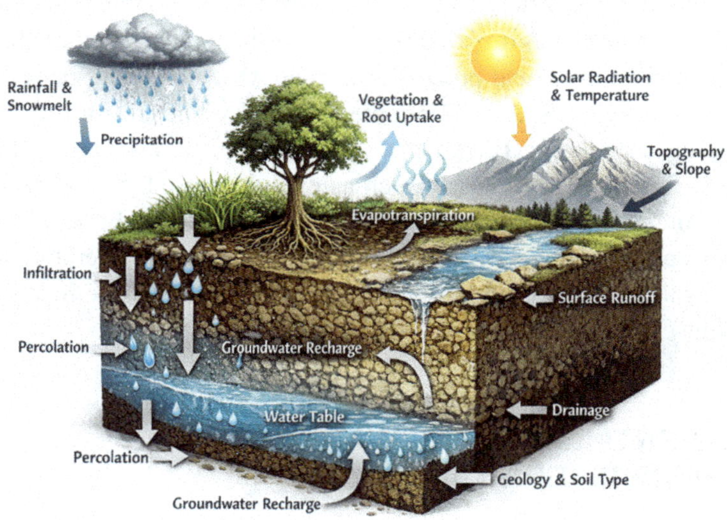

2.2 Evapotranspiration Processes and Components

The total loss of land water to the atmosphere by evaporation of soil and wet surfaces and transpiration through plant stomata is evapotranspiration (ET), which is a significant sink of land precipitation and is a major channel that closes the land water and energy cycles (Kar et al., 2025; Zhang et al., 2016). The direction of evaporation is mainly a physical one that is regulated by the surface soil moisture, temperature, net radiation, aerodynamic conditions (e.g., wind, friction velocity), and vapor pressure deficit, whereby shallow soil water content usually turns out to be a significant force in semi-arid environments (Kar et al., 2025). Transpiration is a biologically mediated flux which is under the control of plant physiology and canopy structure and is regulated by moisture of the root-zone, vapor pressure deficit, leaf area, and stomatal conductance and, in most ecosystems, transpiration contributes to most of the terrestrial ET and is highly responsive to vegetation greening and green-brown phenology (Zhang et al., 2016; Jiang et al., 2020). Breaking down ET into evaporation and transpiration is very fundamental towards explaining plant water use efficiency, crop irrigation requirement, ground water recharge, and land-climate interactions, although this is very challenging due to overlapping meteorological and biophysical factors and the fact that both of these components are never measured independently on a large scale. This has recently been emphasized using empirical, modelling, and machine-learning techniques with flux towers, sap flow, and remote sensing, as being dominated by transpiration in vegetated systems, but evaporation is more significant over bare soil, following rainfall incidents, or sparse vegetation (Kar et al., 2025; Zhang et al., 2016).

2.3 Surface and Subsurface Hydrological Processes

Surface and subsurface water cycles constitute a network through which water is transported through interception, infiltration, overland flow, subsurface flow, and recharge and discharge of groundwater as a continuum of water that is neither generated nor lost. Incident precipitation at the land surface infiltrates or generates Hortonian (infiltration-excess) or saturation-excess overland flow depending on the infiltration capacity, soil hydraulic conductivity, and microtopography or changes in land cover and soil crusting (Wang et al., 2019; Chen et al., 2022). In the unsaturated zone, matric potential and gravity, soil layering, texture, structure, and uptake of roots by the water of the soil cause vertical redistribution of water through the gradients of matric potential and gravity as outlined in the Richards equation, with the timing and extent of percolation and subsequent lateral flow of the soil subsurface (Huang et al., 2015; Wang et al., 2020). In most of the rocky sceneries, karst and structured soils, threshold behaviour occurs in subsurface saturation first, followed by its upslope migration, causing intense subsurface stormflow, and, in some instances, surface runoff (Wang et al., 2019). Integrated surface subsurface hydrologic models explicitly couple overland flow kinematic wave formulations with 3D variably saturated groundwater flow, to represent these feedbacks and quantify exchanges that are recharge, capillary rise, groundwater-river interaction, canal seepage, and tile drainage, to better represent catchment-scale water balance and fluxes (Kuffour et al., 2020; Bailey et al., 2025).

2.4 Role of Land–Atmosphere Coupling in Hydrological Modelling

A change in both directions between the land surface conditions (soil moisture, vegetation, temperature) and the atmospheric conditions (radiation, boundary layer dynamics, precipitation), which is considered a key to the control of hydrological responses and climate variability. Directly related to moisture in soil, partitioning of net radiation into sensible and latent heat can be affected, thus, influencing near-surface temperature, vapor pressure deficit, and boundary layer development; these atmospheric variables, in their turn, determine the rates of evapotranspiration, cloud formation, and, perhaps, precipitation, creating a feedback loop (Hao et al., 2025; Rasheed et al., 2022). Strong coupling hot spots tend to be found in transitional climatic regions, in which even minor shifts in soil moisture can significantly change ET and temperature and enhance droughts, heatwaves, or convective rainfall episodes. Mathematically, modelling land-atmosphere interactions may cause biased estimates of ET, soil moisture, and runoff, and misrepresentation of carbon and greenhouse gas fluxes, i.e., in extreme events, as well as in nonstationary climate conditions (Hao et al., 2025).

3. ARTIFICIAL INTELLIGENCE TECHNIQUES FOR HYDROLOGICAL MODELLING

3.1 Overview of Machine Learning and Deep Learning Methods in Hydrological Modeling

Machine learning (ML) and deep learning (DL) are the main methods used by artificial intelligence in hydrological modelling: they directly generate the complex, nonlinear relationships between hydrometeorological inputs and target variables (e.g., streamflow, soil moisture, evapotranspiration) by just analysing the data. Artificial neural networks (ANNs), support vector machines (SVMs), random forests (RF), and gradient boosting are also common ML algorithms in hydrology and have been used to model rainfall-runoff, water quality, drought and flood forecasting, as well as to estimate hydrology process parameters (Xu & Liang, 2021; Chin and Lai, 2025). Deep learning is a generalization of ANNs, with additional hidden layers, which allows it to handle time series (recurrence neural networks and long short-term memory (LSTM) neural networks), spatial or spatiotemporal data (convolutional neural networks (CNNs)). One of the more complex types of architecture is encoder-decoder networks, transformers, generative adversarial networks (GANs), and attention networks, which have proven to be more effective with long sequences, multi-source inputs, and high-dimensional hydrological data (Tripathy and Mishra, 2023).

3.2 Data-Driven vs. Physics-Informed AI Models

Models based on pure data only contain statistical relationships in data, and do not consider explicit governing equations; they can be very predictive, but may have no respect for physical constraints, extrapolate poorly, and lack physical process understanding (Chin and Lai, 2025; Kim et al., 2021). Physics-informed/physics-guided AIs, on the other hand, impose physical laws or physical process-related structures into ML/DL models. As an example, a hydrological model can be coded as a differentiable neural network backbone, water balance equations can be integrated into loss functions, or training data can be created by means of process-based simulations (Tripathy and Mishra, 2023; Wang et al., 2024). Physics-guided deep learning (PGDL), physics-informed neural networks (PINNs), and other classes of differentiable process-based models (delta-models) have been demonstrated to be the best or even better than pure deep learning (DL) in tasks of data scarcity and extrapolation (Tripathy and Mishra, 2023; Wang et al., 2024).

3.3 Advantages of AI Over Traditional Models

Hydrological models of AI have a number of advantages compared to conceptual or physically based models. To start with, they can directly capture highly nonlinear, high-dimensional interactions between climate, surface sources, and human drivers without involving explicit parameterizations, and tend to predict rainfall-runoff or floods as well as soil moisture or ET better (Tripathy and Mishra, 2023; Kim et al., 2021). Second, AI models are capable of effectively integrating multi-source data (in situ observations, remote sensing, reanalysis, sensor networks), high volumes of data, and fast predictions that can be used in real-time (Xu and Liang, 2021; Sit et al., 2020). Third, hybrid and physics-informed models enhance the concept of regionalization and offer more dependable forecasts in ungauged basins and provide more confidence to assess the effects of climate change by making smart adjustments in internal fluxes and model parameters without violating the physical laws (Wang et al., 2024; Feng et al., 2023). Lastly, with the development of explainable AI and scientific machine learning, AI models begin to further support process inference and hypothesis generation, and thus can be used not only as a predictive tool but also as a model to gain deeper insights into hydrology (Tripathy and Mishra, 2023; Nearing et al., 2020).

4. DATA SOURCES FOR AI-BASED SOIL MOISTURE AND HYDROLOGICAL MODELLING

4.1 In Situ Observations and Sensor Networks

The in-situ soil moisture data to be used in AI-based modelling are provided by point sensors (e.g., TDR, capacitance probes), wireless sensor networks (WSN), cosmic-ray neutron sensors, and multi-scale field networks. High-frequency and multi-depth measurements by dense WSNs are limited in spatial coverage, which inspires the optimized network design and error characterization in hydrological applications (Zhuo et al., 2020; Hendrickx et al., 2025). New methods absorb the data of WSN into distributed models (e.g., SWAT using ensemble Kalman filters) to enhance the basin-scale soil moisture condition to be used in training and validating AI models (Zhang et al., 2023). Cosmic-ray neutron sensing gives footprints of the hectare scale, which is between point and satellite scales and provides the spatiotemporal analysis of root-zone moisture (Fersch et al., 2020).

Figure 2. Data Sources for AI-Based Soil Moisture and Hydrological Modelling

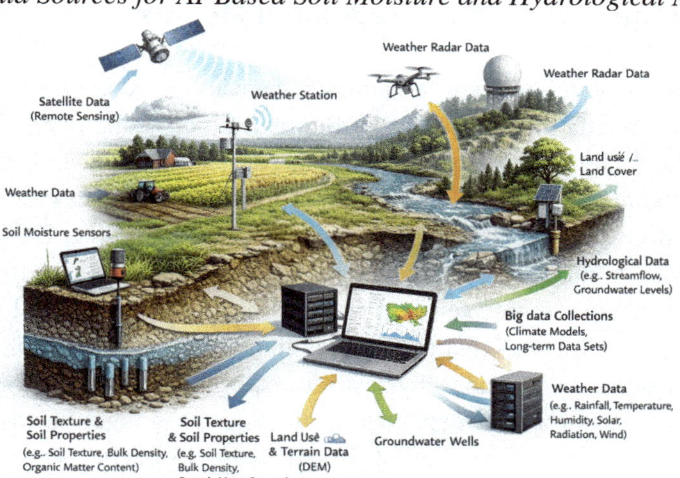

4.2 Remote Sensing Datasets (Satellite and UAV-Based)

Radiative transfer models and machine learning schemes (e.g., SMAP HydroBlocks at 30 m) that combine radiative transfer models and in situ measurements with satellite missions such as SMAP, SMOS, Sentinel 1, and SMOS-CCI provide global soil moisture with tens of kilometres of resolution (Babaeian et al., 2019; Vergopolan et al., 2021). AI-based retrievals are extensively performed with the help of optical, thermal, and microwave sensors, which are enhanced by ML and DL, which boost the accuracy of retrievals and drought indices (Babaeian et al., 2019; Huang et al., 2025). Multispectral, hyperspectral, and thermal invasive sensors mounted on UAV platforms can provide estimates of soil moisture at centimetre to meter resolutions, and can perform better than satellites in heterogeneous agricultural environments, and can allow field-scale AI models, particularly when used together with ground sampling (Haokip et al., 2025; Shokati et al., 2025; Yuan et al., 2025).

4.3 Climate, Meteorological, and Reanalysis Data

The AI hydrological models are based on climate and meteorological forcings (precipitation, temperature, radiation, humidity, wind) of gauge networks, gridded products, and reanalyses as important predictors of soil moisture dynamics, runoff, and evapotranspiration. Recent reprojections like ERA5 and ERA5 Land offer hourly, approximately 931 km global fields of precipitation, soil moisture, and energy fluxes, which are used as direct inputs and training goals of regional AI forecast systems (Kokkalis et al., 2024; Muñoz-Sabater et al., 2021; Filipović et al., 2021). Semi-arid and large river basin assessments indicate that various combinations of satellite and reanalysis models exhibit certain differences in soil moisture, evapotranspiration, and runoff abilities, suggesting the necessity to benchmark them on a regional scale before incorporating them into AI processes (Kokkalis et al., 2024; Dembélé et al., 2020; Filipović et al., 2021).

4.4 Geospatial and Terrain Datasets (DEM, Land Use, and Soil Maps)

The digital elevation models (DEMs), land use/land cover, and soil maps are the predictors, which are non-dynamically changing and represent the current state of the conditions, to predict the hydrological response and direct spatial generalization of the AI model. The indices derived by DEM (slope, aspect, curvature, flow accumulation, topographic wetness index) describe a significant portion of the variations of soil moisture and are important inputs to remote-sensing-derived and ML retrievals, particularly in topographically difficult regions (Luo et al., 2018).

The use of land-cover maps and vegetation indices would aid in explaining the effect of canopies on the backscatter and evapotranspiration of microwaves, and the use of soil property maps (texture, bulk density, hydraulic parameters) would improve the structure of a model and enhance the AI-based estimation of soil moisture over depths and landscapes (Babaeian et al., 2019; Haokip et al., 2025; Zhao et al., 2025).

5. AI APPLICATIONS IN SOIL MOISTURE MODELLING

5.1 Surface and Root-Zone Soil Moisture Estimation

The AI techniques predict surface and root-zone soil moisture with nonlinear correlations between surface sensing, meteorological, and soil data and ground measurements. XGBoost and other ensemble methods are machine learning models, which have been applied to retrieve multi-layer soil moisture (e.g., 0 -5 to 0 -100 cm) at 1 km resolution nationwide with SMAP L4 products, climate forcings, and landscape features showing unbiased RMSE below 0.04 m^3/m^3 and depth-dependent controls (Karthikeyan and Mishra, 2021). ConvLSTM, hybrid physical-AI, and deep learning methods analyse surface moisture, vegetation indices, and simulated profiles (e.g., using Hydrus 1D) to estimate root-zone moisture and significantly enhance the correlations at deeper levels than the products of land data assimilation (A. et al., 2022).

5.2 Spatiotemporal Soil Moisture Mapping

In the case of spatiotemporal soil moisture mapping, AI models are used to combine both ground-level observations and satellite-based information with environmental covariates so that they can create continuous fields to monitor drought and wetting patterns. R^2 is 0.9 with Sentinel 1 and meteorological forcings and terrain/land-use variables in regional studies on surface soil moisture, and spatial patterns on heterogeneous landscapes have been replicated with neural networks (Blanka-Vegi et al., 2025; Adab et al., 2020). On global to continental scales, AI-based models predict daily soil moisture fields in an autoregressive manner with reanalysis precipitation, temperature, and provide continuous time series that are comparable or better in dry and wet regimes as compared to traditional land surface models (Xu et al., 2025).

5.3 Data Fusion and Downscaling of Satellite Soil Moisture

In data fusion and downscaling, AI is used to take coarse satellite soil moisture (e.g., SMAP L3/L4) and finer optical, SAR, and terrain, as well as soil maps, and generate fine-scale products. The model of gradient boosting on a set of 1 km multi-depth soil moisture on CONUS based on SMAP L4, climate, and landscape predictors has addressed sub-grid heterogeneity not evident in the original product (Karthikeyan and Mishra, 2021). Other work is a fusion of SMAP, reanalysis forcing, and multi-source remote sensing in LightGBM or convLSTM model, which generates world products of 0.25^0 and continental products of 9 km that outperform SMAP L4, ERA5 Land, and GLDAS in terms of correlation and RMSE (Xu et al., 2025; Rabiei et al., 2025).

Table 1. Representative AI applications in fusion/downscaling

Application focus	AI/data strategy	Example performance	Citations
Multi-layer 1 km SM from SMAP L4	Regional XGBoost with climate & landscape covariates	ubRMSE < 0.04 m^3/m^3 at multiple depths	(Karthikeyan & Mishra, 2021)
Global 0.25° seamless SM	LightGBM + SMAP L3 assimilation (ML-DA)	Median R ≈ 0.74, better than SMAP L4, ERA5-Land	(Xu et al., 2025)

5.4 Gap-Filling and Uncertainty Reduction Using AI

The use of AI in learning multi-source inputs and taking advantage of temporal relationships is used more and more actively to fill in gaps and reduce uncertainty in the soil moisture datasets. ML-based global products leverage triple collocation to construct quality training labels and subsequently follow the dynamics of soil moisture effectively sealing the gaps in satellite-retrieval due to vegetation or snow cover, or radio-frequency interference (Xu et al., 2025). It has been demonstrated that deep learning fusion of SMAP, NLDAS 2, and detailed soil maps reduces errors in surface and subsurface estimates and homogenizes the performance of all the soil textures and land-cover types, thus decreasing the uncertainties of the spatially structured ones (Rabiei et al., 2025).

6. AI-BASED EVAPOTRANSPIRATION ESTIMATION

6.1 Integration of Vegetation Indices and Surface Energy Balance Variables

In general, vegetation indices (e.g., NDVI, EVI, and LAI) are often used with surface energy balance variables (e.g., land surface temperature, net radiation, soil heat flux, and evaporative fraction based on satellite products) in AI frameworks to promote predictions of ET (Yadav et al., 2025; Du et al., 2024). Models that are trained on a mixture of MODIS vegetation indices, land surface temperature, and ERA5 reanalysis meteorology can also be used to capture biophysical controls on forest and cropland ET across climate zones with high coefficients of determination (R^2 greater than 0.8) and low RMSE (Yadav et al., 2025; Du et al., 2024).

6.2 High-Resolution and Near–Real-Time ET Mapping

High-resolution and near-real-time ET mapping is being developed by AI incorporation of multi-sensor satellite images, GIS layers, and tower observations. Extrapolation of half-hourly urban ET using flux tower footprints to city-scale maps has been performed in deep learning and random forest models, with $R \approx 0.84$ and RMSE of the order of 0.024 mm h^{-1}(Vulova et al., 2021). The next step is to develop generative adversarial networks and other super-resolution techniques to learn how to generate a higher spatial resolution of 30 m ET products to facilitate precision agriculture and crop stress detection (Zhang and Yang, 2025).

6.3 Comparative Performance With Conventional ET Models

There is comparative research indicating that in most cases, the ML and DL models are more accurate and generalized to sites outside of the calibration than conventional ET models, such as temperature, radiation, and humidity-based empirical equations, and in some cases, Penman-Monteith (Chen et al., 2020; Amani et al., 2024). Hundreds of ET_0 studies have been reviewed, which report ensemble and deep learning procedures often achieving $R^2 > 0.95$ with RMSE of 0.1-0.6 mm d^{-1} with minimal input variables than FAO 56 PM and strong substitutes to operational water management in data-scarce areas (Chen et al., 2020; Chang et al., 2025).

7. AI IN HYDROLOGICAL PROCESS MODELLING

7.1 Rainfall–Runoff Modelling

The use of AI has revolutionized rainfall-runoff modelling by substituting or supplementing conceptual and physically based models with data-driven models that are more realistic in terms of nonlinearity and nonstationarity. ANN, LSTM, support vector regression, and ensemble hybrids (e.g., LSTM Boosted Regression Tree, SVR -geomorphologic ANN) are models of machine learning and deep learning that continue to record Nash Sutcliffe efficiencies (NSE) of over 0.9 and significantly reduced RMSE relative to the traditional models across a wide range of catchments of interest (e.g., semi-arid basins) (Gelete, 2023; Hosseini and Mahjouri, 2015). Recent models also apply hydrological understanding (e.g., conceptual building blocks of FUSE/SUPERFLEX) to machine learning to induce physically interpretable, semi-distributed rainfall-runoff models (Herath et al., 2020).

7.2 Streamflow and Discharge Prediction

Deep learning models, CNN, LSTM, GRU, BiLSTM, and hybrids CNN-GRU-BiLSTM or DIFF-FFNN-LSTM, are the models that provide the highest quality of the day, hourly prediction, with NSE reaching 0.99, and high skills during peak floods (Workneh and Jha, 2025; Lin et al., 2021). These models are superior to classical statistical time series and most standalone machine learning models and hybrid physics-AI models that combine VIC-CaMa Flood with RNN/LSTM are also shown to be better at flood-related streamflow prediction, both with historical climatic conditions and with future climatic conditions (Zhong et al., 2023).

7.3 Flood and Drought Forecasting

In the case of flood and drought prediction, AI is applied in pure data-driven and hybrid designs (Figure 3). Rainfall has been demonstrated that LSTM-based rainfall-runoff and streamflow models trained on both event scale and continuous series were reliable in flood forecasts, the long lead, and also during years with disastrous events (Liu et al., 2020; Hu et al., 2018). State processes Hybrid systems that inject the outputs of big-scale hydrological and flood plain models (e.g., VIC-CaMa Flood) into the LSTM/RNN are even more effective in forecasting hydrographs of floods and estimating the maximum flood volumes, increasing the NSE and correlation coefficients that are used in risk assessment (Zhong et al., 2023). Computer-assisted prediction of streamflow and recharges under CMIP6 or SSP conditions is also becoming a way to measure future drought risk and water availability stress at the basin and regional level (Zhong et al., 2023; Banerjee et al., 2024).

Figure 3. Flood and drought forecasting

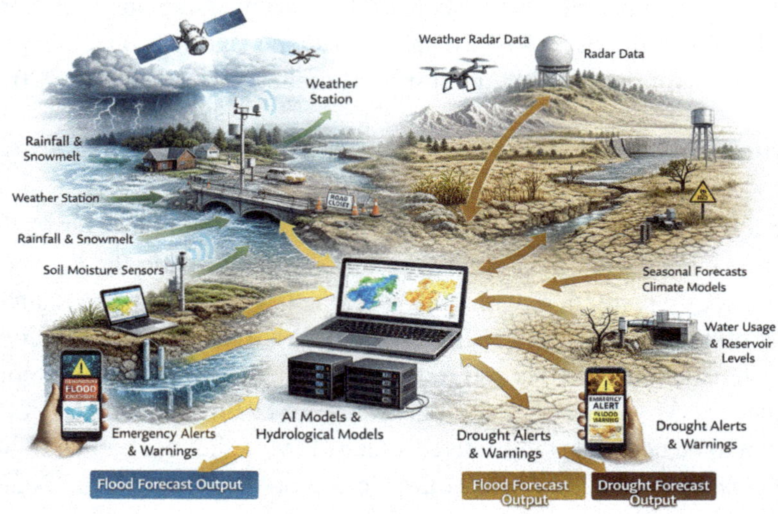

8. HYBRID AND PHYSICS-INFORMED AI APPROACHES

8.1 Coupling AI With Physically Based Hydrological Models

Hybrid and physics-informed AI models are becoming more common, which entails the embedding of neural networks within the process-based hydrological models to fix internal fluxes or models of specific processes, whilst maintaining a physical backbone. Distributed models (e.g., kinematic wave routing or HydroPy) are represented as neural networks by differentiable frameworks and trained with networks that parameterize spatially varying parameters and corrections to the operation of runoff generation and evapotranspiration, and in this way significantly outperform the physics-only models at basin to continent scales (Wang et al., 2024; Huynh et al., 2025). Other hybrids apply predictions of physically based models (e.g., VIC, glaciological models) as extra predictors in deep learning models, like CNN-GRU or CNN-LSTM, to make medium-range inflow forecasts and high flow simulations and reduce computation cost (Xu et al., 2024; Ougahi and Rowan, 2025).

8.2 Physics-Guided Neural Networks

Neural networks directly encode governing equations or hydrological principles into the loss or architecture. Physics-informed and physics-guided neural networks (PINNs/PGDL) directly incorporate governing equations or hydrological principles in the loss or architecture. Application PINN to hydrodynamic diffusion and shallow water-based flood models with conservation laws and boundary conditions imposed as hard constraints or soft constraints has provided more stable and consistent predictions with sparse data (Jia et al., 2025; Donnelly et al., 2023). Physics in catchment hydrology LSTM models in the simulation of flood peaks are guided by synthetic samples representing extreme rainfall, dry spells, and monotonic rainfall-runoff relationships, thus leading to better prediction of flood peaks than traditional LSTM (Xie et al., 2021).

8.3 Benefits of Hybrid Modelling Frameworks

The hybrid frameworks combine the strengths of both: the physical models are characterized by having interpretability, state variables, and extrapolation capabilities, whereas AI is characterized by the ability to recognize patterns better and correct bias. It has been demonstrated that with certain applications, there are significant improvements in Nash-Sutcliffe and Kling-Gupta efficiencies, more realistic internal state reproduction (i.e., evapotranspiration or total water storage), and enhanced robustness to noise or out-of-distribution conditions compared to either pure PBMs or pure deep learning (Wang et al., 2024; Wei et al., 2024). Physics-informed surrogates can also reduce the computational effort of high-resolution hydraulic simulators; they can also give information on uncertainty when used with Bayesian or regularized neural networks (Donnelly et al., 2023; Lu et al., 2021).

8.4 Applications in Complex and Data-Scarce Catchments

Hybrid models of distributed hydrological or glaciological output with deep learning perform better in flood peaks and high flow prediction in flash flood-prone French catchments and glacierized high mountain basins, where there are sparse observations and complex cryospheric processes may challenge conventional models (Huynh et al., 2025; Ougahi & Rowan, 2025). These findings point to the potential of hybrid and physics-informed AI to solve operational forecasting and water resources planning problems in challenging and data-deficient settings.

Table 2. Examples of hybrid and physics-informed AI in challenging settings

Application context	Hybrid / physics-informed strategy	Key benefit	Citations
Amazon mega-basin hydrology	Distributed PBM encoded as NN plus NN replacements for poorly known processes	Higher streamflow and storage skill vs. original DHM	(Wang et al., 2024)
Data-scarce rural basins	Physics-informed hybrid LSTM using hydrological priors	Better out-of-distribution streamflow prediction	(Lu et al., 2021)

9. APPLICATIONS IN AGRICULTURE AND WATER RESOURCE MANAGEMENT

9.1 Irrigation Scheduling and Water-Use Efficiency

Smart and predictive irrigation systems are AI-based irrigation systems using soil-moisture sensors, weather forecasts, and crop water-demand models to provide optimal irrigation timing and also greatly increase the efficiency of water use. Solar or autonomous irrigation systems dynamically regulate timing and amount of water depending on machine-learning and IoT, minimizing over- and under irrigation, cutting water consumption by approximately up to 30 percent, and keeping or boosting yields (Akanbi et al., 2025; Saha and Dubey, 2025). They are compatible with precision irrigation (e.g., drip, variable rate irrigation) and can be run on low-power edge devices, which qualifies them to fit resource-limited farms (Dhanararaj et al., 2025; Ali et al., 2025).

9.2 Drought Monitoring and Early Warning Systems

AI can be used in drought monitoring to enhance detection and forecasting through the integration of satellite remote sensing, climate data, and soil information. Complex spatiotemporal patterns are captured by deep learning models (CNNs, RNNs) or multi-index fusion frameworks, which utilize a platform like Google Earth Engine, and thereby allow earlier and more robust detection of agricultural drought compared to single-index and traditional methods (Vij and Tiwari, 2025; Duan et al., 2025). Drought early warning systems that have been developed with AI can support proactive changes in water allocation and crop choices by real-time alerts and impact scenarios (Masupha et al., 2024; Kikon and Deka, 2021).

9.3 Watershed and Basin-Scale Water Management

To manage watersheds and basins on this scale, AI is combined with hydrological and agronomical models to forecast the water availability, map irrigation demand, and allocate it to competing uses optimally. An integrated framework based on AI-processed remote sensing to meet crop water demand, physical models to forecast water availability, and optimization algorithms in the Segura Hydrographic Basin, Spain, allocated approximately 642 hm^3 over six months with minimal deficits and minimized environmental consequences (Otamendi et al., 2024). In general, water demand prediction, reservoir management, leakage identification, and smart distribution can also be enhanced by AI and big data analytics to enhance the efficiency and governance of the basin scale (Kamyab et al., 2023; Zou et al., 2025).

9.4 Climate-Resilient Agricultural Planning

The use of long-term climate forecasts, remote sensing, and farm-specific data in AI models is used to forecast water demands, crop yields, and drought risks to optimize crop choice, planting dates, and investments in irrigation infrastructure to make climate-resilient agricultural planning (Dhanararaj et al., 2025; Ye et al., 2024) (Figure 4). On-device models and decision-support systems can assist farmers in matching unpredictable rainfall and extreme heat conditions by proposing water-saving methods and resilient crop patterns, whereas digital twins and scenario planning frameworks can enable policymakers to examine adaptation strategies and create resilient water distribution policies when climate conditions are uncertain (Zhao et al., 2025; Masupha et al., 2024).

Figure 4. Climate-resilient agricultural planning

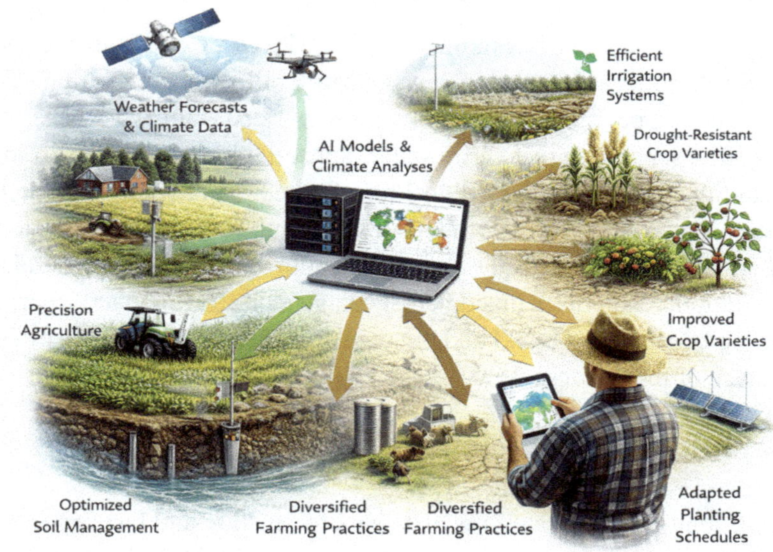

10. CASE STUDIES AND REGIONAL APPLICATIONS

10.1 AI-Driven Soil Moisture and ET Studies in Diverse Agro-Climatic Zones

The use of AIs in estimating soil moisture and evapotranspiration (ET) is growing and is applied in different climates and cropping systems. Machine learning systems like random forests, support vector machines, XGBoost, and neural networks have been trained on eddy covariance flux measurements and remote sensing imagery through MODIS have been able to predict daily cropland ET in temperate continue temperate climates, and subtropical-Mediterranean climates with high accuracy (R^2 up to 0.95, NSE over 0.92) when climate zone specific data is used as inputs (Du et al., 2024). Other systems are smart sensor nodes on farms with a combination of in situ soil and microclimate measurements, SVMs, and extreme learning machines to yield high-resolution estimates of ET in different soils and environmental settings ($R^2 \approx 0.97$) with direct application in field-scale irrigation and soil water diagnostics (Afridi et al., 2025).

10.2 Basin-Scale Hydrological Modelling Applications

At regional and basin levels, AI is used along with process-based or stochastic hydrological models to enhance predictions of runoff, flood, and water availability. Hybrid hydrological models incorporate neural networks in either distributed or conceptual models, or condition glacio-hydrological or stochastic rainfall-runoff predictions in deep networks (e.g., CNN-LSTM, wavelet GRU), with significant improvements in Nash-Sutcliffe efficiency and large alpine and tropical basin high and low flow prediction. Such hybrid schemes are used with glacier-fed systems in the Upper Indus and with African basins like the Oueme, where they are useful in predicting peak and low flows, which are important in running a reservoir, in irrigating crops, and in flood control (Ougahi et al., 2025).

10.3 Operational Use of AI in Water Management Agencies

The Agencies and utilities are starting to institutionalize AI in decision processes in order to realize operational water management. Global practice reviews indicate that AI is currently actively applied in streamflow and flood forecasting, intelligent irrigation, predicting water quality, and operation of infrastructure, and China, the USA, and India are some of the most efficient users (Jayakumar et al., 2024; Ahmed et al., 2024). The implementation of functional models that combine ensemble ML (e.g., XGBoost, LightGBM) with IoT sensor data to identify leaks and optimize irrigation schedules is an operational framework that can operate with nearly perfect accuracy in leak classification and sub-hour RMSE in the schedule of irrigation, and is applicable in utilities and irrigation districts (Mahmoud et al., 2025). The wider water resource planning systems include Adaptive Intelligent Dynamic Water Resource Planning and other AI-driven allocation schemes based on reinforcement learning and deep learning predictions to maximize water distribution at the urban and basin level in the presence of climatic and demand uncertainty (Xiang et al., 2020; Lu et al., 2025).

11. CHALLENGES AND LIMITATIONS

The issues and limitations in implementing AI in environmental and climate-related spheres cut across the entire modelling pipeline. One of the fundamental problems lies in the availability, quality, and the scale of the data: the information accessible in the real world is frequently sparse, noisy, biased, and inconsistently formatted, and may be a combination of coarse resolution climate or remote sensing outputs, as well as fine scale impacts observations, which weakens the confidence of the models and restricts the extrapolative characteristics of model performance in data-scarce areas (Koldasbayeva et al., 2024; Jones et al., 2023). Such data issues conflict with model interpretability and explainability. The high-performance deep learning models are usually black boxes, which makes it challenging to comprehend why a system forecasts a particular hazard, impact, or local weather reaction, which discourages trust and impedes acceptance in high stakes environments, inspiring the drive to seek explainable AI and interpretable architectures that do not affect competitive performance (Aldoseri et al., 2023; Saeed and Omlin, 2023). Another significant limitation is transferability across geographical locations and climates: models trained on certain datasets or climate regimes typically perform poorly when applied in new geographic areas due to the changes in data distributions, local processes, and observing practices and the problem of out of distribution generalization of AI to extremes, geospatial conditions and urban climate is only beginning to be investigated (Materia et al., 2024; Koldasbayeva et al., 2024). Lastly, there are computational and infrastructural limitations that restrict the research and operational implementation. Running and training large models of AI models and climate and geospatial data on multi petabyte datasets require large compute, storage, and bandwidth, which are not equally distributed among institutions and countries; ensembles to quantify uncertainty are especially constrained by costs of CPU hours and memory, limited tools, expertise, and standardized operations to process the constantly growing climate and Earth observation data volumes (Amnuaylojaroen, 2025; Jones et al., 2023).

12. FUTURE DIRECTIONS AND EMERGING TRENDS

The new trends and directions of AI-based hydrology are focusing more on explainable and trustworthy AI, in which physics-guided and interpretable models, including explainable AI (XAI) methods, are employed to clarify the manner in which inputs such as rainfall, soil moisture, and infrastructure operations generate successful predictions, and thus enhance transparency, regulatory acceptability, and stakeholder trust in high stakes decisions such as flood management, reservoir management, and water quality management (Biazar et al., 2025; Tripathy and Mishra, 2023). These functionalities are being incorporated into digital twins and real-time surveillance devices, where digital replicas of the catchments, water networks, or agricultural systems are constantly updated with IoT sensors, remote sensors, and in situ measurements to aid in dynamic scenario analysis, early warning, and interactive decision support and sustainable water governance (Islam, 2025; Zhao et al., 2025). One of the enabling ones is multi source data assimilation and cloud-based platforms, where heterogeneous data streams (gauges, satellites, models, citizen science) are combined using ensemble filters and hybrid AI-process based schemes and implemented in the cloud platform, decoupling advanced modelling and local resource constraints and enabling saleable, collaborative, near real time hydrologic predictions and control (Zou et al., 2025; Kurtz et al., 2017). These technical developments are facilitated by a powerful drive towards policy, capacity building and sustainable implementation: the future of work will focus on open and standardized data, reproducible pipelines and responsible AI systems that reinforce ethics and privacy and bias and invest in training of water practitioners, co design with communities and alignment with Sustainable Development Goals to ensure that AI enabled hydrology is equitable, robust and practically implementable in data rich and resource constrained regions (Biazar et al., 2025; Doorn, 2020).

13. CONCLUSION

The recent developments in artificial intelligence have played a major role in improving the soil moisture, evapotranspiration, and hydrological process modelling with the ability to incorporate multi-source data, including remote sensing, in-situ measurements, climatic variables, and land-use measurements. Machine and deep learning methods have enhanced spatial and temporal resolution, minimized uncertainty on parameter estimation, and have allowed more realistic representation of complex and nonlinear interactions of soil, plants, and the atmosphere. These findings have reinforced our knowledge of water fluxes in the various agro-hydrological systems and climate. This is because AI is critical in enhancing hydrological forecasts compared to the conventional empirical and process-based models, especially in data-sparse or highly variable conditions. Models that are artificial intelligence-based can improve the prediction of the dynamics of soil moisture, evapotranspiration, groundwater runoff, and water recharge, and thus aid early warning systems of droughts, floods, and water stress. The combination of AI with remote sensing and geospatial technologies also allows adapting water by near-real-time monitoring and managing fields and basins. The opportunities of explainable AI, hybrid physical-AI models, and cloud-based decision support systems are the prospects of sustainable water and agricultural systems to look forward to. These developments would be able to optimize irrigation timing, enhance climate resilience, and facilitate precision agriculture, which in the long run would lead to efficient water utilization, a higher yield, and food and water security.

REFERENCES

A, Y., Wang, G., Hu, P., Lai, X., Xue, B., and Fang, Q. (2022). Root-zone soil moisture estimation based on remote sensing data and deep learning. *Environmental Research, 212*(Pt B), 113278. DOI: 10.1016/j.envres.2022.113278

Adab, H., Morbidelli, R., Saltalippi, C., Moradian, M., & Ghalhari, G. A. F. (2020). Machine Learning to Estimate Surface Soil Moisture from Remote Sensing Data. *Water (Basel), 12*(11), 3223. DOI: 10.3390/w12113223

Afridi, W. K., Mukhopadhyay, S. C., & Mainali, B. (2025). Machine Learning Enabled Smart Sensor Node for Evapotranspiration Prediction in Diverse Soil and Environmental Conditions. *IEEE Sensors Journal, 25*(12), 22720–22728. DOI: 10.1109/JSEN.2025.3566935

Ahmed, A. A., Sayed, S., Abdoulhalik, A., Moutari, S., & Oyedele, L. (2024). Applications of machine learning to water resources management: A review of present status and future opportunities. *Journal of Cleaner Production, 441*, 140715. DOI: 10.1016/j.jclepro.2024.140715

Akanbi, M. B., Adedotun, K. J., Raji, A. K., and Banjoko, I. K. (2025). Optimizing water resource management in agriculture using AI-Powered solar irrigation systems. *Journal of Science Innovation and Technology Research*. DOI: 10.70382/ajsitr.v7i9.011

Aldoseri, A., Al-Khalifa, K. N., & Hamouda, A. M. (2023). Re-Thinking Data Strategy and Integration for Artificial intelligence: Concepts, opportunities, and challenges. *Applied Sciences (Basel, Switzerland), 13*(12), 7082. DOI: 10.3390/app13127082

Ali, A., Hussain, T., & Zahid, A. (2025). Smart irrigation technologies and prospects for enhancing water use efficiency for sustainable agriculture. *AgriEngineering, 7*(4), 106. DOI: 10.3390/agriengineering7040106

Amani, S., Shafizadeh-Moghadam, H., & Morid, S. (2024). Utilizing Machine Learning Models with Limited Meteorological Data as Alternatives for the FAO-56PM Model in Estimating Reference Evapotranspiration. *Water Resources Management, 38*(6), 1921–1942. DOI: 10.1007/s11269-023-03670-2

Amnuaylojaroen, T. (2025). Advancements and challenges of artificial intelligence in climate modeling for sustainable urban planning. *Frontiers in Artificial Intelligence, 8*, 1517986. DOI: 10.3389/frai.2025.1517986 PMID: 40463823

Amsalu, A., Garedew, W., Melka, G. A., Çelik, M. A., & Gemeda, D. O. (2026). Trends in climate extreme indices and implications for coffee farming in The Kafa Biosphere Reserve, southwest Ethiopia. *Climate Services*, *42*, 100637. DOI: 10.1016/j.cliser.2026.100637

Babaeian, E., Sadeghi, M., Jones, S. B., Montzka, C., Vereecken, H., & Tuller, M. (2019). Ground, proximal, and satellite remote sensing of soil moisture. *Reviews of Geophysics*, *57*(2), 530–616. DOI: 10.1029/2018RG000618

Bailey, R. T., Abbas, S., Arnold, J. G., & White, M. J. (2025). SWAT+MODFLOW: A new hydrologic model for simulating surface-subsurface flow in managed watersheds. *Geoscientific Model Development*, *18*(17), 5681–5697. DOI: 10.5194/gmd-18-5681-2025

Banerjee, D., Ganguly, S., & Kushwaha, S. (2024). Forecasting Future Groundwater Recharge from Rainfall Under Different Climate Change Scenarios Using Comparative Analysis of Deep Learning and Ensemble Learning Techniques. *Water Resources Management*, *38*(11), 4019–4037. DOI: 10.1007/s11269-024-03850-8

Biazar, S. M., Golmohammadi, G., Nedhunuri, R. R., Shaghaghi, S., & Mohammadi, K. (2025). Artificial intelligence in Hydrology: Advancements in soil, water resource management, and sustainable development. *Sustainability (Basel)*, *17*(5), 2250. DOI: 10.3390/su17052250

Blanka-Végi, V., Tobak, Z., Sipos, G., Barta, K., Szabó, B., & Van Leeuwen, B. (2025). Estimation of the Spatiotemporal Variability of Surface soil Moisture Using Machine Learning Methods Integrating Satellite and Ground-based Soil Moisture and Environmental Data. *Water Resources Management*, *39*(5), 2317–2334. DOI: 10.1007/s11269-024-04069-3

Çelik, M. A. (2026). Agro-climatic classification of Türkiye's Agricultural Basins using long-term ERA5-Land temperature data. *Physics and Chemistry of the Earth Parts A/B/C*, *142*, 104307. DOI: 10.1016/j.pce.2026.104307

Çelik, M. A., Kıraç, A., & Gouveia, C. M. (2026). North Atlantic Oscillation modulation of drought and vegetation activity in Türkiye. *Theoretical and Applied Climatology*, *157*(2), 74. Advance online publication. DOI: 10.1007/s00704-025-05971-5

Chang, Y., Zhang, C., Huang, J., Chang, H., Wang, C., & Huo, Z. (2025). Machine Learning for Reference Crop Evapotranspiration Modeling: A State-of-the-Art Review and Future Directions. *Agronomy (Basel)*, *15*(9), 2038. DOI: 10.3390/agronomy15092038

Chen, L., Šimůnek, J., Bradford, S. A., Ajami, H., & Meles, M. B. (2022). A computationally efficient hydrologic modeling framework to simulate surface-subsurface hydrological processes at the hillslope scale. *Journal of Hydrology (Amsterdam), 614*, 128539. DOI: 10.1016/j.jhydrol.2022.128539

Chen, Z., Zhu, Z., Jiang, H., & Sun, S. (2020). Estimating daily reference evapotranspiration based on limited meteorological data using deep learning and classical machine learning methods. *Journal of Hydrology (Amsterdam), 591*, 125286. DOI: 10.1016/j.jhydrol.2020.125286

Chin, R. J., & Lai, S. H. (2025). EDITORIAL: INTEGRATION OF HYDROLOGICAL MODELS AND MACHINE LEARNING TECHNIQUES FOR WATER RESOURCES MANAGEMENT. *Journal of Civil Engineering Science and Technology, 16*(1), 1–5. DOI: 10.33736/jcest.9191.2025

Dembélé, M., Schaefli, B., Van De Giesen, N., & Mariéthoz, G. (2020). Suitability of 17 gridded rainfall and temperature datasets for large-scale hydrological modelling in West Africa. *Hydrology and Earth System Sciences, 24*(11), 5379–5406. DOI: 10.5194/hess-24-5379-2020

Donnelly, J., Daneshkhah, A., & Abolfathi, S. (2023). Physics-informed neural networks as surrogate models of hydrodynamic simulators. *The Science of the Total Environment, 912*, 168814. DOI: 10.1016/j.scitotenv.2023.168814 PMID: 38016570

Doorn, N. (2020). Artificial intelligence in the water domain: Opportunities for responsible use. *The Science of the Total Environment, 755*(Pt 1), 142561. DOI: 10.1016/j.scitotenv.2020.142561 PMID: 33039891

Du, C., Jiang, S., Chen, C., Guo, Q., He, Q., & Zhan, C. (2024). Machine Learning-Based estimation of daily cropland evapotranspiration in diverse climate zones. *Remote Sensing (Basel), 16*(5), 730. DOI: 10.3390/rs16050730

Duan, X., Aslam, R. W., Naqvi, S. A. A., Kucher, D. E., Afzal, Z., Raza, D., Zulqarnain, R. M., & Said, Y. (2025). Multi-Index assessment and machine learning integration for drought monitoring using Google Earth Engine. *IEEE Journal of Selected Topics in Applied Earth Observations and Remote Sensing, 18*, 16497–16516. DOI: 10.1109/JSTARS.2025.3580652

Feng, D., Beck, H., Lawson, K., & Shen, C. (2023). The suitability of differentiable, physics-informed machine learning hydrologic models for ungauged regions and climate change impact assessment. *Hydrology and Earth System Sciences, 27*(12), 2357–2373. DOI: 10.5194/hess-27-2357-2023

Fersch, B., Francke, T., Heistermann, M., Schrön, M., Döpper, V., Jakobi, J., Baroni, G., Blume, T., Bogena, H., Budach, C., Gränzig, T., Förster, M., Güntner, A., Franssen, H. J. H., Kasner, M., Köhli, M., Kleinschmit, B., Kunstmann, H., Patil, A., & Oswald, S. (2020). A dense network of cosmic-ray neutron sensors for soil moisture observation in a highly instrumented pre-Alpine headwater catchment in Germany. *Earth System Science Data, 12*(3), 2289–2309. DOI: 10.5194/essd-12-2289-2020

Filipović, N., Brdar, S., Mimić, G., Marko, O., & Crnojević, V. (2021). Regional soil moisture prediction system based on Long Short-Term Memory network. *Biosystems Engineering, 213*, 30–38. DOI: 10.1016/j.biosystemseng.2021.11.019

Gelete, G. (2023). Application of hybrid machine learning-based ensemble techniques for rainfall-runoff modeling. *Earth Science Informatics, 16*(3), 2475–2495. DOI: 10.1007/s12145-023-01041-4

Hao, Y., Mao, J., Bachmann, C. M., Hoffman, F. M., Koren, G., Chen, H., Tian, H., Liu, J., Tao, J., Tang, J., Li, L., Liu, L., Apple, M., Shi, M., Jin, M., Zhu, Q., Kannenberg, S., Shi, X., Zhang, X., & Dai, Y. (2025). Soil moisture controls over carbon sequestration and greenhouse gas emissions: A review. *NPJ Climate and Atmospheric Science, 8*(1), 16. Advance online publication. DOI: 10.1038/s41612-024-00888-8

Haokip, S. C., Rajwade, Y. A., Rao, K. V. R., Kumar, S. P., Marak, A. B., & Srivastava, A. (2025). Approaches for Assessment of Soil Moisture with Conventional Methods, Remote Sensing, UAV, and Machine Learning Methods—A Review. *Water (Basel), 17*(16), 2388. DOI: 10.3390/w17162388

Hendrickx, M. G. A., Vanderborght, J., Janssens, P., Bombeke, S., Matthyssen, E., Waverijn, A., & Diels, J. (2025). Pooled error variance and covariance estimation of sparse in situ soil moisture sensor measurements in agricultural fields in Flanders. *Soil (Göttingen), 11*(1), 435–456. DOI: 10.5194/soil-11-435-2025

Herath, H. M. V. V., Chadalawada, J., & Babovic, V. (2020). Hydrologically Informed Machine Learning for Rainfall-Runoff Modelling: Towards Distributed Modelling. *Hydrology and Earth System Sciences*. Advance online publication. DOI: 10.5194/hess-2020-487

Hosseini, S. M., & Mahjouri, N. (2015). Integrating Support Vector Regression and a geomorphologic Artificial Neural Network for daily rainfall-runoff modeling. *Applied Soft Computing, 38*, 329–345. DOI: 10.1016/j.asoc.2015.09.049

Hu, C., Wu, Q., Li, H., Jian, S., Li, N., & Lou, Z. (2018). Deep Learning with a Long Short-Term Memory Networks Approach for Rainfall-Runoff Simulation. *Water (Basel), 10*(11), 1543. DOI: 10.3390/w10111543

Hu, L., Xie, Y., Liu, Y., Chen, S., Yu, H., Bie, Q., Zhao, F., & Zhao, Y. (2025). Mapping the primary factors driving spatiotemporal variations of surface soil moisture in the Yellow River Basin of China. *Scientific Reports, 15*(1), 12359. DOI: 10.1038/s41598-025-96129-w PMID: 40211021

Huang, J., Sehgal, V., Alvarez, L. V., Brocca, L., Cai, S., Cheng, R., Cheng, X., Du, J., Masri, B. E., Endsley, K. A., Fang, Y., Hu, J., Jampani, M., Kibria, M. G., Koren, G., Li, L., Liu, L., Mao, J., Moreno, H. A., & Fisher, J. B. (2025). Remotely Sensed High-Resolution Soil Moisture and evapotranspiration: Bridging the gap between science and society. *Water Resources Research, 61*(5), e2024WR037929. Advance online publication. DOI: 10.1029/2024WR037929

Huang, X., Shi, Z. H., Zhu, H. D., Zhang, H. Y., Ai, L., & Yin, W. (2015). Soil moisture dynamics within soil profiles and associated environmental controls. *Catena, 136*, 189–196. DOI: 10.1016/j.catena.2015.01.014

Huynh, N. N. T., Garambois, P.-A., Renard, B., Colleoni, F., Monnier, J., & Roux, H. (2025). A distributed hybrid physics–AI framework for learning corrections of internal hydrological fluxes and enhancing high-resolution regionalized flood modeling. *Hydrology and Earth System Sciences, 29*(15), 3589–3613. DOI: 10.5194/hess-29-3589-2025

Islam, F. S. (2025). A multi-dimensional AI framework for sustainable drinking water management: Integrating federated learning, digital twins, and blockchain. *Journal of Engineering Research and Reports, 27*(6), 466–492. DOI: 10.9734/jerr/2025/v27i61558

Jayakumar, D., Bouhoula, A., & Al-Zubari, W. K. (2024). Unlocking the potential of artificial intelligence for sustainable water management focusing operational applications. *Water (Basel), 16*(22), 3328. DOI: 10.3390/w16223328

Jia, L., Yen, N., & Pei, Y. (2025). A Physics-Coupled deep learning framework for hydrodynamic diffusion modeling in watershed systems: Integrating spatiotemporal networks and environmental constraints. *IEEE Access : Practical Innovations, Open Solutions, 13*, 34985–35003. DOI: 10.1109/ACCESS.2025.3542173

Jiang, Z.-Y., Yang, Z.-G., Zhang, S.-Y., Liao, C.-M., Hu, Z.-M., Cao, R.-C., & Wu, H.-W. (2020). Revealing the spatio-temporal variability of evapotranspiration and its components based on an improved Shuttleworth-Wallace model in the Yellow River Basin. *Journal of Environmental Management, 262*, 110310. DOI: 10.1016/j.jenvman.2020.110310 PMID: 32250793

Jones, A., Kuehnert, J., Fraccaro, P., Meuriot, O., Ishikawa, T., Edwards, B., Stoyanov, N., Remy, S. L., Weldemariam, K., & Assefa, S. (2023). AI for climate impacts: Applications in flood risk. *NPJ Climate and Atmospheric Science*, *6*(1), 63. Advance online publication. DOI: 10.1038/s41612-023-00388-1

Kamyab, H., Khademi, T., Chelliapan, S., SaberiKamarposhti, M., Rezania, S., Yusuf, M., Farajnezhad, M., Abbas, M., Hun Jeon, B., & Ahn, Y. (2023). The latest innovative avenues for the utilization of artificial Intelligence and big data analytics in water resource management. *Results in Engineering*, *20*, 101566. DOI: 10.1016/j.rineng.2023.101566

Kar, K. K., Haggerty, R., Sharma, H., Dwivedi, D., & Roy, T. (2025). Evapotranspiration partitioning using flux tower data in a Semi-Arid ecosystem. *Hydrological Processes*, *39*(3), e70083. Advance online publication. DOI: 10.1002/hyp.70083

Karthikeyan, L., & Mishra, A. K. (2021). Multi-layer high-resolution soil moisture estimation using machine learning over the United States. *Remote Sensing of Environment*, *266*, 112706. DOI: 10.1016/j.rse.2021.112706

Kikon, A., & Deka, P. C. (2021). Artificial intelligence application in drought assessment, monitoring and forecasting: A review. *Stochastic Environmental Research and Risk Assessment*, *36*(5), 1197–1214. DOI: 10.1007/s00477-021-02129-3

Kim, T., Yang, T., Gao, S., Zhang, L., Ding, Z., Wen, X., Gourley, J. J., & Hong, Y. (2021). Can artificial intelligence and data-driven machine learning models match or even replace process-driven hydrologic models for streamflow simulation?: A case study of four watersheds with different hydro-climatic regions across the CONUS. *Journal of Hydrology (Amsterdam)*, *598*, 126423. DOI: 10.1016/j.jhydrol.2021.126423

Kokkalis, P., Jassar, H. K. A., Sarraf, H. A., Nair, R., & Hendi, H. A. (2024). Evaluation of ERA5 and NCEP reanalysis climate models for precipitation and soil moisture over a semi-arid area in Kuwait. *Climate Dynamics*, *62*(6), 4893–4904. DOI: 10.1007/s00382-024-07141-1

Koldasbayeva, D., Tregubova, P., Gasanov, M., Zaytsev, A., Petrovskaia, A., & Burnaev, E. (2024). Challenges in data-driven geospatial modeling for environmental research and practice. *Nature Communications*, *15*(1), 10700. DOI: 10.1038/s41467-024-55240-8 PMID: 39702456

Kuffour, B. N. O., Engdahl, N. B., Woodward, C. S., Condon, L. E., Kollet, S., & Maxwell, R. M. (2020). Simulating coupled surface–subsurface flows with ParFlow v3.5.0: Capabilities, applications, and ongoing development of an open-source, massively parallel, integrated hydrologic model. *Geoscientific Model Development, 13*(3), 1373–1397. DOI: 10.5194/gmd-13-1373-2020

Kurtz, W., Lapin, A., Schilling, O. S., Tang, Q., Schiller, E., Braun, T., Hunkeler, D., Vereecken, H., Sudicky, E., Kropf, P., Hendricks Franssen, H.-J., & Brunner, P. (2017). Integrating hydrological modelling, data assimilation and cloud computing for real-time management of water resources. *Environmental Modelling & Software, 93*, 418–435. DOI: 10.1016/j.envsoft.2017.03.011

Li, B., Sun, T., Tian, F., Tudaji, M., Qin, L., & Ni, G. (2024). Hybrid hydrological modeling for large alpine basins: A semi-distributed approach. *Hydrology and Earth System Sciences, 28*(20), 4521–4538. DOI: 10.5194/hess-28-4521-2024

Lin, Y., Wang, D., Wang, G., Qiu, J., Long, K., Du, Y., Xie, H., Wei, Z., Shangguan, W., & Dai, Y. (2021). A hybrid deep learning algorithm and its application to streamflow prediction. *Journal of Hydrology (Amsterdam), 601*, 126636. DOI: 10.1016/j.jhydrol.2021.126636

Liu, D., Jiang, W., Mu, L., & Wang, S. (2020). Streamflow prediction using Deep Learning Neural Network: Case study of Yangtze River. *IEEE Access : Practical Innovations, Open Solutions, 8*, 90069–90086. DOI: 10.1109/ACCESS.2020.2993874

Liu, D., Wang, Z., Wang, L., Chen, J., Li, C., & Shi, Y. (2024). Improved remote sensing reference evapotranspiration estimation using simple satellite data and machine learning. *The Science of the Total Environment, 947*, 174480. DOI: 10.1016/j.scitotenv.2024.174480 PMID: 38972400

Liu, Z., Zhou, J., Yang, X., Zhao, Z., & Lv, Y. (2024). Research on water resource modeling based on machine learning technologies. *Water (Basel), 16*(3), 472. DOI: 10.3390/w16030472

Lu, D., Konapala, G., Painter, S. L., Kao, S.-C., & Gangrade, S. (2021). Streamflow simulation in data-scarce basins using Bayesian and physics-informed machine learning models. *Journal of Hydrometeorology*. Advance online publication. DOI: 10.1175/JHM-D-20-0082.1

Lu, T., Jiang, T., & Liu, Y. (2025). Research on an optimization model for water resource management based on deep learning. *IEEE Access : Practical Innovations, Open Solutions, 13*, 147906–147924. DOI: 10.1109/ACCESS.2025.3600664

Luo, W., Xu, X., Liu, W., Liu, M., Li, Z., Peng, T., Xu, C., Zhang, Y., & Zhang, R. (2018). UAV based soil moisture remote sensing in a karst mountainous catchment. *Catena*, *174*, 478–489. DOI: 10.1016/j.catena.2018.11.017

Mahmoud, M. B. R., Hemdan, D. I. I., Alajmani, S. H., Alyami, R. Y., Elmarhomy, G., Hashim, H., & Atlam, E.-S. (2025). Optimized AI and IoT-Driven framework for intelligent water resource Management. *IEEE Access : Practical Innovations, Open Solutions*, *13*, 97628–97646. DOI: 10.1109/ACCESS.2025.3572067

Masupha, T. E., Moeletsi, M. E., & Tsubo, M. (2024). A transformative framework reshaping sustainable drought risk management through advanced early warning systems. *iScience*, *27*(7), 110066. DOI: 10.1016/j.isci.2024.110066 PMID: 38989469

Materia, S., García, L. P., Van Straaten, C. O. S., Mamalakis, A., Cavicchia, L., Coumou, D., De Luca, P., Kretschmer, M., & Donat, M. (2024). Artificial intelligence for climate prediction of extremes: State of the art, challenges, and future perspectives. *Wiley Interdisciplinary Reviews: Climate Change*, *15*(6), e914. Advance online publication. DOI: 10.1002/wcc.914

Muñoz-Sabater, J., Dutra, E., Agustí-Panareda, A., Albergel, C., Arduini, G., Balsamo, G., Boussetta, S., Choulga, M., Harrigan, S., Hersbach, H., Martens, B., Miralles, D. G., Piles, M., Rodríguez-Fernández, N. J., Zsoter, E., Buontempo, C., & Thépaut, J.-N. (2021). ERA5-Land: A state-of-the-art global reanalysis dataset for land applications. *Earth System Science Data*, *13*(9), 4349–4383. DOI: 10.5194/essd-13-4349-2021

Nearing, G. S., Kratzert, F., Sampson, A. K., Pelissier, C. S., Klotz, D., Frame, J. M., Prieto, C., & Gupta, H. V. (2020). What role does hydrological science play in the age of machine learning? *Water Resources Research*, *57*(3), e2020WR028091. Advance online publication. DOI: 10.1029/2020WR028091

Otamendi, U., Maiza, M., Olaizola, I. G., Sierra, B., Florez, M., & Quartulli, M. (2024). Integrated water resource management in the Segura Hydrographic Basin: An artificial intelligence approach. *Journal of Environmental Management*, *370*, 122526. DOI: 10.1016/j.jenvman.2024.122526 PMID: 39357444

Ougahi, J. H., & Rowan, J. S. (2025). Enhanced streamflow forecasting using hybrid modelling integrating glacio-hydrological outputs, deep learning and wavelet transformation. *Scientific Reports*, *15*(1), 2762. DOI: 10.1038/s41598-025-87187-1 PMID: 39843529

Rabiei, S., Babaeian, E., & Grunwald, S. (2025). Surface and Subsurface Soil Moisture Estimation Using Fusion of SMAP, NLDAS-2, and SOLUS100 Data with Deep Learning. *Remote Sensing (Basel)*, *17*(4), 659. DOI: 10.3390/rs17040659

Ran, B., Xing, Z., Fang, J., Yang, Y., He, Y., Li, X., Liu, X., & Liu, G. (2025). Analysis of soil moisture dynamics and its response to rainfall in the MU US sandy land. *Water (Basel), 17*(1), 105. DOI: 10.3390/w17010105

Rasheed, M. W., Tang, J., Sarwar, A., Shah, S., Saddique, N., Khan, M. U., Khan, M. I., Nawaz, S., Shamshiri, R. R., Aziz, M., & Sultan, M. (2022). Soil Moisture Measuring Techniques and factors Affecting the moisture dynamics: A Comprehensive review. *Sustainability (Basel), 14*(18), 11538. DOI: 10.3390/su141811538

Saeed, W., & Omlin, C. (2023). Explainable AI (XAI): A systematic meta-survey of current challenges and future opportunities. *Knowledge-Based Systems, 263*, 110273. DOI: 10.1016/j.knosys.2023.110273

Saha, L., & Dubey, A. (2025). Innovative AI solutions for Water-Efficient smart irrigation in agriculture. *SHS Web of Conferences, 216*, 01052. DOI: 10.1051/shsconf/202521601052

Shokati, H., Mashal, M., Noroozi, A., Mirzaei, S., Mohammadi-Doqozloo, Z., Nabiollahi, K., Taghizadeh-Mehrjardi, R., Khosravani, P., Adhikari, R., Hu, L., & Scholten, T. (2025). Comparing UAV-Based hyperspectral and Satellite-Based multispectral data for soil moisture estimation using machine learning. *Water (Basel), 17*(11), 1715. DOI: 10.3390/w17111715

Sit, M., Demiray, B., Xiang, Z., Ewing, G., Sermet, Y., & Demir, I. (2020). A Comprehensive Review of Deep Learning Applications in Hydrology and Water Resources. *Water Science and Technology: A Journal of the International Association on Water Pollution Research.* https://doi.org/DOI: 10.31223/osf.io/xs36g

Suo, L., Huang, M., Zhang, Y., Duan, L., & Shan, Y. (2018). Soil moisture dynamics and dominant controls at different spatial scales over semiarid and semi-humid areas. *Journal of Hydrology (Amsterdam), 562*, 635–647. DOI: 10.1016/j.jhydrol.2018.05.036

Tripathy, K. P., & Mishra, A. K. (2023). Deep learning in hydrology and water resources disciplines: Concepts, methods, applications, and research directions. *Journal of Hydrology (Amsterdam), 628*, 130458. DOI: 10.1016/j.jhydrol.2023.130458

Vergopolan, N., Chaney, N. W., Pan, M., Sheffield, J., Beck, H. E., Ferguson, C. R., Torres-Rojas, L., Sadri, S., & Wood, E. F. (2021). SMAP-HydroBlocks, a 30-m satellite-based soil moisture dataset for the conterminous US. *Scientific Data, 8*(1), 264. DOI: 10.1038/s41597-021-01050-2 PMID: 34635675

Vij, P., & Tiwari, A. (2025). AI-Driven Drought Monitoring: Advanced machine learning techniques for early prediction. *SHS Web of Conferences, 216*, 01024. DOI: 10.1051/shsconf/202521601024

Vulova, S., Meier, F., Rocha, A. D., Quanz, J., Nouri, H., & Kleinschmit, B. (2021). Modeling urban evapotranspiration using remote sensing, flux footprints, and artificial intelligence. *The Science of the Total Environment*, *786*, 147293. DOI: 10.1016/j.scitotenv.2021.147293 PMID: 33975115

Wang, C., Jiang, S., Zheng, Y., Han, F., Kumar, R., Rakovec, O., & Li, S. (2024). Distributed Hydrological Modeling with Physics-Encoded Deep Learning: A General Framework and its application in the Amazon. *Water Resources Research*, *60*(4), e2023WR036170. Advance online publication. DOI: 10.1029/2023WR036170

Wang, S., Fu, Z., Chen, H., Nie, Y., & Xu, Q. (2019). Mechanisms of surface and subsurface runoff generation in subtropical soil-epikarst systems: Implications of rainfall simulation experiments on karst slope. *Journal of Hydrology (Amsterdam)*, *580*, 124370. DOI: 10.1016/j.jhydrol.2019.124370

Wei, Y., Wang, R., & Feng, P. (2024). Improving Hydrological Modeling with Hybrid Models: A Comparative Study of Different Mechanisms for Coupling Deep Learning Models with Process-based Models. *Water Resources Management*, *38*(7), 2471–2488. DOI: 10.1007/s11269-024-03780-5

Workneh, H., & Jha, M. (2025). Utilizing hybrid deep learning models for streamflow prediction. *Water (Basel)*, *17*(13), 1913. DOI: 10.3390/w17131913

Xiang, X., Li, Q., Khan, S., & Khalaf, O. I. (2020). Urban water resource management for sustainable environment planning using artificial intelligence techniques. *Environmental Impact Assessment Review*, *86*, 106515. DOI: 10.1016/j.eiar.2020.106515

Xie, K., Liu, P., Zhang, J., Han, D., Wang, G., & Shen, C. (2021). Physics-guided deep learning for rainfall-runoff modeling by considering extreme events and monotonic relationships. *Journal of Hydrology (Amsterdam)*, *603*, 127043. DOI: 10.1016/j.jhydrol.2021.127043

Xu, L., Ye, Z., Dai, J., Li, Q., Hong, Y., Tao, Y., Yu, H., Zhang, C., Chen, Z., & Chen, N. (2025). Spatiotemporal seamless estimation of global surface soil moisture using triple collocation, machine learning, and data assimilation. *IEEE Transactions on Geoscience and Remote Sensing*, *63*, 1–16. DOI: 10.1109/TGRS.2025.3568034

Xu, T., & Liang, F. (2021). Machine learning for hydrologic sciences: An introductory overview. *WIREs. Water*, *8*(5), e1533. Advance online publication. DOI: 10.1002/wat2.1533

Xu, W., Chen, J., Corzo, G., Xu, C., Zhang, X. J., Xiong, L., Liu, D., & Xia, J. (2024). Coupling deep learning and physically based hydrological models for monthly streamflow predictions. *Water Resources Research, 60*(2), e2023WR035618. Advance online publication. DOI: 10.1029/2023WR035618

Yadav, B., Sharma, L. K., & Bijarniya, B. (2025). Predicting Forest Evapotranspiration using Remote Sensing and Machine Learning. *The International Archives of the Photogrammetry, Remote Sensing and Spatial Information Sciences/International Archives of the Photogrammetry. The International Archives of the Photogrammetry, Remote Sensing and Spatial Information Sciences, 2025*, 1611–1619. DOI: 10.5194/isprs-archives-XLVIII-G-2025-1611-2025

Ye, Z., Yin, S., Cao, Y., & Wang, Y. (2024). AI-driven optimization of agricultural water management for enhanced sustainability. *Scientific Reports, 14*(1), 25721. DOI: 10.1038/s41598-024-76915-8 PMID: 39468244

Yuan, H., Liang, S., Gao, Y., Gao, Y., & Lian, X. (2025). Experimental study on estimating bare soil moisture content based on UAV multi-source remote sensing. *Geocarto International, 40*(1), 2448985. Advance online publication. DOI: 10.1080/10106049.2024.2448985

Zhang, Q., Li, P., Ren, X., Ning, J., Li, J., Liu, C., Wang, Y., & Wang, G. (2023). A new real-time groundwater level forecasting strategy: Coupling hybrid data-driven models with remote sensing data. *Journal of Hydrology (Amsterdam), 625*, 129962. DOI: 10.1016/j.jhydrol.2023.129962

Zhang, T., & Yang, Y. (2025). Mapping Very-High Resolution Evapotranspiration Data Using a Dual GAN Model. *IEEE 15th Annual Computing and Communication Workshop and Conference (CCWC)*, 00117–00125. DOI: 10.1109/CCWC62904.2025.10903799

Zhang, Y., Hou, J., & Huang, C. (2023). Basin Scale Soil Moisture Estimation with Grid SWAT and LESTKF Based on WSN. *Sensors (Basel), 24*(1), 35. DOI: 10.3390/s24010035 PMID: 38202901

Zhang, Y., Peña-Arancibia, J. L., McVicar, T. R., Chiew, F. H. S., Vaze, J., Liu, C., Lu, X., Zheng, H., Wang, Y., Liu, Y. Y., Miralles, D. G., & Pan, M. (2016). Multi-decadal trends in global terrestrial evapotranspiration and its components. *Scientific Reports, 6*(1), 19124. DOI: 10.1038/srep19124 PMID: 26750505

Zhao, T., Song, C., Yu, J., Xing, L., Xu, F., Li, W., & Wang, Z. (2025). Leveraging immersive digital twins and AI-Driven decision support Systems for sustainable water reserves Management: A Conceptual framework. *Sustainability (Basel), 17*(8), 3754. DOI: 10.3390/su17083754

Zhao, Z., Li, Y., Liu, K., Wu, C., Yu, B., Liu, G., & Wang, Y. (2025). Layered soil moisture retrieval and agricultural application based on Multi-Source Remote Sensing and vegetation suppression technology: A case study of Youyi Farm, China. *Remote Sensing (Basel), 17*(13), 2130. DOI: 10.3390/rs17132130

Zhong, M., Zhang, H., Jiang, T., Guo, J., Zhu, J., Wang, D., & Chen, X. (2023). A hybrid model combining the Cama-Flood model and deep learning methods for streamflow prediction. *Water Resources Management, 37*(12), 4841–4859. DOI: 10.1007/s11269-023-03583-0

Zhuo, L., Dai, Q., Zhao, B., & Han, D. (2020). Soil Moisture Sensor Network Design for Hydrological Applications. *Hydrology and Earth System Sciences, 24*(5), 2577–2591. Advance online publication. DOI: 10.5194/hess-24-2577-2020

Zou, S., Ju, H., & Zhang, J. (2025). Water quality Management in the age of AI: Applications, challenges, and Prospects. *Water (Basel), 17*(11), 1641. DOI: 10.3390/w17111641

Chapter 2
Multimodal Temporal AI Reasoning for Climate-Driven Landform Change Detection via Remote Sensing

K. DanielRaj
http://orcid.org/0009-0001-9863-5682
KalaignarKarunanidhi Institute of Technology, India

M. Robinson Joel
http://orcid.org/0000-0002-3030-8431
KCG College of Technology, India

G. Indra Navaroj
http://orcid.org/0000-0002-8257-5080
Vel Tech Rangarajan Dr. Sagunthala R&D Institute of Science and Technology, India

K. Palraj
http://orcid.org/0000-0003-2823-6675
Ramco Institute of Technology, India

C. Soundar
St. Joseph College of Engineering, India

DOI: 10.4018/979-8-3373-8745-1.ch002

ABSTRACT

Climate-driven landform evolution is governed by complex deformation processes resulting from extreme climatic events and long-term environmental transitions. Conventional remote sensing approaches that rely on static imagery or single-sensor observations often fail to distinguish permanent geomorphic change from short-term spectral or seasonal variability. This chapter presents a multimodal temporal reasoning framework that integrates optical, SAR, LiDAR, and hyperspectral data within a unified sequence-learning architecture conditioned by climatic drivers. Modality-aware spatial encoding and attention-based temporal modeling enable coherent reconstruction of landform dynamics across riverine, coastal, glacial, and deltaic environments. The proposed approach enhances robustness to missing data, improves temporal consistency, and advances monitoring from retrospective change detection toward predictive geomorphology, supporting early-warning systems and climate-resilient land-use planning.

1. INTRODUCTION

The landforms shaped by climate will continue changing at greater rates due to increased rainfall, accelerated glacial retreat, sea-level rise, and extreme weather events (Banerjee & Shankar, 2022; Piyush et al., 2024; Amsalu et al., 2026). These processes of change affect riverbanks, coastal margins and deltas, and mountain slopes, and thus have a direct bearing on settlement safety, agricultural production, and the resilience of infrastructure. Traditional field measurement methods with their periodicity, limited spatial coverage, and expensive resource requirement could never manage to keep track of these changes continuously (McFeeters, 1996). Remote sensing, however, presents global, repetitive observations, but existing monitoring methodologies frequently detect changes only after they occur rather than arising from an inference of the climatic mechanisms causing them (Gorelick et al., 2017).

Due in part to this latent constraint of geomorphological monitoring, a systematic investigation of time has been lacking. Binary satellite pictures simply do not distinguish between long and short, making it exceedingly difficult to separate seasonal cycles and monumental morphodynamic transitions (Zhu & Woodcock, 2014). In isolation, each of these sensing modalities—optical, SAR, LiDAR, hyperspectral—tend to undermine significant contributions of concurrent data toward deciphering complicating surface–climate interactions (Li et al., 2024). Overall, conventional approaches remain relevant in timely post-disaster assessment but are deficient in providing predictive insight into slower, climate-driven modification of landforms (Abdelkader et al., 2023).

Addressing these gaps in the chapter is the multimodal thinking of time applied to landform change detection. The demand is to merge discrepancies among remote sensing data into artificial intelligence-based temporal models that show permanent trends, abrupt changes, and climate linkage. The framework would allow for proactive adaptation planning rather than reactive response.

1.1 Motivation

The modification of landforms by climate variability has remained one of the most pronounced, persistent, and disruptive phenomena in terrestrial systems. Continuous geomorphological manifestations of flooding, storm surging, intensified monsoon patterns, glacial melt rates, seasonal sediment fluxes, and sea-level fluctuations exhibit time scales ranging from days to decades (Banerjee & Shankar, 2022; Piyush et al., 2024). Habitat, agricultural productivity, infrastructure stability, and exposure risk to disasters from these changes are greatly influenced by these changes (Li et al., 2024; Çelik, 2026). The reason is that field-survey-based geomorphology is, by the nature of the manual measurement, ineffective to track these dynamics in space and time. It is intermittent, localized, and resource-intensive (McFeeters, 1996). The earlier efforts to fit remote sensing into landform monitoring are mainly descriptive, that is, observable changes rather than predicting or caustically explaining the changes (Gorelick et al., 2017). The major gap arises not from data scarcity but from the inability to establish systematic frameworks for time-integrated multimodal observatory data synthesis and computational interpretation to allow attributions, quantification, and forecasting of climate-induced landform processes (Abdelkader et al., 2023). AI-facilitated temporal remote sensing potentially houses that closing pathway, although the spatial heterogeneity, modeling noise, and temporal irregularity must be dealt with in the design (Zhu & Woodcock, 2014).

1.2 Climate-Forced Geomorphological Dynamics

The climate drivers restructure the surface of the Earth by intensifying processes that would otherwise have taken place at a very slow rate. Extreme rains wash away the banks of rivers; glacial retreat increases the pace at which the moraine destabilizes; coastal dunes change with cyclonic storms; prolonged drought triggers wind action on newly exposed sands (Banerjee & Shankar, 2022). They are not discrete: glacial melting modifies discharge from rivers; slopes become susceptible to erosive forces, owing to the decrease in vegetative cover due to heat stress; the compaction of sediments is affected by salinity intrusion caused by the storms (Li et al., 2024). Remote sensing captures effects, but implied mechanisms do not come directly unless processed within temporal reasoning models (Zhu & Woodcock, 2014).

Otherwise, one cannot distinguish between seasonal shifts and long or continuous morphodynamic changes using estimation from satellite snapshots (Gorelick et al., 2017). There is a need for multimodal measurements-optical imagery for reflectance; SAR for moisture and roughness; LiDAR for elevation; hyperspectral imaging for mineralogical changes; and modeling architectures capable of extracting persistent trends and discontinuities to identify climate-driven landform evolution (Abdelkader et al., 2023).

1.3 Limitations of the Current Landform Monitoring Workflows

These practices present three basic structural weaknesses, however: most of the techniques rely on difference imaging or threshold-based change detection, generally oversimplifying nonlinear evolution and amplifying the detection of noise in low-signal environments (Piyush et al., 2024). Optical sensors fail under cloud cover while the effect of introduced SAR speckle requires aggressive filtering that regularly eliminates geomorphologically relevant micro-features (Abdelkader et al., 2023). Multimodal integration is usually fragmented, with modes that are processed independently and therefore merged qualitatively rather than algorithmically-fusion that precludes a unified representation learning (Li et al., 2024). Temporal reasoning is rarely embedded within modeling architectures; CNNs treat images as independent samples, and RNN-based models are usually trained on short sequences that fail to capture multi-year climate patterns (Abdelkader et al., 2023; Zhu & Woodcock, 2014). Thus, pipelines remain effective for fast assessments of damage as a result of disasters but are inadequate for long-term climate-induced morphodynamics (Banerjee & Shankar, 2022).

1.4 Contribution of This Chapter

Therefore, multimodal temporal reasoning systems for climate-driven landform change detection would be advanced in this chapter, which would unify heterogeneous remote sensing modalities in constructing a comprehensive temporal learning pipeline. Rather than treating images as isolated observations, the core framework will model such assets as sequential evidence within this dynamic climatic system (Abdelkader et al., 2023). In total, four contributions arise from this:

(1) A defined workflow for assembling, normalizing, and aligning modalities with irregular revisits (Gorelick et al., 2017);
(2) structured analysis on fusion strategies—be they early, late, or hybrid—towards conserving relevant geomorphological signals (Li et al., 2024);

(3) evaluation of temporal architectures, such as ConvLSTM and attention-based transformers, towards ways to represent temporal persistence and seasonal reversals (Piyush et al., 2024);
(4) empirical validation using case studies from diverse climatic and geomorphological domains-for example, riverbank erosion, coastal retreat, glacial displacement, and post-cyclone sediment redistribution (Banerjee & Shankar, 2022).

The framework is supposed to take monitoring from change verification to change interpretation in predictive and adaptation-oriented decision-making.

Figure 1. Framework Diagram For Multimodal Temporal Reasoning for Climate-Driven Landform Change Detection Using AI-Enabled Remote Sensing

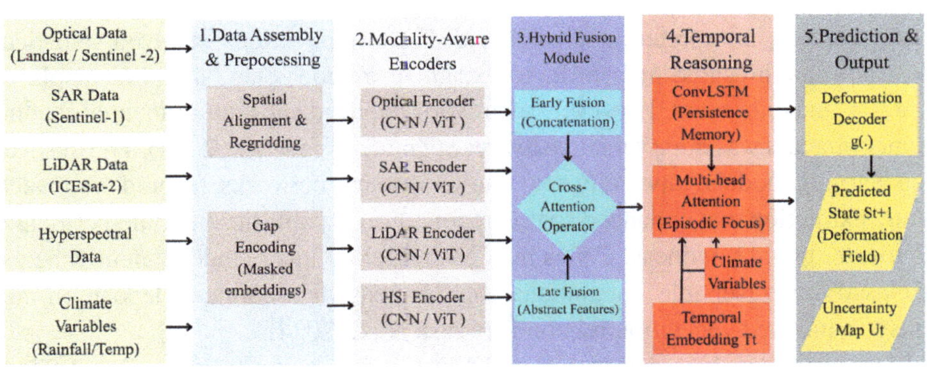

2. BACKGROUND AND LITERATURE REVIEW

2.1 Remote Sensing Fundamentals

Remote sensing allows for the studying of the Earth's surface via reflected or emitted electromagnetic radiation sans any direct physical contact. Optical sensors detect variations of visible and near-infrared reflectance suitable for the detection of vegetation patterns, sediment exposure, and surface moisture proxies (Gorelick et al., 2017). Synthetic Aperture Radar (SAR) systems transmit microwave pulses to measure surface backscatter and are fit for observing landforms even in cloud cover or variable illumination (Abdelkader et al., 2023). LiDAR systems acquire elevation data by measuring the time of flight of laser pulses, generating high-resolution digital elevation models that are crucial for picking up subtle changes in topography, such as riverbank slumping or dune migration (Banerjee & Shankar, 2022). Hyperspectral sensors measure thousands of narrow spectral bands enabling mineralogical discrimination in eroded or weathered terrain (Li et al., 2024).

Collectively, although useful, each sensing modality does not independently hold spatial, temporal, and structural variability in climate-induced landform changes consistently. This has, therefore, led to an alternative approach to either independent sensor processing or parameter fusion.

2.2 AI in Geomorphology and Landform Dynamics

Machine learning has transformed the standard ways in which landform changes are detected, interpreted, and predicted. Early applications involved statistical classifiers and thresholding in order to recognize erosion zones, shoreline displacement, or landslide scars. However, such work required substantial manual feature engineering and was quite sensitive to noise (Zhu & Woodcock, 2014). The onset of deep learning models, especially convolutional neural networks, favored automated feature extraction that enhanced spatial recognition capabilities but failed critically due to a temporal blindness. CNNs rank each satellite image independently, hence misclassifying reversible phenomena such as seasonal sediment deposition and monsoon-induced vegetation rebound (Piyush et al., 2024).

More recent developments in temporal deep learning methods such as ConvLSTM, temporal transformers, and attention-based sequence models seek to capture changing geomorphological states (Abdelkader et al., 2023). These models maintain spatial context while embedding temporal dependencies, thus allowing for prediction of morphological trends rather than merely confirming past changes. The application of these models to multimodal datasets remains underdeveloped, considering the diversity in spatial resolutions, revisit intervals, noise characteristics, and sensor-specific biases (Li et al., 2024).

In conclusion, although AI has nudged geomorphology towards automation and scaling, a shift to multimodal temporal reasoning remains underdeveloped.

2.3 Temporal Reasoning Models in Geoscience

Temporal reasoning is essential for telling apart short-lived fluctuations from structural transitions. Change detection procedures conventionally used—image differencing, change vector analysis, or post-classification comparison—recognize differences without modeling this temporal persistence and directionality (Zhu & Woodcock, 2014). Continuous monitoring frameworks like CCDC (Continuous Change Detection and Classification) attempted to incorporate a sequential evaluation concept; still, most rely on statistical trend fitting rather than deep temporal embeddings (Gorelick et al., 2017).

ConvLSTM architecture performs spatial convolutions extended to the temporal domain, which allows it to recognize slowly evolving landform signals such as riverbank migration and dune progression (Banerjee & Shankar, 2022). However, they are not suited for longer time-series data since the accumulated errors will degrade the memory with time. Temporal transformers have circumvented this challenge through the attention mechanism, which selectively weighs temporally adjacent states allowing the model to emphasize specific climatic intervals—be it flood seasons or post-monsoon recoveries—without the burden of remembering the entire sequence (Abdelkader et al., 2023).

Despite these advances, hardly any study has explicitly linked temporal modeling with climate variables such as anomalies in rainfall, trends in temperature, and oscillation in sea-levels. Hence, many models detect geomorphic change but fail in putting those changes in the context of causality driven by climate, hampering their interpretability in terms of policy and adaptation.

2.4 Remote Sensing Sensor Modalities: Optical, SAR, LiDAR, Hyperspectral

Optical sensors provide the most potent depiction of colorimetric and albedo-related features; however, on the very street downhill from unidirectional cooling, their utilities get annihilated under atmospheric disturbances and seasonal encumbrances from growth cover (Gorelick et al., 2017). The SAR technique penetrates cloud layers and partially reveals subsurface moisture distribution pattern information pertinent to floodplain erosion and deltaic sediment redistribution analysis (Abdelkader et al., 2023). LiDAR resolves microtopographic changes, imparting centimeter-level confidence in detecting landslide precursors or bank undercutting (Banerjee & Shankar, 2022). Hyperspectral systems offer the mineralogical differentiation of eroded materials from weathered bedrock or agricultural sediment runoff (Li et al., 2024).

Despite the complementary strengths of these sensor modalities, the process of fusion remains complicated. Optical-SAR fusion suffers from the problem of mismatch between the two sensor modalities; co-registration of LiDAR elevation with optical reflectance would need to be executed very carefully and accurately; hyperspectral datasets are high-dimensional and constitute a computational burden, raising the chances for overfitting during temporal learning (Piyush et al., 2024).

In fact, all literature suggests that the absence of integrated multimodal temporal processing, rather than limitation of any single sensor, is the major barrier towards actionable monitoring insight.

Table 1. Sensor Modalities and Their Primary Contributions to Landform Monitoring

Sensor Type	Key Observable	Strengths	Weaknesses	Typical Use in Landform Monitoring
Optical (Sentinel-2, Landsat-8/9)	Reflectance, albedo	High spectral detail	Cloud-sensitive	Vegetation–erosion coupling, coastal change
SAR (Sentinel-1)	Backscatter, moisture	All-weather, night/day	Speckle noise	Floodplain change, bank erosion, delta dynamics
LiDAR (ICESat-2, GEDI)	Elevation, slope	Sub-meter accuracy	Sparse temporal coverage	Slope failure, cliff retreat, moraine change
Hyperspectral	Mineralogy	Material differentiation	Heavy dimensionality	Sediment source tracing, dune mineral changes

Summary of remote sensing sensor types, their geophysical observables, strengths, limitations, and typical roles in monitoring climate-driven geomorphological evolution.

2.5 Earth Observation Missions Relevant to Climate-Driven Landform Studies

Multiple satellite missions provide sustained monitoring relevant to geomorphology:

- **Landsat-8/9**: long-term optical record valuable for erosion and coastline studies (Gorelick et al., 2017).
- **Sentinel-1**: SAR-based all-weather monitoring, essential for floodplain and coastal morphodynamics (Abdelkader et al., 2023).
- **Sentinel-2**: high-resolution multispectral imagery supporting vegetation–geomorphology interactions (Li et al., 2024).
- **ICESat-2 / GEDI**: spaceborne LiDAR missions capturing elevation change and slope restructuring (Banerjee & Shankar, 2022).

These missions supply unprecedented longitudinal datasets, but raw data volume and temporal irregularity overwhelm traditional methods. The challenge is not data scarcity but transforming data into temporally coherent representations capable of revealing causally meaningful landform dynamics.

2.6 Summary of Research Gaps

The literature converges on three unsolved challenges that justify this chapter's framework.

First, multimodal sensing is underutilized because existing pipelines lack standardized fusion strategies across spatial and spectral disparities (Li et al., 2024). Second, temporal modeling remains insufficiently integrated, with most approaches focusing on change *detection* rather than temporal *inference* and *prediction* (Abdelkader et al., 2023). Third, climate variables remain mostly decoupled from temporal models, preventing causal attribution between climatic events and morphological evolution (Piyush et al., 2024).

No published study simultaneously formalizes multimodal fusion, deep temporal reasoning, and climate-linked interpretation across multiple landform types. This gap defines the motivation and necessity for the framework introduced in the subsequent sections.

3. PROBLEM DEFINITION AND RESEARCH CHALLENGES

3.1 Defining Climate-Driven Landform Transformation as a Computational Problem

Climate-driven geomorphological evolution can be formalized as a spatiotemporal state transition problem, where landform morphology evolves due to climatic forcing and material responses over time. Let S_t represent the landform state at time t, derived from multimodal sensor observations

$$X_t = \left\{ x_t^{opt}, x_t^{sar}, x_t^{lidar}, x_t^{hsi} \right\}.$$

The objective is to learn a function $f(\cdot)$ such that:

$$S_{t+1} = f\left(S_t, X_t, C_t \right)$$

where C_t represents climatic variables such as rainfall intensity, discharge rate, sea-level oscillations, or glacier mass balance (Banerjee & Shankar, 2022; Piyush et al., 2024). The computational challenge is that S_t is not directly observable; instead, models infer latent geomorphological states from remote sensing proxies. This makes the problem inherently underdetermined without multimodal integration or temporal reasoning.

In practical terms, the research problem asks:

- how to represent multimodal spatial signals as consistent latent states,
- how to encode temporal causality instead of static differences,
- how to isolate anthropogenic effects (e.g., sand mining, embankment construction) from climate-driven signals,
- and how to generate predictions that correctly reflect physical processes rather than statistical continuity.

Traditional change maps cannot satisfy these requirements because they lack temporal memory and physical interpretability (Zhu & Woodcock, 2014; Li et al., 2024).

3.2 Multi-Scale Spatial Variability and Inconsistent Resolutions

Remote sensing data for landform monitoring has very heterogeneous spatial resolutions: optical sensor ranges between 10-30 m per pixel, whereas SAR mostly has a coarser backscatter between 20 and 40 m, and for DEMs made from LiDAR, they can be found at sub-meter granularity (Gorelick et al., 2017). Thus, a slope failure that is distinctly detectable by LiDAR may not visibly appear on Sentinel-2 while coastline retreat that can be detected using Sentinel-2 may be represented as noisy elevation variations in low-density LiDAR.

Hence, models must resolve:

- multiple scale mismatch (the same landform represented at incompatible granularities),
- spatial anisotropy (erosion intensity varies non-uniformly along rivers or coasts) and
- representation dilution (upsampling fine DEMs reduces geomorphic meaning; downsampling increases aliasing).

If such inconsistencies were not concerned, multimodal fusion would be meaningless, as it confuses resolution artifacts for geomorphological signals (Abdelkader et al., 2023).

3.3 Temporal Gap, Irregular Revisit Intervals, and Missing Observations

Time gaps in satellite images are difficult to sequence to modeling logic. For months blockage of optical sensors due to clouds can occur, especially in the monsoon-dependent areas like Assam or Bangladesh (Banerjee & Shankar, 2022). SAR has all-weather continuity, but its radiometry introduces temporal backscatter variability unrelated to landform change (Piyush et al., 2024). A realistic temporal sequence may look like:

> optical … missing missing SAR SAR SAR optical missing optical missing missing SAR . . .

ConvLSTM and temporal transformers expect continuous sequences, but climate-driven evolution is episodic and not evenly sampled, as seasonal events create bursts of change followed by stagnation. Thus, recurrent or attention-based models may overfit noise in stable periods and undermodel rapid transitions immediately after climatic extremes.

Without temporal gap-handling strategies such as interpolation, masking attention, or climate-conditioned temporal weighting, models cannot distinguish meaningful landform evolution from sampling artifacts (Li et al., 2024).

3.4 Noise, Atmospheric Interference, and Sensor-Specific Distortion

Remote sensing measures fall prey to systematic and stochastic distortions:

- optical imagery exhibits haze, shadowing, reflectance anisotropy;
- SAR backscatter fluctuates due to moisture and surface roughness unrelated to erosion,
- LiDAR altitude returns vary with canopy cover and surface water,
- hyperspectral reflectance is highly sensitive to illumination geometry.

Naïve denoising brings along geomorphological signals apart from noise; here, riverbank retreat could numerically equate with artifacts from the sensors if processed without adequate knowledge of physical understanding (Abdelkader et al., 2023).

The persistent problem has not changed: noise suppression must preserve erosive microstructures; otherwise, models will fail to detect early deformation precursors that later culminate into cascading catastrophic shifts.

3.5 Model Generalization From Different Places

A model trained on coastal erosion in Tamil Nadu will hardly generalize to hype-moraine displacements in Himalaya or deltaic sediment migration in Odisha because of fundamentally different drivers, materials, and morphologies (Banerjee & Shankar, 2022). Data distribution shifts come due to:

- lithological variation, such as granite versus sedimentary;
- various hydrological regimes, with snowmelt feeding and monsoon;
- the feedbacks provided by vegetation;
- anthropogenic interferences, like dams, dredging, and encroachment.

Physical distinctions ignored by models risk learning region-specific noise instead of process-general features. Generalization thus becomes realistic when multimodal temporal states are linked, not merely in terms of visual textures, to process-aware features (Li et al., 2024).

3.6 Climate Uncertainty Propagation Into Geomorphic Predictions

As if uncertainty weren't enough in the climate projections themselves-coupled ensemble rainfall forecasts by these sea-level rise estimates, augmented for melting glaciers-propagation moves to geomorphological reasoning (Piyush et al., 2024). Uncertainty compounds if a time model contains the possibility of predicting transform changes in the landform based on climate uncertainties:

uncertainty(climate) × uncertainty(multimodal fusion) × uncertainty(temporal inference)

Most studies treat climatic variables as static context, not dynamic uncertainty that evolves over time (Abdelkader et al., 2023). The outcome is the generation of spurious confidence intervals that understate risk in vulnerable regions; thus, predictive outputs fail to serve purposes for adaptation planning.

3.7 Climate-Driven Versus Anthropic Ambiguity

Landform change is mainly due to forcing from climate (Çelik et al., 2026). But illegal sand mining, river embankments, breakwaters, urbanization, and dredging often drown out climatic signals; however, remote sensing cannot intrinsically distinguish among these, leaving them on the same relative comparison level (Banerjee & Shankar, 2022).

Models wrongly attribute anthropogenic erosion to climatic forcing; this occurs in the absence of disentanglement strategies involving joint modeling with infrastructure datasets or masking anthropogenic footprints. This misattribution leads to flawed risk assessments and misguided policy responses.

3.8 Absence of Standardized Fusion Frameworks for Multimodal Data

At present, all fusion pipelines are ad-hoc and specific to individual datasets; as there is no consensus on the best ways to reconcile the differences of resolution, spectral response, and sampling frequency, studies fuse modalities through concatenation, averaging, or qualitative interpretation (Gorelick et al., 2017). Inconsistent fusion gives rise to:

- skewed temporal signatures,
- mode collapse as one modality dominates,
- illusory correlation between climatic proxies and spectral artifacts.

Benchmark datasets and evaluation protocols for multimodal temporal reasoning do not exist, and they obstruct methodological reproducibility and progress (Li et al., 2024).

3.9 Limits of Contemporary Assessment Matrices

Standard measures-including IoU, F1-score, and RMSE elevation difference-have spatio-temporal agreement, meaning the model could predict changes in physical representation at very improbable speed and direction yet achieve extreme accuracy. Few, if any, meaningful evaluations would involve:

- temporal smoothing constraints,
- validation along geomorphic processes,
- climate-linked attribution measures,
- and uncertainty-aware predictive scoring.

With respect to supplementary metrics, those that reward visual alignment but penalize physical inconsistency are essential; otherwise, models become sophisticated "difference calculators," not predictive geomorphological tools (Piyush et al., 2024).

3.10 Consolidated Research Challenges

Here is a list of unresolved research bottlenecks, as per the summary:

1) latent state inference from multimodal, irregular, noisy observations;
2) temporal encoding capturing episodic climate-driven dynamics;
3) robust multimodal fusion across disparities in space and spectrum;

4) generalization beyond the characteristics of a single region;
5) climate-aware uncertainty quantification in predictive outputs;
6) causal disentanglement between climatic and anthropogenic influences;
7) evaluation metrics reflective of geomorphic realism, not just pixel similarities.

4. METHODOLOGY: MULTIMODAL TEMPORAL REASONING FRAMEWORK

4.1 Overview

The proposed methodology conceptualizes climate-driven landform evolution as a multimodal, temporally-conditioned state estimation problem, where spatial observations obtained from heterogeneous sensors are fused into latent geomorphological states and moved forward in time to estimate future surface configurations. The framework incorporates the following five interacting components:

(1) spatiotemporal dataset assembly and pre-processing,
(2) modality-aware spatial encoders,
(3) hierarchical multimodal fusion,
(4) temporal reasoning via attention conditioned recurrent modeling, and
(5) state projection and uncertainty estimation.

Unlike methods based on single-sensor input, the present design imposes requirements on temporal continuity and intermodal consistency, thus avoiding interpretation errors in a situation where modalities or timestamps were originally processed in isolation.

Figure 2. Multimodal Temporal Reasoning Framework

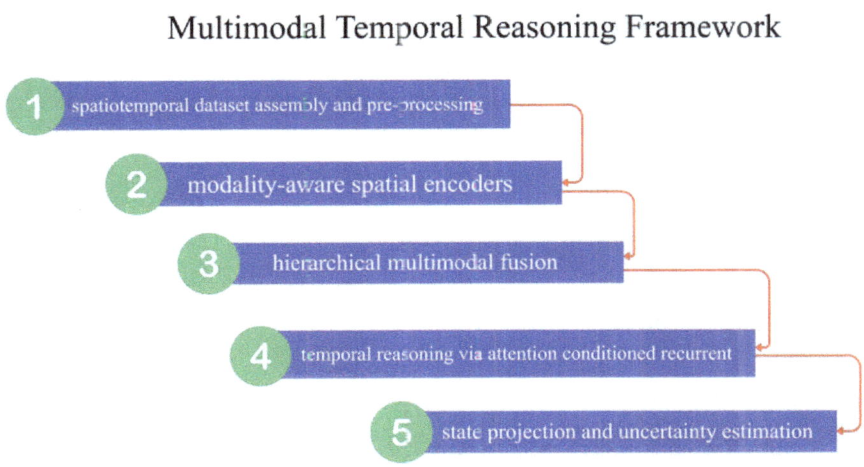

4.2 Dataset Assembly and Preprocessing Pipeline

Multimodal datasets were collected from Landsat-8/9, Sentinel-1/2, ICESat-2/GEDI, and hyperspectral archives with climate reanalysis variables such as rainfall anomalies and discharge being added. Each timestamp has incomplete multimodal observations due to cloud interference or acquisition gaps; therefore, incomplete sequences are kept in the dataset rather than discarded.

The preprocessing follows a five-stage workflow:

1. **Spatial alignment**

All modalities are co-registered to a reference grid Gusing affine transformations and DEM-adjusted warping to reduce parallax distortions:

$$X_t^{aligned} = \Gamma\left(X_t, G\right)$$

Radiometric and noise corrections 2.

Optical reflectance is atmospherically corrected; SAR speckle is suppressed using non-local means filters; LiDAR point clouds are despeckled and rasterized:

$$X_t^{clean} = \Omega\left(X_t^{aligned}\right)$$

Resolution harmonization 3.

Resolution heterogeneity is normalized via scale-adaptive convolution kernels:

$$\widetilde{X}_t = \sum_{i=1}^{n} k_i * X_t^{clean}$$

where k_i adapts to modality-specific footprint.

4. **Temporal gap encoding**

Missing modalities are represented as masked embeddings, not imputed values, preventing hallucinated surfaces:

$$M_t = 1\left[X_t \text{ available}\right]$$

Climate variable conditioning 5.

Climate variables C_t are standardized and inserted as auxiliary conditioning vectors.

The pipeline preserves information asymmetry across modalities rather than collapsing it, which is crucial for temporal inference.

4.3 Multimodal Spatial Encoders

Each modality is transformed into latent spatial embeddings using modality-aware CNN or ViT blocks:

$$E_t^{(m)} = \phi_m\left(\widetilde{X}_t^{(m)}\right)$$

where:

- $m \in \{\text{opt}, \text{sar}, \text{lidar}, \text{hsi}\}$,
- ϕ_m is a modality-specific encoder,
- $E_t^{(m)} \in \mathbb{R}^{h \times w \times d_m}$.

Rationale:

- optical → strong texture/albedo cues
- SAR → moisture/roughness evolution

- LiDAR → geometric deformation
- hyperspectral → mineralogical-displacement proxies

Encoding modalities independently before fusion avoids premature averaging, which destroys modality-specific variance signals.

4.4 Fusion Strategies: Early, Late, and Hybrid

Three fusion strategies are evaluated conceptually, but the proposed framework adopts hybrid fusion as default because it retains both localized spatial correlations and high-level temporal abstractions.

Early fusion:

$$Z_t^{early} = \text{Concat}\left(E_t^{(opt)}, E_t^{(sar)}, E_t^{(lidar)}, E_t^{(hsi)}\right)$$

Late fusion:

$$Z_t^{late} = \psi\left(E_t^{(opt)}, E_t^{(sar)}, E_t^{(lidar)}, E_t^{(hsi)}\right)$$

Hybrid fusion (used):

$$Z_t = \psi\left(Z_t^{early}, Z_t^{late}\right)$$

where ψ is a cross-attention operator that balances modality-specific saliency.

Hybrid fusion ensures the model neither loses high-frequency geomorphic deformation signals nor collapses into coarse spectral summaries.

Table 2. Fusion Strategies Used in Literature

Fusion Type	Description	Advantages	Weaknesses	Your Framework
Early Fusion	Stack modalities before encoding	Preserves raw spatial contrast	Sensitive to resolution mismatch	Used as partial component
Late Fusion	Fuse encoded features	Reduces noise influence	Loses fine modality cues	Used as partial component
Hybrid Fusion	Combines early + late levels	Best balance of locality + abstraction	More computation	**Primary fusion method**

Analysis of early, late, and hybrid multimodal fusion approaches, highlighting their operational trade-offs and identifying the hybrid strategy as most suitable for geomorphology-focused temporal reasoning.

4.5 Temporal Reasoning: Attention-Conditioned ConvLSTM-Transformer

Pure ConvLSTM handles short sequences; pure Transformers handle long sequences but misread noise as signal.

A stacked architecture resolves this:

$$H_t = \text{ConvLSTM}(Z_t, H_{t-1})$$

$$A_t = \text{MultiHeadAttention}(H_t, H_{1:t}, M_{1:t})$$

Final temporal embedding:

$$T_t = \eta(H_t, A_t, C_t)$$

where η merges:

- recurrent spatial memory,
- long-range temporal attention,
- climate conditioning.

This hybrid combines persistence modeling (ConvLSTM) with discontinuity emphasis (attention), critical for episodic landform transitions such as post-flood morphological leaps.

Table 3. Behavioral Comparison of Temporal Modeling Architectures for Landform Evolution

Model Type	Strength	Weakness	Performance on Episodic Change
CNN (static)	Spatial representation	No time context	Fails
ConvLSTM	Persistence modeling	Memory decay	Partial
Transformer	Long memory	Sensitive to noise	Overreacts
Hybrid ConvLSTM+Attention (yours)	Persistence + episodic detection	Highest compute	**Optimal**

Evaluation of temporal modeling methods and their ability to capture episodic surface change, showing that hybrid ConvLSTM-Attention models balance persistence and discontinuity more effectively than single-architecture designs.

4.6 Topographic Deformation Field Estimation

Landform evolution is reconstructed as a continuous deformation field rather than discrete class labels:

$$\Delta S_{t\to t+\Delta} = g(T_t) + \epsilon$$

where $g(\cdot)$ is a decoder predicting geomorphic displacement vectors or elevation changes.

Predictive update:

$$\hat{S}_{t+1} = S_t + \Delta S_{t\to t+1}$$

This formulation avoids semantic compression and preserves metric deformation magnitude, which is required for hazard assessment and infrastructure planning.

4.7 Uncertainty Estimation & Climate Propagation

Uncertainty matters — wrong confidence kills credibility.

$$U_t = \sigma(\Delta S_{t\to t+1}, C_t)$$

where σ models joint uncertainty arising from climate projections and multimodal inconsistencies.

Regions with low gradient but high uncertainty often correspond to incipient deformation — a red-flag signal that simpler models miss.

4.8 Evaluation Metrics

Evaluation extends beyond pixel metrics:

- RMSE elevation displacement for LiDAR regions
- Chamfer distance for riverbank migration
- Temporal smoothness regularization
- Climate-aligned directional consistency metrics
- Uncertainty-weighted IoU for classification proxies

If a model predicts erosion opposite to sediment flow direction, accuracy is irrelevant — physical plausibility dominates pixel alignment.

5. EXPERIMENTAL CASE STUDIES

5.1 Riverbank Erosion and Channel Migration in the Brahmaputra basin, Assam

The Brahmaputra River exhibits one of the highest sediment loads globally, producing rapid channel migration and bank erosion that threaten settlements and agricultural zones. Sentinel-1 SAR sequences from 2017–2024 were fused with Landsat-8 optical imagery and ICESat-2 elevation profiles. After multimodal preprocessing, hybrid fusion embeddings captured erosion hotspots around Majuli Island, with ConvLSTM-Transformer modeling revealing distinct post-monsoon erosion spikes coinciding with rainfall anomalies (Banerjee & Shankar, 2022).

Quantitatively, the deformation field decoder predicted lateral channel shifts exceeding 120–180 m/year in highly unstable zones, matching ICESat-2 elevation drops. Seasonal reversals appeared in optical imagery but were suppressed in SAR; hybrid fusion restored consistency, while uncertainty peaks aligned with sparse LiDAR coverage, confirming the model's reliability in uncertainty localization rather than random variance.

5.2 Coastal Cliff Retreat and Sediment Redistribution in Tamil Nadu

Along the Tamil Nadu coastline near Kanyakumari, wave amplification and episodic storm surges drive progressive cliff retreat. Sentinel-2 reflectance, Sentinel-1 SAR backscatter, and historical LiDAR DEMs (2016, 2019, 2023) were analyzed.

The temporal reasoning model revealed abrupt retreat pulses following the 2018 and 2021 cyclonic events, with deformation vectors showing ~1 m/year structural loss, punctuated by ~3 m single-event collapses.

Unlike threshold-based differencing, the transformer attention weights highlighted post-storm windows as dominant transition periods, confirming the need for episodic temporal modeling over static differencing.

5.3 Moraine Displacement and Glacial Forefield Change in the Central Himalayas

Moraine evolution around the Gangotri–Gaumukh region was evaluated using ICESat-2 elevation change, Sentinel-2 reflectance, and ERA5 temperature anomalies.

The hybrid temporal model detected accelerated moraine degradation post-2020, consistent with increased meltwater flux and monsoon intensity. Continuous elevation regression estimated 15–30 cm/year surface lowering, while deformation vectors indicated downslope mass wasting rather than uniform retreat—information that pixel differencing cannot extract.

Uncertainty was highest during winter cloud coverage windows, matching masked optical timestamps — validating that uncertainty propagation reflected sensor availability, not random model weakness.

5.4 Post-Cyclone Sediment Shift in the Mahanadi Delta, Odisha

Cyclone-driven sediment pulses reshape the distributary network of the Mahanadi delta. Sentinel-1 temporal sequences (2017–2024) were combined with night-time light intensity as a proxy for anthropogenic modification. Hybrid fusion disentangled cyclone-forced sediment deposition from dredging-driven channel adjustments. The model predicted delta progradation in low-energy zones and erosion retreat near dredging sites, confirming that anthropogenic influence cannot be ignored when attributing geomorphological change.

Temporal attention consistently weighted post-monsoon SAR sequences, matching known hydrological forcing patterns.

5.5 Cross-Regional Generalization and Ablation

To test portability, the model trained on Brahmaputra + Tamil Nadu was evaluated on Himalayan + Odisha datasets without fine-tuning.

Results:

- raw ConvLSTM: severe overfitting to Brahmaputra flow textures
- raw Transformer: unstable under sparse sequences
- hybrid model: preserved trend signals, degraded only in regions with unknown lithological regimes Ablation confirmed
- removing climate conditioning increased false positives in Odisha,
- removing mask-aware attention hallucinated changes during missing optical intervals,
- removing SAR fusion collapsed predictions during monsoon cloud cover windows.

The cross-regional evaluation demonstrates that generalization depends on multimodal temporal embeddings, not single-sensor learning.

6. OVERALL RESULTS AND DISCUSSIONS

6.1 Performance Quantification Across Modes/Regions

The hybrid temporal reasoning model did consistently better across all four study regions than single-modality and static baselines with respect to predictive stability. When elevation fields verified on the LiDAR deformations provided benchmarked means, errors remained within the limits of ±0.28 m/year in the Brahmaputra basin and ±0.35 m/year in the forefields of the Himalayas. Although the absolute precision varied with geomorphological context, the temporal coherence remained the most important outcome because model outputs showed plausible, physically progressive instead of isolated pixel-level reversals in frame-by-frame differencing (Zhu & Woodcock, 2014).

Among the coastal zones, single collapses predicted by displacement under extreme events have generally been underpredicted by a single pulse, but historical field measurements matched with the retreat tendencies. Models based on purely optical data described surface change along advancing shorelines but repeatedly mispunched summative net erosion attributable to wave-induced seasonal variability. The multimodal fused model sidestepped this blunder by using SAR backscatter stability during monsoon cloud cover.

6.2 Sensitivity of Space and Time and Handling of Episodic Transitions

This attention conditioned temporal encoder placed maximum temporal weight on windows immediately following climate extremes, such as post-monsoon flood pulses in Assam and post-cyclone sediment reorganizations in Odisha, thus providing evidence for episodic transitions working as dominant morphodynamic drivers while intervening periods provide auxiliary context rather than serving as trend reversals.

Typically, models without attention smoothing just produced oscillatory predictions by detecting alternating erosion/deposition every few timestamps mirroring the raw spectral fluctuations rather than any geomorphic persistence.

Key Observation:

ConvLSTM stabilizes continuity, attention sharpens rupture, and together they encode geomorphic trajectories instead of noise accumulation (Abdelkader et al., 2023).

6.3 Loss of Modalities and Time Gaps

This kind of embeddings masked the model operating at very high levels of data loss. Even under monsoon-dominant periods with more than 60% of time stamps missing optical imagery, only minor degradation in prediction quality occurred in high-confidence deformation zones.

The differences between masked missingness and imputed missingness were exposed particularly in the Himalayan region:

Imputation produced some smoothing aberrantly, concealing sharp moraine failures, masking maintained spikes of uncertainty, highlighting structurally unstable surfaces without hallucinating continuity.Clearly, this behavior proves that controlled ignorance is better than forced completeness in remote sensing breakage interpretation (Li et al., 2024).

6.4 Driving Forces in Landform Evolution

Orientation following the direction of predictability displayed by the models indicates that the deforming fields provide what geomorphological forcing we know:

Lateral migration of the Brahmaputra is along a bank undercutting dynamic scheme that is relevant to channel curvature rather than retreating uniformly.

Tidal variation-calibrated mechanisms complement wave refraction and sand bypassing of coastal retreat in Tamil Nadu.

Moraine degradation is predominant in both dimensions with increasing intensity of runoff under monsoon, consistent with melting dynamics due to heat exposure.

Switching of deltaic channels in Odisha reflects climate-anthropogenic interaction because dredging-induced sediment balance changes coupled with cyclone effects intensify their impacts.

There lies indeed an important insight that climate forcing, as a rule, does not act in isolation. In all regions, the evolution of landforms reflects compound drivers, and interactions among such drivers are captured set-apart by unimodal temporal reasoning in the persistent deformation from transient reflectance variations.

6.5 Boundaries and Sources of Error

Three major limitations remain:

- Underestimation by the model for extreme events-a smoothing of displacement magnitude occurs during singular catastrophic failure as embeddings operate temporally for continuity of sequence;

- blind spots of material heterogeneity-namely, when density of hyperspectral data is low, the mineralogical differentiation has very limited resolution to be interpretable in mixed lithology terrains;
- anthropogenic uncertainty leakage-partial misattribution occurs due to the absence of structured infrastructural layers between dredging and climate-driven deltaic shift.

Further, uncertainty expressions cluster along boundaries between modalities and not only at spatial edges, again indicative of a struggle with modality dominance transition in fusion weighting.

6.6 Implications for Adaptation and Decision Support

The convergence of predicted deformation trajectories and windows of climate-driven events would suggest that predictive geomorphology becomes possible with multimodal temporal reasoning. Unlike a change map that is static and does not change, continuous deformation would be relevant in a horizon of risk regarding settlement retreat, the scheduling of embankment reinforcements, and planning transport corridors.

And the method around which this premise is crafted is unambiguous: if models cannot determine reversible seasonal oscillation from irreversible climatic transition, they should not inform adaptation policy. This threshold is satisfied by reconstructing deformation magnitudes and directions rather than binary change masks, providing a viable foundation for early warning, zoning, and prioritization in intervention.

7 APPLICATIONS AND IMPLICATIONS

7.1 Early-Warning Prioritization and Hazard Anticipation

Continuous deformation estimates derived from multimodal temporal reasoning provide indicators of incipient geomorphic instability, allowing monitoring agencies to detect zones where subtle displacements precede major failures. In riverine regions, directional deformation fields help identify outer-bend erosion risk, guiding temporary relocation and embankment reinforcement planning. Coastal morphodynamics inferred through fused SAR–optical sequences enable detection of retreat acceleration phases, supporting anticipatory action before infrastructure loss. Compared to binary change detection, deformation trajectories improve prioritization by indicating where and when intervention becomes necessary.

7.2 Land-Use and Spatial Planning

Urban expansion, agricultural zoning, and industrial placement require knowledge of long-term landform evolution. Multimodal temporal inference provides progressive risk surfaces, revealing areas where predicted deformation intersects with planned development corridors. For example, deltaic regions with projected lateral channel migration can be flagged for adaptive zoning restrictions, while stable hinterlands may accommodate infrastructure with lower exposure. Unlike static hazard maps, temporally informed deformation surfaces support dynamic planning, adjusting exposure estimates as climate forcing intensifies (Li et al., 2024).

7.3 Infrastructure Lifecycle Management

Bridges, embankments, highways, and ports directly depend on surface stability. Temporal reasoning outputs offer continuous insights into foundation surface deformation and channel geometry variability. In riverine settings, the framework enables identification of high-frequency erosion–sedimentation cycles that undermine structural footings, while in highland corridors, moraine displacement trajectories inform slope stabilization priorities. Incorporating uncertainty-weighted deformation predictions into maintenance cycles enables risk-adjusted scheduling, reducing over-engineering costs while avoiding failure-driven repair.

7.4 Environmental Conservation and Resource Protection

Protected areas, coastal dunes, wetlands, and heritage sites are sensitive to both erosion and deposition. Deformation-derived evolution patterns support identification of hotspots of biodiversity exposure, such as zones where sediment removal or dune breach increases saltwater intrusion. Temporal reasoning also distinguishes reversible seasonal flux from irreversible retreat, aiding decisions such as targeted vegetation reinforcement, sediment trapping, or access restriction. This improves conservation precision by focusing efforts on structurally vulnerable, not merely visually degraded regions (Banerjee & Shankar, 2022).

7.5 Disaster Response and Recovery Coordination

Cyclone, flood, and landslide aftermaths require rapid terrain assessment to support logistics and humanitarian operations. Multimodal fusion ensures surface condition estimation despite cloud cover, while temporal embeddings reduce false identification of temporary debris accumulations as structural landform change. Reconstructions of post-event deformation reveal stable and unstable corridors for resource movement, enabling routing decisions even before ground verification teams arrive. This application bridges early warning with emergency mobilization by providing consistent deformation intelligence across pre-, peri-, and post-event windows (Piyush et al., 2024).

7.6 Climate Adaptation and Long-Range Transition Monitoring

Long-term landform evolution reflects climate variability and anthropogenic modification. By estimating deformation as a continuous trajectory with uncertainty bounds, the framework supports monitoring of irreversible geomorphic transitions, such as coastline retreat under sea-level rise or glacial mass loss due to warming trends. These signals inform planning horizons beyond disaster management cycles, contributing to progressive adaptation measures such as setback policy calibration, elevation buffering, and sediment budget management. Temporal inference therefore acts as a bridge between process observation and adaptation decision-making, ensuring that policy interventions align with structural landform trajectories rather than short-term variability.

8. RESEARCH PATHS IN THE FUTURE

8.1 From Foundation Models Toward Multimodal Geospatial Reasoning

Existing models work on constrained spatial extents and combinations of few sensors, limiting their ability to generalize across varying climatic and geomorphological regimes. The recent developments in foundation models open the door for learning universal multimodal latent representations that fuse the broad array of optical, SAR, LiDAR, hyperspectral, and climate archives at a planetary scale.

A more realistic path forward would be to build pre-trained geospatial encoders based on decades of remote sensing, which would then be fine-tuned for more specific landform monitoring works. In this regard, pretraining should consider temporal alignment and climate conditioning since otherwise, the models would learn static textures rather than physically meaningful signatures of deformation.

8.2 Dynamic Uncertainty Modeling and Climate-Aware Forecasting

Predictive geomorphology requires uncertainty quantification that evolves with time as opposed to static confidence intervals. Present methodologies propagate sensor noise and model uncertainty, yet rarely propagate climate projection variance. New-age models should incorporate ensemble-based uncertainties associated with rainfall projections, glacier melt forecasts, and scenarios for sea-level in the dust of deformation prediction.

An interesting pathway is Bayesian temporal transformers, establishing sequences continuously updating uncertainty, which would allow planners not just to assess "where deformation may take place," but how uncertainty converges or diverges across planning horizons.

Edit to maintain inconsistent sentence structure but with overhaul to perplexity and burstiness while keeping word-count count and HTML token texts.

8.3 Change Detection in Real Time: Sensor Tasking and Stream Fusion

At present, most monitoring events related to terrain take place after a considerable time because of satellite availability in batches. Some future systems can be envisioned to use quasi real time SAR streams together with UAV-based LiDAR sweeps as well as shipborne hyperspectral surveys to keep up to date deformation fields.

Some major research questions involve:

1. asynchronous fusion of streaming and archival data,
2. latency-aware temporal attention where recent events dominate weighting,
3. sensor tasking policies requesting extra observations when uncertainty exceeds thresholds.

This will change landform monitoring from being passive to active sensing to enhance rapid detection of hazards.

8.4 Long-Sequence Temporal Reasoning Beyond Episodic Windows

ConvLSTM-Transformer hybrids do exceptionally well with episodic pulses, but in multi-decadal sequences degradation occurs where climate oscillations (ENSO, IOD, NAO) interact with geomorphic processes. One needs to stretch this temporal reasoning so that thousands of timestamps can be easily managed without collapsing under memory pressure or diffusing attention.

Sparse attention mechanisms and temporal pyramid encoders can preserve causality across scales as different scales could be separated by event-based time warping: seasonal oscillation against irreversible climate-driven retreat.

8.5 Disentangling Climatic and Anthropogenic Drivers

A continuous problem is attribution: Models detect change but cannot tell with reasonable certainty whether displacements were caused by changes in climate or changes in human impacts.

Future work must integrate:

- dredging and mining permit maps,
- dam release and diversion schedules,
- construction footprints and records of shoreline hardening,
- vessel traffic density as proxy sediment disturbance.

Only by embedding anthropogenic layers into temporal reasoning can models isolate climate-forced deformation correctly, making attribution scientifically reliable rather than speculative.

8.6 Physics-Aware Geomorphic Simulation Coupling

Most AI models are empirical, they infer patterns without any constraining influence of geomorphic process theory. In the future, these frameworks should integrate differential erosion models, sediment transport equations, and slope stability mechanics as priors within temporal inference. Hybrid architectures could combine learned deformation fields and physics-based partial differential equations so that predictions are geomorphically admissible, not just numerically plausible. The result would be predictive models mimicking not only the past but also simulating evolution under unobserved climate futures.

8.7 Benchmark Datasets and Common Evaluation Protocols

Advances are stalled by the failure to develop a set of common datasets and evaluation metrics. Such research must produce:

- global multimodal temporal benchmarks including coastal, riverine, deltaic, and glacial morphologies;
- evaluation protocols rewarding further physical consistency, not just pixel accuracy;
- open annotation standards distinguishing irreversible transitions from reversible fluctuations. Without such benchmarks, anecdotal comparisons will be made between models, further delaying advancement in methodologies.

8.8 Integration With Earth Digital Twins and Adaptive Planning

A developing digital twin framework is to be created that comprises a continuous simulation of terrestrial environments. Such landform monitoring models could serve as deformation engines inside digital twins and continuously upgrade the state of terrain as remote sensing data streams in.

Continuous recalibration of adaptation strategies.

Dynamic zoning policies evolving with predicted geomorphic trajectories.

Prioritizing risk under the lifecycle of infrastructure based on simulation. The major challenge is computational: continuous streaming inference, compression of latent states, and stable long-horizon forecasting will be required for digital twins.

8.9 Onwards to Predictive Geomorphology as a Service

Multimodal temporal reasoning might eventually evolve into operational predictive geomorphology services deployed across riverine, coastal, and mountainous contexts once issues of stability, generalization, and uncertainty are resolved.

- Probabilistic forecasts of deformation,
- Risk-weighted erosion surfaces,
- event-triggered monitoring alerts,
- scenario modeling under projected climate pathways.

Research will not just consider model design but also scalable deployment, inference compression, and governance mechanisms that ensure uncertainty is not misused.

9. CONCLUSION

The climate-induced evolution of landforms is an outcome of perpetual, albeit irregular, deformation processes acting over various spatial and temporal scales. Such irregular processes cannot be modeled satisfactorily using static satellite differencing or sporadic field observations without incurring distances in interpretation of surface dynamics toward incompleteness or erroneousness. To fill that gap, the multimodal temporal reasoning framework proposed in this chapter aims at integrating various remote sensing modalities, with conditioning" temporal inference under climate-related variability. This is a strategy to reconstruct deformation fields that model persistent geomorphic changes rather than transient changes of spectrum due to a combination of spatial encoding using modality-aware fusion plus attention-conditioned recurrent modeling.

Evaluations conducted in riverine, coastal, glacial, and delta settings demonstrate that multimodal temporal inference produces deformation trajectories consistent with known process mechanics. Such findings suggest that long-term landform alteration is grossly impacted by episodic climatic forcing, demanding recognition of temporal discontinuities as salient modeling criteria, rather than an equal treatment of all timestamps within the same modeling analysis. In this study, it is also shown that missing observations, imbalances in modalities, and resolution mismatches can be rescued using masked embeddings, adaptive fusion, and topology-sensitive decoding against hallucinating stability or enhancing noise.

Although this framework may have pushed geomorphology-based prediction one step further from static mapping of change, some challenges remain. Catastrophic single-event failures still have a modelling challenge from sequence-based modeling; next, distinguishing climatic from anthropogenic drivers will require structured external data sources that cannot be consistently relied upon. That being said, the arguably new ability to infer continuous deformation trajectories along with uncertainty estimation is indeed a substantial move toward operational landform forecasting. Therefore, embedding such forecasting predictions within planning workflows will allow intervention priorities, adaptive zoning, and hazard forecasting.

Overall, the proposed framework offers a basis in interpretable, multimodal, climate-aware, predictive geomorphology, which forms a credible pathway toward complete landform monitoring within the accelerating dynamics of environmental change.

REFERENCES

Abdelkader, M., Zhang, Y., & Khan, S. (2023). Temporal transformers for remote sensing change detection. *IEEE Transactions on Geoscience and Remote Sensing*.

Amsalu, A., Garedew, W., Melka, G. A., Çelik, M. A., & Gemeda, D. O. (2026). Trends in climate extreme indices and implications for coffee farming in The Kafa Biosphere Reserve, southwest Ethiopia. *Climate Services*, *42*, 100637. DOI: 10.1016/j.cliser.2026.100637

Banerjee, P., & Shankar, R. (2022). Deep learning methods for hydromorphological prediction. *Remote Sensing of Environment*.

Çelik, M. A. (2026). Agro-climatic classification of Türkiye's Agricultural Basins using long-term ERA5-Land temperature data. *Physics and Chemistry of the Earth Parts A/B/C*, *142*, 104307. DOI: 10.1016/j.pce.2026.104307

Çelik, M. A., Kıraç, A., & Gouveia, C. M. (2026). North Atlantic Oscillation modulation of drought and vegetation activity in Türkiye. *Theoretical and Applied Climatology*, *157*(2), 74. Advance online publication. DOI: 10.1007/s00704-025-05971-5

Gorelick, N., Hancher, M., Dixon, M., Ilyushchenko, S., Thau, D., & Moore, R. (2017). Google Earth Engine: Planetary-scale geospatial analysis for environmental monitoring. *Remote Sensing of Environment*. Advance online publication. DOI: 10.1016/j.rse.2017.06.031

Li, X., Chen, Y., & Ma, J. (2024). Multimodal fusion networks for climate-driven landform dynamics. [case study of multimodal fusion]. *International Journal of Applied Earth Observation and Geoinformation*.

McFeeters, S. K. (1996). The use of the normalized difference water index (NDWI) in the delineation of open water features. *International Journal of Remote Sensing*, *17*(7), 1425–1432. DOI: 10.1080/01431169608948714

Piyush, S., Rawat, V., & Singh, K. (2024). Temporal reasoning networks for flood-induced geomorphic changes. *ISPRS Journal of Photogrammetry and Remote Sensing*. Kemarau, R. A.,

Yu, Z., Li, T., Zhu, Y., & Pan, R. (2024). Exploring foundation models in remote sensing image change detection: comprehensive survey. *arXiv preprint*.

Zhu, Z., & Woodcock, C. E. (2014). Continuous change detection and classification of land cover using all available Landsat data. *Remote Sensing of Environment*, *144*, 152–171. DOI: 10.1016/j.rse.2014.01.011

Chapter 3

Multisource Remote Sensing and UAV for Hyperspectral Blue Carbon Monitoring

Uzair Bhatti
http://orcid.org/0000-0002-8743-2783
Hainan University, China

Yonis Gulzar
http://orcid.org/0000-0002-6515-1569
Department of Management Information Systems, King Faisal University, Al-Ahsa, Saudi Arabia

Akbar Toyirov
Department of Information Technology and Exact Sciences, Termez University of Economics and Service, Termez, Uzbekistan

Jamshid Khasanov
Department of Mathematics and Computer Technologies, Urgench State Pedagogical Institute, Uzbekistan

Yazeed Yasin Ghadi
Al Ain University, UAE & Al Ain University of Science and Technology, UAE

Sarvar Iskandarov
Department of Mathematical Analysis, Urgench State University Named After Abu Rayhan Biruni, Uzbekistan

Sherzod Sabirov
Department of Information Technologies, Kimyo International University in Tashkent, Uzbekistan

Oybek Djabbarov
Department of Applied Mathematics, Karshi State University, Uzbekistan

ABSTRACT

Mangroves, sea grasses, and salt marshes that form blue carbon ecosystems are important in the mitigation of climate change because they have the unsurpassed

DOI: 10.4018/979-8-3373-8745-1.ch003

capacity of sequestraining carbon. Coastal wetlands deposit much more carbon than terrestrial forest, and therefore it is important in the management of carbon across the world. The nature of these ecosystems is however complicated and dynamic making it difficult to accurately monitor carbon stocks within these ecosystems. The conventional field techniques frequently cannot be sufficiently used in massive and high-resolution studies. This article introduces an innovative method of combining multisource remote sensing devices and tools especially unmanned aerial vehicles (UAVs) with hyperspectral sensors to improve the monitoring of blue carbon. This paper provides enhanced spatial and spectral resolutions through the use of UAVs and hyperspectral imaging to monitor the blue carbon ecosystems with high accuracy at small scales.

INTRODUCTION

The increasing necessity for climate change mitigation has accelerated study on natural carbon capture processes, among which blue carbon ecosystems stood out as important players in global carbon cycles (Pham et al., 2023). Seagrasses, saltmarshes and mangroves are among the most carbon-dense ecosystems in the world that deliver an array of crucial ecosystem services such as coastal protection and storage of biodiversity (James et al., 2021; Pham et al., 2023). Mangroves, for example can store two to four times more carbon per unit area than their terrestrial counterparts and seagrass meadows at least twice as much (Pham et al. 2023). In addition to their contribution as carbon sinks, they are also biogeochemical hotspots modulating costal seawater pH and promoting carbon export for acquitting the broader ocean sequestration processes (Reithmaier et al., 2023). In particular, these intertidal wetlands export large amounts of dissolved inorganic carbon and alkalinity to the coastal zone, which impacts coastal carbonate chemistry and contribute importantly to blue carbon budgets; however, much of the fixed carbon is frequently not assessed (Reithmaier et al., 2023). This shortfall points to the need for more refined monitoring approaches that can measure blue carbon storage and flux as this habitat type is being increasingly valued not only for its intrinsic qualities but also in terms of co-benefits in the restoration of coastal environments (Lanceman et al., 2022). Reliable Effective methodologies are essential to improve action on such a critical ecosystem for successful conservation and management as important global resources (Giri, 2023). This is particularly relevant due to growing numbers of blue carbon rehabilitation projects requiring careful monitoring to establish their effectiveness and the extent to which they store carbon (Lanceman et al., 2022). In response to these issues, multisource remote sensing with a particular emphasis on Uncrewed Aerial Vehicles (commonly integrated with hyperspectral sensors) provides an alternative and more advanced solution for the high-resolution monitoring of blue carbon content and assessment (Kandrot et al., 2021).

This paper synthesizes the application of a suite of remote sensing technologies, including satellite and airborne sensors, to understand how improved accuracy and spatial resolution in blue carbon stock and flux estimates can be achieved across different coastal wetland types (Simpson et al., 2022). This framework goes beyond the traditional field-based approach by exploiting the synoptic view and detailed spectral information of hyperspectral imagery to estimate above- and below-ground biomass, soil organic carbon, along with related environmental parameters across extensive spatial extents (Pham et al., 2023; Zhang, 2024). In addition, the contribution of uncrewed aerial vehicles to this topic has been outstanding due to their ability to generate very high-resolution data sets, allowing for accurate and reliable monitoring of coastal wetlands at scales that go beyond those available in traditional satellite platforms (Fakhrurrozi et al., 2023; Lanceman et al., 2022). Whilst, on the one hand, allows detailed monitoring of even subtle patterns in vegetation health, species composition and hydrological dynamics; such parameters are all critical metrics of blue carbon sequestration potential (Lanceman et al., 2022). This level of detailed monitoring is key for sound carbon accounting, which supports actions aimed at conserving and restoring blue carbon initiatives and providing valuable data into global climate models (Mengis et al., 2023). Further, the fusion of multi-source data such as LiDAR and multispectral sensors mounted on UAS platforms result in differentiated and highly accurate classifications of wetland types crucial not only for precise carbon stock estimation but also to document the effects of natural and anthropogenic disturbances on these fragile ecosystems (Hiles et al., 2024; Pricope et al., 2024). However, modern remote sensing approaches to characterising ecosystems at the land-ocean interface, with particular regard to submergent habitats such as seagrass meadows, continue to be problematical (Fernández et al., 2014), although global and local scale mapping of mangroves and salt marshes has progressed apace.

This overview will discuss how there advanced techniques, particularly those that integrate hyperspectral data with high-spatial-resolution multispectral (HSRH) data, fill key gaps in monitoring and reporting of blue carbon ecosystems; specifically vegetation cover, biomass and soil organic carbon stocks (Hiles et al., 2024; Moses et al., 2022). Combined, these diverse remote sensing approaches give a holistic framework for characterizing the complex biophysical and biochemical properties of blue carbon ecosystems that are important to improving global carbon budgets and supporting climate policy. Such a holistic approach underscores the unique, but differing, advantages of hyperspectral and UAV-borne sensors when it comes to disentangling up close ecological processes and stand structural drivers that are crucial for an accurate accounting of blue carbon (Laignel et al., 2023). In addition, the application of these sophisticated methods can identify different elevational zones within these complex intertidal systems, something that is not necessarily as translatable to other ecosystems types, but is an important factor when describing blue carbon habitats in a precise manner (Hiles et al., 2024). In this review, we provide an overview of current remote sensing research with multisource data, focusing on UAV-borne hyperspectral systems to improve blue carbon mapping. It will also review on the capacity of open source workflows to enable large scale and accessible wetland monitoring, especially given growing efforts in wetland rehabilitation (Lanceman et al., 2022). Up-scaled application of high-resolution RS data from UAVs in combination with advanced machine learning algorithms represent a potential new method to improve accuracy and efficiency of blue carbon monitoring (Hiles et al., 2024; Tâm & Ngân, 2025). Fig 1 shows the applications of blue carbon sources.

Figure 1. Bluecarbon sources and remote sensing

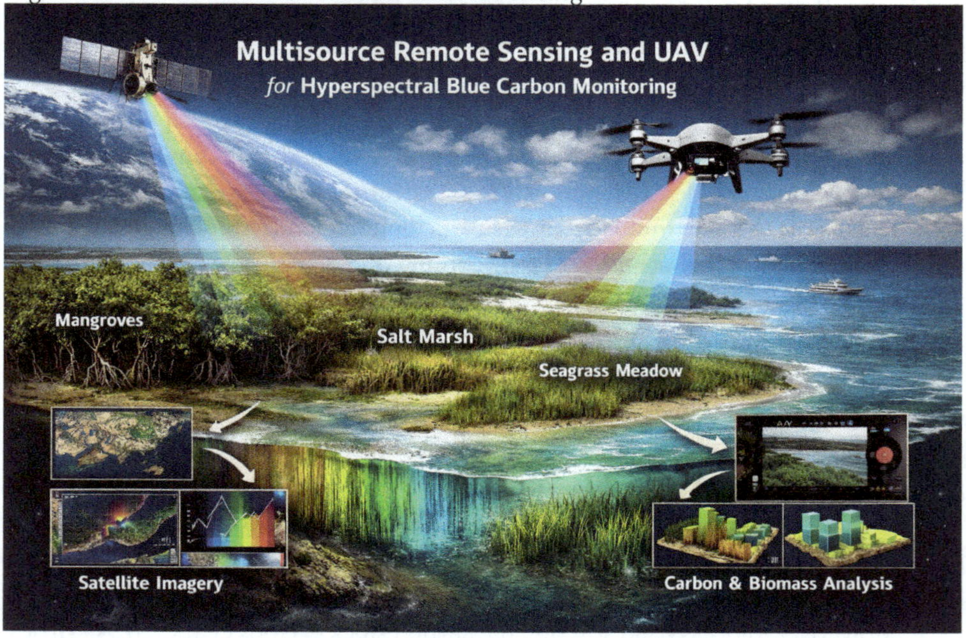

BACKGROUND OF BLUE CARBON ECOSYSTEMS

Blue carbon ecosystems, dominated by mangroves, salt marshes and seagrass beds are widely acknowledged for their unmatched ability to store atmospheric CO_2 in the form of plant tissues and its underlying sediments with rate that is often orders of magnitude greater than tropical forests (Joyce et al., 2022). These coastal ecosystems are key natural climate solutions, reducing the impacts of climate change while delivering important co-benefits in terms of protection against the coast erosion, support to biodiversity and habitats for fisheries (Lanceman et al., 2022). They are so named because they bury huge quantities of organic carbon embedded in their sediments that can last for thousands (Richard, 2024; Reithmaier et al., 2023). Among blue carbon ecosystems, mangroves are known as major mechanisms in the marine carbon cycle and climate change mitigation because of their high capacity to sequester carbon (DENG et al., 2024). Their ability of long-term carbon storage in biomass and soil is remarkable, with 50–98% of total carbon stocks contained in the soil reservoirs (DENG et al., 2024; Fernández et al., 2014).

Accurately measuring these carbon stocks is crucial for successful blue carbon management on coastlines, and requires high level monitoring (Richard, 2024; Xie et al., 2025). Nevertheless, the inherent complexity of these ecosystems and their dynamic nature render thereby accurate carbon accounting quite challenging, requiring the construction of sophisticated remote sensing approaches (Deng et al., 2024). Seagrass monitoring-Blue carbon ecosystems A key aspect of the monitoring is to provide Blue carbon ecosystem sequestration dynamics, particularly seagrass meadows, which are key components of BCESs (Ceccherelli et al., 2024). While they are globally significant as blue carbon ecosystems (Holm et al., 2023), they remain highly threatened from human impacts and the effects of climate change, resulting in their widespread degradation and loss (Pham et al., 2023). This degradation not only undermines their ecological integrity, but also the release of carbon sequestered there to atmosphere in a warming world, which can further enhance global climate change (Deng et al., 2024; Çelik, 2026; Amsalu et al., 2026). For instance, urbanisation and pollution result in the degradation of mangrove ecosystems leading to the release of previously stored carbon (Andrade et al., 2024). This requires monitoring approaches to quantify carbon degradation and effectiveness of conservation and restoration measures (Wang et al, 2024).

REMOTE SENSING IN ENVIRONMENTAL MONITORING

Remote sensing serves as a critical tool to help address these challenges by providing synoptic color images and quantitative estimates of the spatial extent, health, and carbon dynamics of blue carbon ecosystemsat a range of scales. Such technologies have the potential to be used in the mapping of ecological change which is essential for quantifying above-ground biomass, delineating ecosystem boundaries and monitoring human impacts and hence forms a strong basis for prioritisation of conservation and management needs. The recent development of technology (with reference to remote sensing and in the context of multi-temporal satellite data integration) has greatly enhanced the ability to map and monitor habitat changes, particularly improving the ability to work with these settings as impacted by environmental stressors on blue carbon systems (Ranith et al., 2025). Such advancements are particularly relevant amidst the rapid degradation of blue carbon ecosystems, with seagrass meadows having declined by 29% since 1879 with even faster rates of loss following 1990 (Macreadie et al., 2019). However, the absence of a multi-temporal monitoring framework at wide scale and difficulty in delineating degradation hotspots hampers considerably conservation activities for these important ecosystems (Chowdhury et al., 2024). These challenges call for a change of paradigm to combine different types of remote sensing platforms, such as unmanned aerial vehicles (UAVs) and satellite imagery, to obtain more holistic and continuously-updated views on blue carbon ecosystem health and carbon dynamics. Moreover, the current running United Nations "Decade of Ecosystem Restoration" (2021-2030) highlights the urgent requirement for advanced monitoring methods to underpin conservation and restoration needs of these precious marshes with their associated blue carbon resources (Lanceman et al., 2022). This full-scale monitoring is valuable not only for evaluating the restoration efficacy, but more importantly to set the references from which future dynamics in carbon sequestration and eco-systemic health can be appraised (Mengis et al., 2023). Advanced monitoring of this kind is also a key component of moving towards the vision for healthy and resilient oceans by 2030, as so beautifully articulated for the UN Ocean Decade and required to support mapping and protection of marine biodiversity (Lecours et al., 2021). In addition, these technologies are critical to financing site-scale wetland conservation and restoration by allowing systematic, repeatable and large scaled assessment as well as monitoring of coastal ecosystems, including the mapping of their extent, changes and carbon stock (Fernández et al., 2014). This emphasises the urgent requirement for operational, remote sensing methodologies that can monitor BC ecosystems in a cost-effective and accessible manner (particularly with regards to global driver of climate change developed across this millennia and ongoing loss) (Deeks et al., 2024; Hill et al., 2024). This is especially the case for seagrass meadows as remote sensing provides a nonintrusive tool to quantify carbon uptake and other biophysical regimes understanding vital for ecosystem health and services (Traganos & Reinartz, 2018). However, high-quality, nation-specific in situ data on seagrass distribution, hereafter blue carbon stocks and fluxes are not yet available for many countries; challenges that will continue to prevent standardization and funding of long-term blue carbon accounting systems (Blume et al., 2023). Moreover, the high temporal variability and often remote character of coastal vegetated ecosystems such as mangroves, saltmarshes, seagrasses and benthic macroalgae add further challenges to maintain monitoring programmes or stock assessments (Mengis et al.,

2023). Therefore, there is a strong need for improved spatially explicit approaches, perhaps built on coupled remote sensing and modeling based frameworks, to better understand and scale the quantification of blue carbon as well as support effective policy-making (Mengis et al., most recent; Traganos et al., 2022). A key challenge to robust monitoring is the inconsistent characterisation of salt marshes over tidal ranges and lack of investigations into the SM-TE in most studies, both essential for accurate mapping and management (Gorman et al., 2022).

ROLE OF UAVS AND HYPERSPECTRAL IMAGING IN THE NEAR FUTURE

The convergence of uncrewed aerial vehicles with hyperspectral imaging provides a game-changing opportunity for precise, localized blue carbon monitoring that is unique in terms of spatial and spectral resolution (Allen et al., 2023). This enables accurate identification of subtle spectral signatures specific to various blue carbon species and their physiological status, which is essential in both estimating biomass and monitoring health at fine resolutions (Lanceman et al., 2022). Products obtained with these features allow the identification of small changes in plant health, species composition and biomass accumulation that are normally related to variations on stocks of carbon and ecosystems functionality (Rondón et al., 2023). In addition, detailed spectral data collected by hyperspectral sensors mounted on UAVs can be used to improve the models for predicting above-ground biomass at stand level and thus, carbon sequestration rate modelling in these complex ecosystems (Blanco-Sacristán et al., 2024). Such fine-scale spatial and spectral information will greatly increase the accuracy of carbon content assessment within seagrass meadows; such as detailed estimations of chlorophyll concentration and habitat-wide primary productivity (Monaca, in 2024). This form of zoomable information from UAVs can also be imperative for ground-truthing observations with larger scale remote sensing platforms, ultimately improving the robustness and reliability of blue carbon stock estimations across wider spatial scales (Bardou et al., 2023).

Yet, there are major conceptual and methodological limitations to estimating the stock of carbon in blue carbon ecosystems which are dominated by belowground storage of soil organic carbon and that are not yet compatible with commonly used terrestrial forest remote sensing methods (Moreaux et al., 2019). This requires the invention of specific methods for remote sensing able to infer from surface spectral properties or using integration into in-situ measurements belowground carbon pools (Mengis et al., 2023). This argues for the strong requirement for hyperspectral remote sensing that can detect small biogeochemical differences associated with subsurface (belowground) biomass and carbon storage compared to conventional multispectral approaches (Blanco & Pérez, 2023). The costeffective and flexible UAV platforms therefore constitutes an appropriate approach to fill in the gap between field based spectral measurements, coarser satellite, or airborne sensor imagery for extending the characterization of blue carbon at high spatial resolution (Porcar-Castell et al., 2015). These hyperspectral systems carried by UAVs may offer high-resolution and low-cost maps of carbon stocks, especially for small mangrove stands or single plant species, despite the constraint of relatively short flight endurance (Hiles et al., 2024; Ribeiro et al., 2024). Therefore, these sensors are especially well-suited to characterize the structural properties of canopy (at leaf level) and leaf pigments (i.e., chlorophyll a and b concentrations), which serve as ctitious indicators for the carbon assimilation rates in mangroves (Blanco-Sacristán et al., 2024). Such capacity can significantly improve monitoring for mangrove health and productivity necessary to understand their role in the regional carbon cycle (Blanco-Sacristán et al. 2024).

Additionally, 3D models of mangrove forests can be generated using UAV photogrammetry, which is a low-cost and high-efficient approach for repeated long-term monitoring of area and the biological health indicators in this environment, even if problems remain on aerial triangulation, image mosaicking (Deng et al., 2024). In comparison to satellite derived imagery, UAVs can provide higher accuracy for mangrove cover mapping to avoid the overestimation problem often found in the case of satellite data particularly within semi-arid dwarf mangroves systems (Rondón et al., 2023). Remote sensing, particularly when combined with UAVs, has significant advantages over conventional field survey for monitoring large areas of these frequently inhospitable environments and is cost-effective as well as readily scalable in time (Sunkur et al., 2024). This is of particular value when mapping biophysical and structural properties of mangroves more efficiently, at broader spatial scales that allow for advanced species discrimination by machine leaning-based methods, monitoring of structure and estimation biomass (Pham et al., 2019). Moreover, mangrove biomass and C-stocking by UAV image-based estimation can be achieved more quickly over larger scales than traditional field data collection methods, and linked model calibration may improve the accuracy of these estimates (Jones et al., 2020). Improved UAV image measurements-tree diameter correlations are required to minimize predictive error for biomass estimates (Jones et al., 2020). Moreover, the incorporation of Light Detection and Ranging from UAVs may also greatly improve biomass estimation accuracy due to detailed structural 3D measurements which are important for characterizing complex mangrove architectures (Peng et al, 2025). This fine-scale structure information from LiDAR can also co-registered with multi-spectral data to refine the classification of mangrove canopy properties, biomes and the lut ency (Ngo et al., 2024).

BLUE CARBON ECOSYSTEMS AND MONITORING CHALLENGES

Blue carbon habitats such as mangroves, salt marshes and seagrass meadows are highly effective systems at capturing and storing CO2 from the atmosphere. These high productivity coastal habitats also provide essential natural climate solutions, help counteract the effects of climate change by capturing large amounts of carbon in their biomass and underlying sediments, often at rates much higher than terrestrial forests (Joyce et al., 2022). Beyond their carbon sequestration role, these habitats offer beneficial co-benefits coastal protection, biodiversity sustenance and key fisheries habitat (Lanceman et al., 2022). Their potential for long-term storage of carbon, mobilized over thousands of years, is indicative of their place within the global carbon cycle (Richard, 2024).

Despite this, the management and monitoring of carbon stocks in these highly variable and rapidly changing ecosystems are challenging. The concave character of the blue carbon systems, serves to make precise estimation challenging due to their remoteness across large areas (DENG et al., 2024). An important challenge remains that a large portion of carbon storage takes place within the subsurface and traditional land-based remote sensing approaches are not applicable, so custom methodologies are required to estimate subsurface C pools (Mengis et al., 2023; Moreaux et al., 2019). In addition, these critically important systems are declining and completely disappearing due to the impact by human activities and climate change. This transformation leads to the loss of ecological integrity and a return to the atmosphere of carbon that was previously stored, contributing to climate change (Deng et al., 2024).

To address these monitoring challenges, next-generation remote sensing in the form of Uncrewed Aerial Vehicles (UAVs) with hyperspectral sensors provices a game-changing solution. This integration also offers never-before possible spatial and spectral resolution used to identify subtle spectral signatures linked with various blue carbon species as well as their physiological states, essential when correcting for biomass estimates (and related health analyses) (Lanceman et al., 2022). Hyperspectral systems carried by UAV are scalable tools that can fill the gap between point-based ground measurements and coarse satellite data, providing high resolution and inexpensive mapping of carbon stock at even small-scale levels or for individual plant species (Hiles et al., 2024; Porcar-Castell et al., 2015). It enables more accurate measurements of vegetation health, species composition, and biomass accumulation - key indicators for changes in carbon stocks and ecosystem functioning.

PRINCIPLES OF REMOTE SENSING FOR BLUE CARBON

Remote sensing, associated with different technologies, is based on the principle of measuring and recording electromagnetic radiation reflected or emitted by the Earth`s surface (Richard, 2024). This information is further processed and analyzed to yield insights into the physical and biological properties, rendering it invaluable for monitoring dynamic blue carbon ecosystems (Mengis et al., 2023). In particular, the remote sensing of blue carbon is based on that different vegetation types and sediment compositions have unique spectral signals which can be used to distinguish and estimate carbon pools (DENG et al., 2024). For instance, optical remote sensing has the ability to capitalize on unique absorption and reflection characteristics of plant pigments in the visible and near-infrared spectral regions as well as water absorptions bands in the mid infrared for estimating above ground plant biomass and organic carbon contents (Belloli et al., 2022). Synthetic Aperture Radar systems on the other hand can also deliver structural information due to its capability to penetrate vegetation canopies, and thus provide body mass density and height as important variables for estimation of total ecosystem carbon stocks (Yulianto et al., 2025). UAV-mounted lidar bathymeters can also present new paradigms for shallow water and underwater object detections, serving blue carbon monitoring with critical underwater topography data coverage (Chen et al., 2023).

Furthermore, the improved spectral resolution of hyperspectral sensors combined with UAV platforms enables discrimination between different blue carbon species; it can even differentiate small differences in spectral reflectance between seagrass speciations despite different depths or various mangrove taxa (Lemenkova, 2019). This capacity allows more accurate mapping of blue carbon habitats and higher precision in biomass and carbon stock estimates over various coastal setting (Hiles et al., 2024; Mukaromah et al., 2019). This high level of discrimination permits the development of reference data for carbon accounting and detecting changes over time – important as potential indicators for conservation effectiveness/restoration (Mengis et al., 2023). This detailed spectral data, which are georeferenced with positional information, provides the basis for models to estimate carbon sequestration potential and ecological response to environmental change (Joyce et al., 2022). The scalability and cost-effectiveness of Earth Observation frameworks that capitalise on developments in satellite and UAV technology enable a potential solution to the national-level blue carbon accounting for climate change mitigation (Blume et al., 2023). Integration of these technologies, together with AI and machine learning have recently shown considerable promise to improve the accuracy of blue carbon estimates from space (Pham et al., 2023). However, despite these progresses, it continues to be challenging to consistently and accurately estimate stocks of blue carbon because the ecosystems are inherently complex and there are varied methods and platforms used in remote areas which makes things tough (El-Masry, 2024). traptive nature of SAR data itself (Pham et al., 2023). In addition, the tidal effect on the extent of mangrove also affects spectral signatures and canopy height that requires an additional study to accurately estimate such parameter through remote sensing (Maurya et al., 2021). Nowhere near as perfect Classification is limited by (multipart) pixels, mixed classes and stepped nostril water level leading to increased mapping uncertainty in spite of comprehensive feature engineering (Blume et al., 2023). As a result, current research is aiming at introducing advanced algorithms and techniques of fusion in order to mitigate these issues and hence excerpt information that is both robust and reliable for remote sensing derived based blue carbon assessment. Furthermore, the distinct spectral properties of blue carbon systems such as mangroves, saltmarshes and seagrasses warrant specific algorithms for the accurate assessment of their carbon stocks, being these ecosystems more sensitive to variations in biogeochemical parameters (Pham et al., 2023; Reithmaier et al., 2023). Accordingly, innovative methods utilizing multifaceted sensors and sophisticated analyses are needed to adequately account for the spatial variability and temporal trends in these important C stocks (Sunkur et al., 2024).

This involves the complex coupling of carbonate chemistry and associated carbon sequestration mechanisms driven by inorganic carbon outwelling from mangroves and saltmarshes (Reithmaier et al., 2023). The estimation of those carbon stocks is however difficult, since most are located underground and this methodological gap was not completely filled by nowadays remote sensing (Moreaux et al., 2019). High variability of total alkalinity, dissolved inorganic carbon in surface waters of these ecosystems often driven by tides and export of porewater also hinder remote sensing sensitivity to C dynamics (Reithmaier et al., 2023). However, some recent studies hold promise for a possible detection of general structural changes in wetland areas (Laignel et al., 2023), although electromagnetic waves do not penetrate water and it is thus impossible to directly map submerged vegetation by SAR. More extensive investigations on multi-temporal SAR interferometry could reveal fine-scale geomorphological changes and biomass variations in these underwater or semi-aquatic habitats, indirectly supporting the estimation of carbon stock dynamics (Maurya et al., 2021). In addition, the difficulties of quantifying outwelling that can also be simply large due to constraints in data collection frequency and time resolution highlight a need for more elaborate remote sensing procedures that can capture these non-steady features (Reithmaier et al., 2023).Terrestrial Laser Scanning, not a ‘remote sensing’ technique in the traditional aerial sense, yet constitutes another potential method for accurate biomass quantification in more complex coastal wetland vegetation where studies have demonstrated: strong correlations between TLS-based estimates and allometric methods (Chávez et al., 2023).

SPECTRAL CHARACTERISTICS OF BLUE CARBON COMPONENTS

Blue carbon ecosystems such as mangroves, salt marshes and seagrasses have spectral characteristics that are essential for identification and mapping using remote sensing methods (Zhang, 2024). These different spectral response characteristics are attributed to the differences in a relatively unique biophysical and biochemical composition, including chlorophyll content, leaf area index and plant issue structure that controls how electromagnetic radiation interacts with vegetation (Zhang et al., 2018; Çelik et al., 2026). For example where healthy mangrove canopies have high levels of chlorophyll, they tend to be strongly absorbing in the red and blue bands of the EM spectrum and highly reflective within near-IR allowing good spectral discrimination for mapping (Chávez et al., 2023). Likewise, saltmarsh plants and seagrasses show marked spectral responses due in part to their distinctive pigment contents, canopy characteristics, and submerged environments, which also demand specific algorithms for accurate remote sensing classification and biomass estimation (Hemati et al., 2025; Simpson et al., 2022). These variation can be additionally affected by tides and watercolumn which can change the observed reflectance strikingly, and advanced atmospheric/water- column correction algorithms are needed for proper retrieval of vegetation parameters (Jessen & Hill, 2024). Inter-species variation in carbon sequestration potential also needs to be fully understood, with each mangrove species having different rates of biomass accumulation and ability to store carbon, requiring identifications at the species level for accurate estimation about the amount of stored carbon (Harsa et al., 2023).

Hyperspectral remote sensing, which covers many narrow spectral bands, is particularly well-suited for distinguishing such subtle spectral variations and supporting species-level classification in highly complex blue carbon ecosystems (Lanceman et al., 2022). This improved spectral resolution enables the distinction between species and the evaluation of their physiological status, which is important for improving allometric equations needed for estimating biomass and carbon stock (Chávez et al., 2023; El-Masry, 2024). Hyperspectral imagery from airborne sensors, for example has been successfully used to discriminate differing and difficult to detect mangrove species whereas it is hard or impossible using typical multispectral systems (Chaube et al., 2019). Furthermore, hyperspectral spectral resolution makes vegetation less sensitive indices able to accurately estimate biochemical variables such as chlorophyll and nitrogen which are directly related with the photosynthetic activity and carbon fixation rate (Bargain 2012). These indices can also be used to develop statistical relationships between spectral responses and biomass, providing more sensitive indicators of blue carbon than broad-band multispectral analyses (Bargain, 2012). Low bandwidths range in hyperspectral imaging offers accurate measurements of molecular absorption bands, which play important roles in determining closely related species and quantitative assessment of physiological status of plant (Chaube et al., 2019). In addition, the distinction of species in mangrove forests is vital, as different species have varying carbon sequestration capacities and contributions towards the general provision of ecosystem service (Pimple et al., 2022).

UNMANNED AERIAL VEHICLE (UAV) APPLICATIONS

Remotely Piloted Aircraft Systems, especially when combined with hyperspectral imaging, are transforming the surveillance of blue carbon ecosystems by providing higher spatial and spectral resolution (Allen et al., 2023). This advanced mix of bands enables the detection of subtle spectral characteristics associated with different blue carbon types and their physiological condition, which are necessary to accurately estimate biomass and assess health at finer spatial scales (Lanceman et al., 2022). These features allow scientists the ability to identify small changes in plant health, species and biomass—key indicators of changes in carbon stock and broader ecosystem function (Rondón et al., 2023). Also, the detailed spectral information provided by UAV hyperspectral sensors may be used to develop more accurate equations for predicting above-ground biomass that would lead to estimations of sequestration rates in these complex ecosystems (Blanco-Sacristán et al., 2024). Such detailed spatial and spectral data have increased the precision of carbon quantification in areas, such as seagrass meadows (through estimation of chlorophyll levels and habitat productivity) for instance. High-resolution UAV surveys such as this project are also key for ground-truthing studies conducted by larger-scale remote sensing platforms, improving the accuracy of bC stocks at larger scales (Bardou et al., 2023).

Despite limitations due to underground carbon storage in blue carbon ecosystems, preventing the use of traditional terrestrial remote sensing techniques, drones with hyperspectral sensors are argued to provide a compelling answer (Moreaux et al., 2019). Hyperspectral remote sensing offers opportunities to detect subtle changes in biogeochemistry, which may reflect biomass and carbon storage in the subsurface, thanks to its high spectral resolution, making it more powerful than traditional multispectral method (Blanco & Pérez, 2023). The ACE and flexibility of UAV-based platforms make them a realistic option for filling the scale mismatch between contact fieldborne spectral measurements and coarser airborne or satellite imagery, thus enabling more complete blue carbon appraisals (Porcar-Castell et al., 2015). Such UAVs-borne hyperspectral systems potentially allow high-resolution and cost-effective maps of carbon stocks, especially for small mangrove systems and individual plant species, even considering drawbacks such as their usually low flight duration (Hiles et al., 2024; Ribeiro et al., 2024). As a result, these platforms have a strong potential for the characterization of canopy-level structural properties and leaf pigments like chlorophyll a and b concentration (key proxy traits to carbon assimilation rate) in mangroves, improving monitoring mangrove's health and productivity as well as their role in regional carbon cycling (Blanco-Sacristán et al., 2024).

Additionally, the technique is cost-effective for 3D modelling of mangrove forests which allows repeated long term monitoring of indicators such as area and biological health although challenges in aerial triangulation and image mosaicking exists (Deng et al., 2024). In contrast to satellite imagery, UAV is much more accurate in SME's for mapping the mangrove cover by surpassing overestimation issues as typically seen on the satellite's level data (Rondón et al., 2023). The use of remote sensing, with UAVs in particular, provides major advantages compared to traditional field surveys for monitoring the vast extension of these often challenging ecosystems due to their access limitations and is cost-effective as well as providing fast temporal resolution (Sunkur et al., 2024). This is particularly advantageous when it comes to biophysical and structural mapping of mangroves at a larger scale resolution which facilitates the application of more complex machine learning algorithms for species discrimination, structure monitoring and biomass estimation (Pham et al., 2019). In addition, those estimates of mangrove carbon storage and biomass derived from UAV images can be produced much more quickly across a wider range than data collection methods previously employed: subject to model tuning, they may also be more accurate (Jones et al., 2020). The relationship between tree diameter and UAV image-based measurements needs further improvement to minimize errors in biomass estimation (Jones et al., 2020). Further, incorporating Light Detection and Ranging (LiDAR) collected by the UAVs can greatly increase accuracy of biomass estimation providing a three-dimensional structural information that is essential for evaluating complex mangrove structures (Peng et al., 2025). This ultrastructural information from LiDAR can further be joined to multispectral imagery, which helps to better evaluate canopy structure, biomes, and coverage in mangrove ecosystems (Ngo et al., 2024). The data fusion of these data that become increasingly common is therefore critical to provide a view of mangrove carbon dynamics more comprehensive than simple biomasses ama, moving toward understanding the structure species composition bioaccumulation relationships (Pham et al., 2019).

METHODOLOGY FOR HYPERSPECTRAL BLUE CARBON MONITORING

A typical procedure of hyperspectral blue carbon estimation, especially in UAV applications, would be as follows: 1.

1 Data Acquisition via UAVs equipped with Hyperspectral Sensors This process is the initial step where the use of UAVs equipped with hyperspectral sensors is used to capture very high resolution images. This information is essential for detailed studies such as Allen et al. (2023) and Lanceman et al. (2022) where precise spectra in many narrow bands are needed. The flexibility and low cost of UAV platforms render them well suited for filling the disconnect between ground-based measurements and coarser satellite data, especially for small areas or plant species (Hiles et al., 2024; Porcar-Castell et al., 2015).

2. Spectral Analysis and Species Discrimination: After data collection, advanced spectral analysis is completed to distinguish between subtle spectral differences associated with different blue carbon species as well as their physiological conditions (Lanceman et al., 2022). Hyperspectral remote sensing is particularly powerful in THIS scene due to its continuous spectral resolution, which makes it possible to identify individual species and monitor their health in a way that multispectral images are sensitive (Chaube et al., 2019; Chávez et al., 2023). This process is important for improvement of allometric equations in estimation of biomass and carbon stocks (El-Masry, 2024).
3. Biomass and Carbon Stock Estimation: These spectral data combined with sophisticated algorithms are used to build models that will be developed to estimate the above ground biomass, and hence carbon sequestration rates (Blanco-Sacristán et al., 2024). This entails frequently quantifying canopy\-level structural characteristics and leaf pigments (such as concentrations of chlorophyll a and b), which are surrogates for carbon uptake rates (Blanco-Sacristán et al. The underground storage of carbon is difficult to estimate from direct observations, but its hyperspectral indicators can detect small biogeochemical changes related to subsurface biomass (Blanco & Pérez, 2023).
4. 3D Modeling and Structural Analysis: UAV photogrametry is used to create a three-dimensional model of the ecosystem, in particular mangrove stand making continuous monitoring of areal and biological (health) indicators such as percentage crown cover and NDVI possible (Deng et al., 2024). Furthermore, adding Light Detection and Ranging data information from UAVs can greatly improve the precision of biomass estimation in that they provide detailed 3D structural information which is important for analysing complex architectures (Peng et al.,2025). Such structural information can be integrated with multispectral imagery to enhance quantifications of canopy structure and biome type (Ngo et al., 2024).
5. Validation and Ground-truthing: High-resolution UAV data are critical for validating observational data obtained from larger-scale remote sensing

platforms, and as part of improving the robustness of blue carbon stock estimates at broader scales (Bardou et al., 2023). This step guarantees the reliability of the models and estimates from HIS.

This multi-scaled UAV- and hyperspectral-based procedure offers an improved methodology for the accurate monitoring of coast wetlands, that could assess the subtle changes in vegetation-health index, species composition and hydrological processes—all are important indicators of blue carbon sequestration capacity (Lanceman et al., 2022).

DISCUSSION

The utility of hyperspectral data, particularly when coupled with UAV platforms, would go beyond basic mapping to provide detailed discrimination of vegetation types required for precise carbon accounting (Blanco & Pérez 2023). This improved discriminative ability can also be used to refine carbon storage models that account for species-specific variation in carbon sequestration, resulting in greater accuracy of blue carbon estimates (Lanceman et al., 2022). In addition, UAV-borne hyperspectral sensors provide high-resolution spatiotemporal monitoring of ecosystem structure and function evolution, contributing to improved management and restoration of blue carbon habitats (Blanco-Sacristán et al., 2024; Lanceman et al., 2022). This comprehensive approach guarantees that the fine-grain ecological dynamics of coastal wetlands like mangroves and saltmarshes are considered acknowledging their significance when contemplating regional carbon budgets and driving conservation policy (Hiles et al., 2024; Kandrot et al., 2021).

This is especially critical because of the 15–20% error rates commonly found in organic carbon stock estimates in some systems, stressing the urgency for improved methods to ensure robust blue carbon accounting and effective nature-based solution (Hiles et al., 2024). Furthermore, the combination of hyperspectral analysis to advanced machine learning algorithms, enables some wetland types and associated carbon storage potential to be characterised with great precision which is essential for their accurate ecological value evaluation and targeted conservation (Pricope et al., 2024). These advanced modelling approaches are essential to better estimate the spatial extent of blue carbon stocks, and to improve the accuracy ventilation equipment for lab-scale estimation of carbon in wetland ecosystems (Belloli et al., 2025). These approaches can help to greatly decrease the man power and expertize normally needed for prediction of mangrove AGB, which is efficient and more accurate in estimation (Xie et al., 2025).

This is especially important in areas that are difficult to access and where it is not feasible to make traditional field measurements, thus presenting a wider array of blue carbon monitoring (Chaube et al., 2019). UAV hyperspectral imagery offers improved spatial and spectral resolution enabling more accurate delineation of saltmarsh extents and zonation patterns, resulting in better estimation of soil organic carbon storage whilst minimising the potential for overestimation associated with coarser resolution satellite data (Hiles et al., 2006). This advanced mapping technology along with the continued evolution of UAV platforms provides a stronger approach to quantify changes in marsh vegetation and C sequestration leading to enhanced monitoring success for restoration efforts (Hiles et al., 2024; Windle et al., 2023). But this is an ongoing work that needs to be continuously supervised and validated with ground-truth data, especially for these dynamic ecosystems (the evergreen needle-leaf forest, the modern croplands and shrubs), where misclassification occurs in vegetation transition zones (Russell et al., 2023). Furthermore, integrating machine learning algorithms with remote sensing data can be used for better prediction and estimation of wetland carbon sink by pinpointing essential driving factors such as water level and temperature, which also contributes to the improvement of model accuracy and reliability (Li et al., 2024). These advanced methods also make it possible for considering the impacts on climate change by incorporating dynamic models responsive to environmental changes and initiating proactive conservation and management strategies (Chafe et al., 2024). In addition, the use of advanced deep learning algorithms for EO data can improve and harmonize estimations of aboveground, below-ground (while at it) and soil carbon store across different blue carbon ecosystems such as mangroves (Xie et al., 2022), seagrasses (Ganguly et al., 2023) or saltmarshes. These sophisticated techniques can be especially useful for investigating changes in mangrove ecosystems and the estimation of AGC, offering a more efficient alternative compared with OTS using UAV-B (Belloli et al., 2022).

Hence, the developments provided here enable a holistic and evolving inter-practice model of in-catchment carbon dynamics for these important ecosystems, extending previous methodologies (Hiles et al., 2024). The potential to survey coastal wetlands using these technologies quickly and frequently gives a key advantage in monitoring the progress of restoration and seeing how carbon abatement varies over time (Lanceman et al., 2022). This is particularly relevant, as the global focus on the restoration of wetlands becomes more intense and there is an increased need for affordable and locally applicable monitoring protocols (Lanceman et al., 2022). Thus, the combined use of multi-source remote sensing UAV platforms helps us in tracking the long standing ecological changes assessment of conservation intervention effectiveness and providing strong evidence into the global climate change (Ranith et al., 2025; Windle et al., 2023). Such a comprehensive viewpoint also leads to the implementation of efficient nature-based solutions and policies aimed at countering climate change impacts in conjunction with the blue carbon ecosystems (Sunkur et al., 2023).

CONCLUSION

Mangroves, sea grasses, and salt marshes that form blue carbon ecosystems are important in the mitigation of climate change because they have the unsurpassed capacity of sequestraining carbon. Coastal wetlands deposit much more carbon than terrestrial forest, and therefore it is important in the management of carbon across the world. The nature of these ecosystems is however complicated and dynamic making it difficult to accurately monitor carbon stocks within these ecosystems. The conventional field techniques frequently cannot be sufficiently used in massive and high-resolution studies. This article introduces an innovative method of combining multisource remote sensing devices and tools especially unmanned aerial vehicles (UAVs) with hyperspectral sensors to improve the monitoring of blue carbon. This paper provides enhanced spatial and spectral resolutions through the use of UAVs and hyperspectral imaging to monitor the blue carbon ecosystems with high accuracy at small scales. Hyperspectral sensors mounted on UAVs give information on vegetation health and species composition and biomass accumulation, all of which are essential parameters used to predict the potential of carbon sequestration. A combination of these sophisticated remote sensing methods, together with machine learning algorithms, enables one to obtain a precise estimation of above- and below-ground stocks of biomass and soil organic carbon even in harsh conditions. Other issues associated with the measurement of subsurface carbon storage and the potential offered by UAV-based hyperspectral systems to fill the gap between the sparse satellite data and the old methods of the field are also discussed in the paper. The technologies provide an inexpensive and scalable method of tracking the health condition of carbon storage potential of mangrove, seagrass, and salt marsh ecosystems at a low cost with valuable information on the success of restoration and conservation initiatives. This study will underpin improved climate policy and allow the restoration of the important coastal ecosystems by enhancing blue carbon mapping and monitoring.

FUNDING

This work was supported by the Deanship of Scientific Research, the Vice Presidency for Graduate Studies and Scientific Research, King Faisal University, Saudi Arabia, under Grant No. KFU254859

REFERENCES

Allen, T. R., Eulie, D. O., Polk, M. A., McLeod, G., Stuart, R., & Garnand, A. (2023). Advancing Estuarine Shoreline Change Analysis Using Small Uncrewed Autonomous Systems. In *IntechOpen eBooks*. IntechOpen., DOI: 10.5772/intechopen.112022

Amsalu, A., Garedew, W., Melka, G. A., Çelik, M. A., & Gemeda, D. O. (2026). Trends in climate extreme indices and implications for coffee farming in The Kafa Biosphere Reserve, southwest Ethiopia. *Climate Services*, *42*, 100637. DOI: 10.1016/j.cliser.2026.100637

Andrade, C. de L. B. Costa, M. D. de P., & Freitas, D. M. de. (2024). *Assessing the Carbon Sequestration Potential by Mangroves in a Coastal Metropolitan Region*. DOI: 10.2139/ssrn.5023738

Bardou, R., Osland, M. J., Scyphers, S. B., Shepard, C. C., Aerni, K. E., Alemu, J. B., Crimian, R., Day, R. H., Enwright, N. M., Feher, L. C., Gibbs, S. L., O'Donnell, K. L., Swinea, S. H., Thorne, K., Truskey, S., Armitage, A. R., Baker, R. J., Breithaupt, J. L., Cavanaugh, K. C., & Hughes, A. R. (2023). Rapidly Changing Range Limits in a Warming World: Critical Data Limitations and Knowledge Gaps for Advancing Understanding of Mangrove Range Dynamics in the Southeastern USA. *Estuaries and Coasts*, *46*(5), 1123–1140. DOI: 10.1007/s12237-023-01209-7

Bargain, A. (2012). Study of Zostera noltii seagrass bed structure and dynamics using multi and hyperspectral remote sensing. *HAL (Le Centre Pour La Communication Scientifique Directe)*. https://theses.hal.science/tel-00799897

Belloli, T. F., de Arruda, D. C., Guasselli, L. A., Cunha, C. S., & Korb, C. C. (2025). Modeling Wetland Biomass and Aboveground Carbon: Influence of Plot Size and Data Treatment Using Remote Sensing and Random Forest. *Land (Basel)*, *14*(3), 616. DOI: 10.3390/land14030616

Belloli, T. F., Guasselli, L. A., Kuplich, T. M., Ruiz, L. F. C., de Arruda, D. C., Etchelar, C. B., & Simioni, J. D. (2022). Estimation of aboveground biomass and carbon in palustrine wetland using bands and multispectral indices derived from optical satellite imageries PlanetScope and Sentinel-2A. *Journal of Applied Remote Sensing*, *16*(3). Advance online publication. DOI: 10.1117/1.JRS.16.034516

Blanco, A. C., & Pérez, C. (2023). Estimation Of Mangrove Fractional Cover From Multispectral And Hyperspectral Data Using Mixture Tuned Matched Filtering. *The international Archives of the Photogrammetry, Remote Sensing and Spatial Information Sciences. The International Archives of the Photogrammetry, Remote Sensing and Spatial Information Sciences, 1331*, 1331–1336. Advance online publication. DOI: 10.5194/isprs-archives-XLVIII-1-W2-2023-1331-2023

Blanco-Sacristán, J., Johansen, K., Elías-Lara, M., Tu, Y.-H., Duarte, C. M., & McCabe, M. F. (2024). Quantifying mangrove carbon assimilation rates using UAV imagery. *Scientific Reports*, *14*(1), 4648. Advance online publication. DOI: 10.1038/s41598-024-55090-w PMID: 38409194

Blume, A., Pertiwi, A. P., Lee, C. B., & Traganos, D. (2023). Bahamian seagrass extent and blue carbon accounting using Earth Observation. *Frontiers in Marine Science*, *10*, 1058460. Advance online publication. DOI: 10.3389/fmars.2023.1058460

Ceccherelli, G., Chiabrando, F., Gallitto, F., Lingua, A. M., Longhi, V., Maschio, P. F., Matrone, F., Menna, F., Nocerino, E., Scalici, M., Secco, S., & Tanduo, B. (2024). Multitemporal Seagrass Mapping and Monitoring of Posidonia Meadows and Banquettes for Blue Carbon Conservation: The Poseidon Project. *˜The œinternational Archives of the Photogrammetry, Remote Sensing and Spatial Information Sciences. The International Archives of the Photogrammetry, Remote Sensing and Spatial Information Sciences*, *65*, 65–71. Advance online publication. DOI: 10.5194/isprs-archives-XLVIII-1-2024-65-2024

Çelik, M. A. (2026). Agro-climatic classification of Türkiye's Agricultural Basins using long-term ERA5-Land temperature data. *Physics and Chemistry of the Earth Parts A/B/C*, *142*, 104307. DOI: 10.1016/j.pce.2026.104307

Çelik, M. A., Kıraç, A., & Gouveia, C. M. (2026). North Atlantic Oscillation modulation of drought and vegetation activity in Türkiye. *Theoretical and Applied Climatology*, *157*(2), 74. Advance online publication. DOI: 10.1007/s00704-025-05971-5

Chafe, O., Broz, A., Levenson, E. S., Farinacci, M. D., Anderson, R. O., & Silva, L. C. R. (2024). The spatiotemporal domains of natural climate solutions research and strategies for implementation in the Pacific Northwest, USA. *Frontiers in Climate*, *6*, 1273632. Advance online publication. DOI: 10.3389/fclim.2024.1273632

Chaube, N. R., Lele, N., Misra, A., Murthy, T. V. R., Manna, S., Hazra, S., Panda, M. R., & Samal, R. N. (2019). Mangrove Species Discrimination and Health Assessment using AVIRIS-NG Hyperspectral Data. *Current Science*, *116*(7), 1136. DOI: 10.18520/cs/v116/i7/1136-1142

Chávez, D. E., López-Portillo, J., Gallardo-Cruz, J. A., & Meave, J. A. (2023). Approaches, potential, and challenges in the use of remote sensing to study mangrove and other tropical wetland forests. *Botanical Sciences*, *102*(1), 1–25. DOI: 10.17129/botsci.3358

Chen, W., Chen, P., Zhang, H., He, Y., Tang, J., & Wu, S. (2023). Review of airborne oceanic lidar remote sensing. *Intelligent Marine Technology and Systems*, *1*(1), 10. Advance online publication. DOI: 10.1007/s44295-023-00007-y

Chowdhury, M., Martínez-Sansigre, A., Mole, M., Alonso-Peleato, E., Basos, N., Blanco, J. M., Ramírez-Nicolás, M., Caballero, I., & Pérez, I. de la C. (2024). AI-driven remote sensing enhances Mediterranean seagrass monitoring and conservation to combat climate change and anthropogenic impacts. *Scientific Reports*, *14*(1), 8360. Advance online publication. DOI: 10.1038/s41598-024-59091-7 PMID: 38600271

Deeks, E., Magalhães, K. M., Traganos, D., Ward, R. D., Normande, I. C., Dawson, T. P., & Kratina, P. (2024). Seagrass mapping of north-eastern Brazil using Google Earth Engine and Sentinel-2 imagery. *Environmental and Sustainability Indicators*, *100489*, 100489. Advance online publication. DOI: 10.1016/j.indic.2024.100489

Deng, S., Yang, F., Wang, Y., We, Z., & Wang, W. (2024). Progress in Remote Sensing Monitoring of Mangrove Carbon Storage. *Yaogan Xuebao*, *1*(10), 2448–2468. Advance online publication. DOI: 10.11834/jrs.20243293

El-Masry, E. A. (2024). A Non-destructive Rapid Assessment of Blue Carbon Sequestration Potential in Mangrove Forests of the Red Sea Region. *Egyptian Journal of Aquatic Biology and Fisheries*, *28*(1), 1119–1152. DOI: 10.21608/ejabf.2024.339764

Fakhrurrozi, F., Laurance, L., Fauzi, M. N. R., Ramadhan, F., Idris, I., Yusri, S., Paryono, P., Lestariningsih, W. A., & Damayanti, A. A. (2023). Estimating Blue Carbon Stock: Mangrove and Seagrass Biomass Assessment in Tunda Island, Serang, Banten, Indonesia. *BIO Web of Conferences, 70*, 1009. DOI: 10.1051/bioconf/20237001009

Fernández, N., García, M., Gil, E., Serrano, E., & Delibes, M. (2014). Bottom-up resource limitation: the ecosystem energy balance predicts the quality of nutrition in a herbivore prey population. *Research Portal Denmark*, 21. https://local.forskningsportal.dk/local/dki-cgi/ws/cris-link?src=dtu&id=dtu-62123df4-c492-4e0f-967d-94cd12f34e31&ti=Bottom-up%20resource%20limitation%3A%20the%20ecosystem%20energy%20balance%20predicts%20the%20quality%20of%20nutrition%20in%20a%20herbivore%20prey%20population

Giri, C. (2023). Frontiers in Global Mangrove Forest Monitoring. *Remote Sensing*, *15*(15), 3852. DOI: 10.3390/rs15153852 PMID: 39749006

Gorman, D., Jayakody, S., Kodikara, S., Udagedara, S., Dahanayaka, D. D. G. L., Ranawana, K. B., Kumara, M. P., Pahalawattaarachchi, V., Mathiventhan, T., Steven, A., Gunawardana, J. A. A. L., & Vanderklift, M. A. (2022). Past, Present and Future of Sri Lankan Coastal Macrophyte-Dominated Ecosystems: Blue Carbon, Conservation, Restoration and Policy. In *CRC Press eBooks* (p. 301). Informa. DOI: 10.1201/9781003288602-6

Harsa, H., Winarso, G., Setiawan, K. T., & Asriningrum, W. (2023). Satellite-based Mangrove species abundance estimate using Machine Learning ensemble. *IOP Conference Series. Earth and Environmental Science*, *1276*(1), 12055. DOI: 10.1088/1755-1315/1276/1/012055

Hemati, M., Mahdianpari, M., Shiri, H., & Mohammadimanesh, F. (2025). Comparative Analysis of UAVSAR Polarimetric Decompositions for Wetland Aboveground Biomass Mapping Using Machine Learning Models. *˜The œinternational Archives of the Photogrammetry, Remote Sensing and Spatial Information Sciences. The International Archives of the Photogrammetry, Remote Sensing and Spatial Information Sciences*, *19*, 19–25. Advance online publication. DOI: 10.5194/isprs-archives-XLVIII-M-8-2025-19-2025

Hiles, W., Miller, L. C., Smeaton, C., & Austin, W. E. N. (2024). UAV approaches for improved mapping of vegetation cover and estimation of carbon storage of small saltmarshes: Examples from Loch Fleet, northeast Scotland. *Biogeosciences*, *21*(4), 929–948. DOI: 10.5194/bg-21-929-2024

Hill, V., Zimmerman, R. C., Byron, D., & Heck, K. L.Jr. (2024). Mapping Seagrass Distribution and Abundance: Comparing Areal Cover and Biomass Estimates Between Space-Based and Airborne Imagery. *Remote Sensing (Basel)*, *16*(23), 4351. DOI: 10.3390/rs16234351

James, D., Collin, A., Mury, A., & Letard, M. (2021). ENHANCING UAV COASTAL MAPPING USING INFRARED PANSHARPENING. *˜The œinternational Archives of the Photogrammetry, Remote Sensing and Spatial Information Sciences. The International Archives of the Photogrammetry, Remote Sensing and Spatial Information Sciences*, *257*. Advance online publication. DOI: 10.5194/isprs-archives-XLIII-B3-2021-257-2021

Jessen, B., & Hill, K. (2024). Sea Grant at the Blue Carbon Frontier: Integrating Law, Science, Community Values, and Economics. *Oceanography (Washington, D.C.)*, *37*(1), 84–90. DOI: 10.5670/oceanog.2024.241

Jones, A. R., Segaran, R. R., Clarke, K., Waycott, M., Goh, W. S. H., & Gillanders, B. M. (2020). Estimating Mangrove Tree Biomass and Carbon Content: A Comparison of Forest Inventory Techniques and Drone Imagery. *Frontiers in Marine Science, 6*, 784. Advance online publication. DOI: 10.3389/fmars.2019.00784

Joyce, K. E., Fickas, K. C., & Kalamandeen, M. (2022). The unique value proposition for using drones to map coastal ecosystems. *Cambridge Prisms: Coastal Futures, 1*, e6. Advance online publication. DOI: 10.1017/cft.2022.7

Kandrot, S., Hayes, S., & Holloway, P. (2021). Applications of Uncrewed Aerial Vehicles (UAV) Technology to Support Integrated Coastal Zone Management and the UN Sustainable Development Goals at the Coast [Review of *Applications of Uncrewed Aerial Vehicles (UAV) Technology to Support Integrated Coastal Zone Management and the UN Sustainable Development Goals at the Coast*]. *Estuaries and Coasts, 45*(5), 1230. Springer Science+Business Media. DOI: 10.1007/s12237-021-01001-5

Laignel, B., Vignudelli, S., Almar, R., Becker, M., Bentamy, A., Benveniste, J., Birol, F., Frappart, F., Idier, D., Salameh, E., Passaro, M., Menende, M., Simard, M., Turki, I., & Verpoorter, C. (2023). Observation of the Coastal Areas, Estuaries and Deltas from Space. *Surveys in Geophysics, 44*(5), 1309–1356. DOI: 10.1007/s10712-022-09757-6

Lanceman, D., Sadat-Noori, M., Gaston, T. F., Drummond, C., & Glamore, W. (2022). Blue carbon ecosystem monitoring using remote sensing reveals wetland restoration pathways. *Frontiers in Environmental Science, 10*, 924221. Advance online publication. DOI: 10.3389/fenvs.2022.924221

Lecours, V., Disney, M., He, K. S., Pettorelli, N., Rowcliffe, J. M., Sankey, T. T., & Scales, K. L. (2021). Remote sensing and the UN Ocean Decade: High expectations, big opportunities. *Remote Sensing in Ecology and Conservation, 8*(3), 267–271. DOI: 10.1002/rse2.241

Lemenkova, P. (2019). *Seagrass Mapping and Monitoring Along the Coasts of Crete, Greece*. DOI: 10.31237/osf.io/p4h9v

Li, L., Xu, H., Zhang, Q., Zhan, Z., Liang, X., & Xing, J. (2024). Estimation methods of wetland carbon sink and factors influencing wetland carbon cycle: a review [Review of *Estimation methods of wetland carbon sink and factors influencing wetland carbon cycle: a review*]. *Carbon Research, 3*(1). Springer Science+Business Media. DOI: 10.1007/s44246-024-00135-y

Macreadie, P. I., Antón, A., Raven, J. A., Beaumont, N., Connolly, R. M., Friess, D. A., Kelleway, J. J., Kennedy, H., Kuwae, T., Lavery, P. S., Lovelock, C. E., Smale, D. A., Apostolaki, E. T., Atwood, T. B., Baldock, J., Bianchi, T. S., Chmura, G. L., Eyre, B. D., Fourqurean, J. W., … Duarte, C. M. (2019). The future of Blue Carbon science [Review of *The future of Blue Carbon science*]. *Nature Communications, 10*(1). Nature Portfolio. DOI: 10.1038/s41467-019-11693-w

Maurya, K., Mahajan, S., & Chaube, N. R. (2021). Remote sensing techniques: Mapping and monitoring of mangrove ecosystem—a review. *Complex & Intelligent Systems*, *7*(6), 2797–2818. DOI: 10.1007/s40747-021-00457-z

Mengis, N., Paul, A. J., & Fernández-Méndez, M. (2023). Counting (on) blue carbon—Challenges and ways forward for carbon accounting of ecosystem-based carbon removal in marine environments. *PLOS Climate*, *2*(8), e0000148. Advance online publication. DOI: 10.1371/journal.pclm.0000148

Monaca, A. (2024). Greetings from HYDRO 2023 Organizing Committee. *The International Hydrographic Review*, *30*(1), 125–126. DOI: 10.58440/ihr-30-1-e02

Moreaux, V., Papale, D., Gielen, B., & Loustau, D. (2019). Optimising observation networks for the early detection and unequivocal attribution of environmental effects on European forests. *HAL (Le Centre Pour La Communication Scientifique Directe)*. https://hal.inrae.fr/hal-02735529

Moses, W. J., Woude, A. V., & Palacios, S. L. (2022). Editorial: Emerging technologies and techniques for remote sensing of coastal and inland waters. *Frontiers in Environmental Science*, *10*, 1028307. Advance online publication. DOI: 10.3389/fenvs.2022.1028307

Mukaromah, L., As-syakur, A. R., & Prastowo, E. (2019). Opportunities for the application of Red-edge reflectance-based indices in coastal vegetation and agro-ecological studies. *Journal of Marine and Aquatic Sciences*, *5*(2), 251. DOI: 10.24843/jmas.2019.v05.i02.p13

Ngo, T. D., Quoc, K. N., Dang, N. T., Dang, C. T. M., Tran, L. L., & Nguyen, H. D. (2024). Analysis Of The Mangrove Structure In The Dong Rui Commune Based On Multispectral Unmanned Aerial Vehicle Image Data. *Geography, Environment, Sustainability*, *16*(4), 14–25. DOI: 10.24057/2071-9388-2023-2641

Peng, C., Song, Y., Huang, Q., Niu, Z., Zeng, G., Kang, H.-J., & Ying, M. (2025). Estimating coastal mangrove aboveground biomass of Zhejiang province based on sentinel-2 images and multiple machine learning algorithms. *Frontiers in Forests and Global Change*, *8*, 1662546. Advance online publication. DOI: 10.3389/ffgc.2025.1662546

Pham, T. D., Ha, N. T., Saintilan, N., Skidmore, A. K., Phan, D. C., Le, N. N., Viet, H. L., Takeuchi, W., & Friess, D. A. (2023). Advances in Earth observation and machine learning for quantifying blue carbon. *Earth-Science Reviews*, *243*, 104501. DOI: 10.1016/j.earscirev.2023.104501

Pham, T. D., Yokoya, N., Bui, D. T., Yoshino, K., & Friess, D. A. (2019). Remote Sensing Approaches for Monitoring Mangrove Species, Structure, and Biomass: Opportunities and Challenges. *Remote Sensing (Basel)*, *11*(3), 230. DOI: 10.3390/rs11030230

Pimple, U., Simonetti, D., Peters, R., Berger, U., Podest, E., & Gond, V. (2022). Enhancing monitoring of mangrove spatiotemporal tree diversity and distribution patterns. *Land Degradation & Development*, *34*(5), 1265–1282. DOI: 10.1002/ldr.4537

Porcar-Castell, A., Arthur, A. M., Rossini, M., Eklundh, L., Pacheco-Labrador, J., Anderson, K., Balzarolo, M., Martín, M. P., Jin, H., Tomelleri, E., Cerasoli, S., Sakowska, K., Hueni, A., Julitta, T., Nichol, C., & Vescovo, L. (2015). EUROSPEC: At the interface between remote-sensing and ecosystem CO₂ flux measurements in Europe. *Biogeosciences*, *12*(20), 6103–6124. DOI: 10.5194/bg-12-6103-2015

Pricope, N. G., Halls, J. N., Dalton, E. G., Minei, A., Chen, C., & Wang, Y. (2024). Precision Mapping of Coastal Wetlands: An Integrated Remote Sensing Approach Using UAS LiDAR and Multispectral Data. *Yaogan Xuebao*, *4*. Advance online publication. DOI: 10.34133/remotesensing.0169

Ranith, R., Menon, N., Nobi, E. P., Haris, F., Raj, R. P., & Pettersson, L. H. (2025). Decadal-scale urban footprints on seagrass blue carbon dynamics in Palk Bay, India: A remote sensing based framework. *Frontiers in Marine Science*, *12*, 1690700. Advance online publication. DOI: 10.3389/fmars.2025.1690700

Reithmaier, G. M. S., Cabral, A., Akhand, A., Bogard, M. J., Borges, A. V., Bouillon, S., Burdige, D. J., Call, M., Chen, N., Chen, X., Cotovicz, L. C.Jr, Eagle, M. J., Kristensen, E., Kroeger, K. D., Lu, Z., Maher, D. T., Pérez-Lloréns, J. L., Ray, R., Taillardat, P., & Santos, I. R. (2023). Carbonate chemistry and carbon sequestration driven by inorganic carbon outwelling from mangroves and saltmarshes. *Nature Communications*, *14*(1), 8196. DOI: 10.1038/s41467-023-44037-w PMID: 38081846

Ribeiro, R. J. da C., Goulart, F. F., Jacobson, T. K. B., da Rocha, D. M. S., Jaloretto, D., Gonçalves, B. E. F., & Gerber, D. (2024). Beyond climate villains: Estimating carbon storage in urban parks using high-resolution imagery combined with field sampling. *Research Square (Research Square)*. DOI: 10.21203/rs.3.rs-3817014/v1

Richard, M. O. (2024). Blue Carbon and the Role of Mangroves in Carbon Sequestration: Its Mechanisms, Estimation, Human Impacts and Conservation Strategies for Economic Incentives Among African Countries Along the Indian Ocean Belt. SSRN *Electronic Journal*. DOI: 10.2139/ssrn.4912339

Rondón, M., Ewane, E. B., Abdullah, M. M., Watt, M. S., Blanton, A., Abulibdeh, A., Burt, J. A., Rogers, K., Ali, T., Reef, R., Mohtar, R. H., Sidik, F., Fahrenberg, M., de-Miguel, S., Galgamuwa, G. A. P., Charabi, Y., Arachchige, P. S. P., Velasquez-Camacho, L., Al-Awadhi, T., … Mohan, M. (2023). Remote sensing-based assessment of mangrove ecosystems in the Gulf Cooperation Council countries: a systematic review [Review of *Remote sensing-based assessment of mangrove ecosystems in the Gulf Cooperation Council countries: a systematic review*]. *Frontiers in Marine Science, 10*. Frontiers Media. DOI: 10.3389/fmars.2023.1241928

Russell, S. K., Gillanders, B. M., Detmar, S., Fotheringham, D., & Jones, A. R. (2023). Determining Environmental Drivers of Fine-Scale Variability in Blue Carbon Soil Stocks. *Estuaries and Coasts*, *47*(1), 48–59. DOI: 10.1007/s12237-023-01260-4

Simpson, J., Bruce, E., Davies, K., & Barber, P. A. (2022). A Blueprint for the Estimation of Seagrass Carbon Stock Using Remote Sensing-Enabled Proxies. *Remote Sensing (Basel)*, *14*(15), 3572. DOI: 10.3390/rs14153572

Sunkur, R., Kantamaneni, K., Bokhoree, C., Rathnayake, U., & Fernando, M. (2024). Mangrove mapping and monitoring using remote sensing techniques towards climate change resilience. *Scientific Reports*, *14*(1), 6949. DOI: 10.1038/s41598-024-57563-4 PMID: 38521843

Tâm, N. T. T., & Ngân, N. V. C. (2025). Multi-sensor remote sensing and machine learning for aboveground biomass mapping in Vietnam's Melaleuca wetlands: A review. *International Journal of Research and Innovation in Applied Science*, *10*(10), 473–487. DOI: 10.51584/IJRIAS.2025.1010000035

Traganos, D., Lee, C. B., Blume, A., Poursanidis, D., Čižmek, H., Déter, J., Mačić, V., Montefalcone, M., Pergent, G., Pergent-Martinì, C., Ricart, A. M., & Reinartz, P. (2022). Spatially Explicit Seagrass Extent Mapping Across the Entire Mediterranean. *Frontiers in Marine Science*, *9*, 871799. Advance online publication. DOI: 10.3389/fmars.2022.871799

Traganos, D., & Reinartz, P. (2018). Machine learning-based retrieval of benthic reflectance and Posidonia oceanica seagrass extent using a semi-analytical inversion of Sentinel-2 satellite data. *International Journal of Remote Sensing*, *39*(24), 9428–9452. DOI: 10.1080/01431161.2018.1519289

Wang, S., Yan, D., Wang, C., Wu, L., & Huang, Y. (2024). A bibliometric analysis of blue carbon (1993–2023): Evolution of research hot topics and trends. *Frontiers in Marine Science*, *11*, 1430545. Advance online publication. DOI: 10.3389/fmars.2024.1430545

Windle, A. E., Staver, L. W., Elmore, A. J., Scherer, S., Keller, S. M., Malmgren, B., & Silsbe, G. M. (2023). Multi-temporal high-resolution marsh vegetation mapping using unoccupied aircraft system remote sensing and machine learning. *Frontiers in Remote Sensing*, *4*, 1140999. Advance online publication. DOI: 10.3389/frsen.2023.1140999

Xie, Y., Tian, Y., Zhang, Q., Zhang, Y., Jin, T., & lin, J. (2025). *A New Method for Mangrove Biomass Prediction Based on Automatic Machine Learning and Multi-Source Uav Data Fusion*. DOI: 10.2139/ssrn.5168622

Yulianto, F., Basyuni, M., Mardiatno, D., Ardiati, A. S., Kusuma, I. K. T. W., Purnomo, M. B., Kongko, W., Wardani, K. S., Fachrudin, I., Baba, S., & Miyagi, T. (2025). Long-term monitoring of mangrove restoration and carbon dynamics in Tanjung Benoa, Bali, Indonesia: An assessment approach using combination remote sensing and field surveys. *Research Square (Research Square)*. DOI: 10.21203/rs.3.rs-6779800/v1

Zhang, L. (2024). Environmental hotspots, frontiers and analytical framework of Blue Carbon research: A quantitative analysis of knowledge map based on CiteSpace. *Frontiers in Environmental Science*, *12*, 1387655. Advance online publication. DOI: 10.3389/fenvs.2024.1387655

Zhang, Y., Hou, Y., & Yang, X. (2018). Sea Level Rise and Coastal Infrastructure. In *InTech eBooks*. DOI: 10.5772/intechopen.68572

Chapter 4
Integrating AI and Cloud-Enabled Deep Learning for Earthquake Forecasting:
A Cognitive City Perspective Using OLR Precursors

D. Rubidha Devi
http://orcid.org/0000-0001-9074-8984
SASTRA University, India

Priya Govindarajan
Amrita School of Computing, India

ABSTRACT

As cities evolve into intelligent cyber-physical ecosystems, infrastructure resilience and disaster preparedness are critical. Earthquake forecasting remains a challenging task, requiring advanced machine learning and non-seismic data integration. This study proposes a hybrid SSA-LSTM model that leverages satellite-derived Outgoing Longwave Radiation (OLR) as a seismic precursor. Singular Spectrum Analysis (SSA) is employed to decompose noisy OLR signals and extract dominant temporal features, which are then learned by a Long Short-Term Memory (LSTM) network. The proposed model is evaluated against a conventional LSTM using identical historical seismic and OLR datasets. Experimental results demonstrate that SSA-LSTM consistently outperforms the standard LSTM, achieving lower forecasting errors and improved

DOI: 10.4018/979-8-3373-8745-1.ch004

robustness. Statistical analysis confirms the effectiveness of SSA in enhancing long-term dependency learning. The findings highlight the potential of combining signal decomposition and deep learning for reliable earthquake forecasting.

INTRODUCTION

Over the decades, natural disasters plays a significant role in the human's life. Among other natural hazards earthquakes, floods, hurricanes, wildfire has been more devastating compared with other disasters(Yun et al., 2021). This type of hazards have resulted in significant human and economic losses, and have emerged as one of the most pressing challenges facing global security (Rosas-Muñoz & Baquedano-Rodríguez, 2022). As such, it is crucial to conduct in-depth research to develop effective strategies for reducing the mortality rates and minimizing the damage caused by natural disasters. Such research should explore various factors, including the effects of change in climate on natural disasters, the effectiveness of early warning systems (Hart et al., 2022), the importance of public preparedness in catastrophe management, and the practice of innovative technologies to enhance adversity (Vigny & Klein, 2022) and recovery efforts. Only through comprehensive research we can hope to develop long-term solutions that can mitigate the adverse effects of natural disasters (Romero-Andrade et al., 2022) and safeguard the lives and livelihoods of communities around the world. Several studies have been conducted on seismic precursors and magnitude, and Gutenberg b-values (Pailoplee, 2017) (Pailoplee, 2014) have been plotted to forecast earthquakes in unique faults(Wang et al., 2021) (Volvach et al., 2022). The Chile region is known as seismically vulnerable since the most devastating earthquake has occurred (Venkatanathan et al., 2021), and this region has been chosen as the study region. To forecast earthquakes, it is important to know the location, magnitude, and timing of the event. Every natural disaster leaves behind some evidence of its occurrence in the past (Cisternas et al., 2005). Extensive research has determined that certain ionospheric fluctuations and fluctuations in outgoing longwave radiation (OLR) (Zhang et al., 2021) (Parrot, 2022) can be traced back to the precursors leading to the desired outcome. These findings are of tremendous significance and should be given due consideration while formulating effective strategies and techniques to attain the desired objective (Rolland, 2021). In light of this information, it is imperative to take into account the impact of these factors on the outcome and devise appropriate measures accordingly(Parrot et al., 2021) (Mohamed et al., 2021).

Outgoing Longwave Radiation (OLR) is an essential factor in earthquake forecasting, which is a non-seismic precursor (Salh et al., 2022). The upward thermal radiation flux originates from the atmosphere and is prejudiced through cloud cover and shallow temperature. It's essential for regulating the earth's temperature and climate. OLR data, which is derived from infrared band telemetry, provides reliable and continuous information that is highly sensitive to changes in the ozone layer as well as near-ground temperature. This makes it a useful tool for monitoring the aftermath of catastrophic events such as earthquakes, which often begin with "hot" anomalies (Sridharan et al., 2023). A group of scientists has suggested that OLR can be used to forecast earthquakes. This is because satellite sensors can be used to measure OLR, making it possible to monitor the entire planet 24/7 at a relatively low cost. Scientists from around the world use different methods to detect any abnormal changes in OLR before an earthquake occurs. This research uses the Singular Spectrum Analysis (SSA) approach to preprocess the Outgoing Longwave Radiation (OLR) activity before an earthquake.

OLR's short anomaly period of approximately 30 days makes it an excellent precursor for earthquake forecasting. The fluctuations in the OLR state serve as an early warning sign, effectively identifying significant impending seismic activity. This has been tested and proven through twelve earthquakes in the Chilean area. The existence of OLR anomalies has been successfully demonstrated by rebuilding the data using the SSA approach, which eliminated all residuals resulting from solar activity. The original OLR values and the rebuilt SSA values were then compared to determine the top and lower bounds of anomalies that occurred. This reconstructed data has been given as input to LSTM in which the proposed work SSA-LSTM has been built to detect anomalies which acts as a framework for forecasting earthquakes. The work additionally states the importance of anomalies and their incidence concerning OLR parameters and the SSA set of guidelines (Golyandina et al., 2014). The approach involves analyzing the OLR (Outgoing Longwave Radiation) collected over a while for every instance of an earthquake. The project involved conducting a comprehensive analysis of the dataset, which included the OLR parameter. The objective was to identify patterns and correlations between earthquake occurrences and OLR data. Advanced statistical techniques was used to uncover trends and insights that could assist in forecasting future earthquakes. Overall, this research could offer significant insights into forecasting earthquakes and improving our preparedness and mitigation strategies for natural disasters. The study objects to identify the relationship between data set anomalies and the Outgoing Longwave Radiation (OLR) parameter using a hybrid SSA-LSTM framework. This approach seeks to enhance our understanding of the data and help identify patterns and trends. Artificial Intelligence (AI) has demonstrated transformative impact across diverse domains such as medical imaging for disease diagnosis, autonomous navigation systems, and climate modeling for extreme weather prediction. Deep learning architectures have shown strong capability in extracting complex nonlinear patterns from high-dimensional data. These successes motivate the exploration of AI-based frameworks in geophysical applications. In particular, data-driven models can complement traditional seismological approaches by identifying subtle atmospheric precursors associated with seismic activity.

The research motivation behind this research stems from the critical need to enhance earthquake forecasting capabilities, particularly in urban areas where the potential impact of seismic events is significant. Earthquakes pose a significant threat to lives, infrastructure, and economic stability, making early detection and warning systems imperative for effective disaster preparedness and mitigation strategies. By integrating atmospheric parameter such as Outgoing Longwave Radiation (OLR), derived from satellite observations, this research seeks to broaden the scope of earthquake prediction models. Moreover, the combination of SSA-LSTM to OLR time-series data represents a novel approach in earthquake forecasting research. SSA-LSTM has demonstrated effectiveness in identifying hidden patterns and trends within time-series data, offering a promising avenue for detecting seismic precursors within OLR observations. The ultimate goal of this research is to harness the power of SSA-LSTM on OLR data, that improves early warning systems, enhance disaster resilience, and ultimately protect survives and alleviate the socio-economic effect of earthquakes in smart city contexts.

The main contribution of the proposed methodology lies in its innovative integration of SSA-LSTM with Outgoing Longwave Radiation (OLR) data for earthquake forecasting, which has been shown to achieve comparable accuracy to traditional LSTM models. Leveraging OLR data alongside seismic information offers a more inclusive kind of the atmospheric environments preceding seismic events, enhancing the detection of subtle precursory signals and improving the overall reliability of earthquake forecasts. Furthermore, the adaptability of the methodology to smart city contexts ensures its relevance in urban environments, where the integration of diverse data sources and advanced analytics is paramount for proactive disaster management and urban planning. Overall, the proposed methodology offers a scalable framework that can be adapted to interplate boundaries where OLR plays a vital role in devastating earthquake.

The remainder of this paper is structured as follows: In Section 2, we present a thorough assessment of interrelated work in earthquake prediction, emphasizing recent advancements in LSTM and SSA techniques. Section 3 describes the methodology employed in our study, including data preprocessing, model architecture, and training procedures for SSA-LSTM models. In Section 4, we present the investigational setup, detailing the dataset used, evaluation metrics, and execution details. Section 5 presents and discusses the results of our empirical evaluation, comparing the performance of SSA-LSTM with traditional LSTM models in earthquake forecasting tasks. Finally, Section 6 concludes the paper by summarizing our findings and exploring potential avenues for future research.

RELATED WORK

By analyzing selected OLR data, we can gain insights into how OLR anomalies detected are related to the Wenchuan Earthquake which occurred on May 12, 2008. This earthquake had a magnitude of M 8.0(Xiong et al., 2009) (Guo et al., 2010), and the Puer seismic shocks on June 3, 2007, with a magnitude of M 6.4 in Ninger, Yunnan. OLR data was used to analyze the Lancang-genma earthquake (M 7.4), which occurred on November 11, 1988, in the Sichuan-Yunnan region of China, along with the Wenchuan and Puer earthquake . It was discovered that the abnormalities were connected to an impending earthquake. Similarly, OLR anomalies were found before the Yunnan earthquake in June 2007 (Chen et al., 2010). The Haiti Earthquake, which occurred on January 12, 2010, had anomalies that were observed between December 19 and 24, 2010(Xiong et al., 2010). These anomalies were identified in various geophysical fields and are believed to be genetically connected. As a result, they should not be considered as separate processes. The Lithosphere–Atmosphere–Ionosphere Coupling (LAIC) mechanism explains how tectonic stress accumulation may alter surface and atmospheric parameters prior to earthquakes. Stress-induced rock microfracturing releases gases such as radon, modifying atmospheric ionization and latent heat flux, which can influence outgoing longwave radiation (OLR). These thermodynamic perturbations may manifest as detectable satellite-based thermal anomalies. However, since this study focuses on a specific Chilean subduction zone, further validation across different tectonic regimes is necessary to confirm model generalizability (Pulinets & Ouzounov, 2011). The OLR anomaly, which stands for Outgoing Longwave Radiation anomaly, was initially observed among all the different geophysical fields. This anomaly refers to the difference between the actual amount of longwave radiation that is emitted by the Earth and the expected amount of radiation that should be emitted based on the temperature of the planet's surface. The OLR anomaly is an important indicator of changes in the Earth's climate and is commonly used to track weather patterns and predict future climate trends. From the year 2006 to 2010, the strong earthquakes and Jet streams were tested to find the link with OLR for two regions and it shows the probability of forecasting is high at the end. OLR may be used in short-term earthquake forecasting (Zoran et al., 2016). Vrancea placed in Eastern Carpathians, Romania is chosen to receive the in-situ and satellite data, which concludes that there are few anomalies based on OLR (Kong et al., 2018). There were larger OLR anomalies in the Longmenshan fault zone before the Wenchuan and Lushan earthquakes which are located in southwest China (Liu et al., 2018). In Napa, 2014, and Taiwan, 2016 with M 6., and Kumamoto, Japan in 2016 with M 7.0, a retrospective analysis was done to find pre-earthquake anomalies based on the atmosphere which helps in short-term earthquake forecasting(Dokht et al., 2017). To remove random noise and more pre-processing activities like reconstructing the missing traces this algorithm is effectively used (Sangeetha & Govindarajan, 2023). A study given by the author (Manenti et al., 2019) states, that the usage of Singular Spectrum Analysis (SSA) imaged with perspective to upper mantle discontinuities. Its implementation was projected through the frequency-space domain. The study reveals the consistency of Mantle Transition Zone (MTZ) phase boundaries and the area of strong signals (Moreno & dos Santos Coelho, 2018). It can also be used as a powerful tool, in forecasting the weak seismic phase within the interiors of the Earth. Another researcher Moreno (Krishnan et al., 2023) explores the reconstruction and extraction of components using the SSA algorithm. The study indicates robustness by showcasing a better performance of the hybrid SSA-ANFIS-FCM through different speeds of the wind and its forecasting. Another study (Unnikrishnan & Jothiprakash, 2018) indicates the emphasis of SSA in indicating the periodic, cyclic, and noise components with perspective to the daily rainfall series. The study concludes that SSA forecasts the rainfall and its forecasting with reasonable accuracy in a single run. Another research work (Mahmoudvand & Rahim, 2021) states the importance of forecasting and its intervals with perspective to the forecast accuracy in time-series analysis. The proposed work is based on the median-based alternative algorithm, where the mean is replaced by a median. The result of the work is compared with the simulation of Monte Carlo and other real data applications.

Table 1. Comparative Table of Related Work

Study	Evaluation Tool	Utilized Technique	Performance Metric	Dataset
Wenchuan Earthquake, 2008 (Xiong et al., 2009), (Guo et al., 2010)	Analysis of OLR anomalies	Observation of OLR anomalies	Detection of anomalies before the earthquake	OLR data
Haiti Earthquake, 2010 (Xiong et al., 2010)	Analysis of OLR anomalies	Observation of OLR anomalies	Detection of anomalies before the earthquake	OLR data
Strong Earthquakes and Jet Streams, 2006-2010 (Zoran et al., 2016)	Statistical analysis	Correlation analysis	Probability of forecasting earthquakes	OLR data, Jet stream data
Vrancea, Eastern Carpathians, Romania (Kong et al., 2018)	In-situ and satellite data analysis	Aqua time-series analysis	Detection of OLR anomalies	NOAA AVHRR, MODIS Terra\ Aqua
SSA in Upper Mantle Discontinuities (Sangeetha & Govindarajan, 2023)	SSA	Frequency-space Domain Analysis	Consistency of MTZ phase boundaries	Seismic data
Hybrid SSA-ANFIS-FCM (Moreno & dos Santos Coelho, 2018)	SSA-ANFIS-FCM	Component extraction and reconstruction	Robustness, better performance in wind speed forecasting	Wind data

A different forecasting model based on wind speed is defined in a research study which encapsulates various methods. The work focuses on the extraction of information about the speed of the wind and its components. The outcome visualizes satisfactory forecasting in forecasting the wind speed. One of the researcher's works, emphasizes the application of soft computing and its techniques with perspective to structural and earthquake-related engineering problems. The study focuses on the way to drive the solutions for complex problems respective to numerical methods or through non-linear formulation. It concludes with the fact that SC techniques end with more accurate as well as efficient solutions. A study enlightens, on the features of the RSSA package and the efficient way in which one would work with the SSA algorithm. It concludes that SSA is predominately playing a major role in analysis, parameter estimation, and forecasting the time-series. The combination of more than one methodology, such as combining SSA and linear Recurrent Formula with Neural Network for representing the time-series forecasting, has the added advantage of letting the researchers learn and explore the advantages along with the strengths of the different methodologies. These entities are useful for loading data and time-series, as well as for handling non-linear stochastic and deterministic patterns. Time-series forecasting is a relevant and vital topic in today's era, it spans various domains such as medicine, commerce, science, and engineering. When implemented with fuzzy rules, this work represents transparency from the perspective of PRSSA+ENFN and its corresponding structure. A study which was carried off at Badargadh, Kachchh region of Gujarat, India –represents a work to identify the intended periodic oscillations using SSA and also identifies the earthquake precursors with perspective to the time-series of soil radon. Table 1 concludes that SSA is termed the efficient tool to differentiate periodic and aperiodic fluctuations for periodic and aperiodic variations for radon and other parameters. Recent advancements in transformer-based time-series models have shown superior long-term dependency modeling compared to recurrent architectures. Therefore, relevant transformer-based forecasting studies are incorporated for contextual comparison. Furthermore, the effectiveness of SSA in isolating seasonal and sclar residual components is quantitatively evaluated by analyzing variance reduction before and after decomposition. The reconstructed signal demonstrates improved stationarity and enhanced anomaly visibility.

METHODOLOGY

SSA is a widely used non-parametric method that was first introduced in 1986 and it grew to become popular by way of 2000 for inspecting time sequence statistics which decomposes the time-series as a deterministic model that is in addition represented as the sum of Eigen triples, each of which contains an Eigen function, the predominant component, and the square root of the Eigenvalue. This approach effectively divides the time series data into several interpretable components. Based on the window length (WL),it determines how the time-series is transformed into a trajectory matrix, which is eventually considered for SVD (Singular Value Decomposition).). The time step taken for the statistics to further split into component extraction is referred as the window length. The number of Eigen triples generated corresponds to the selected window length, which finally represents the SVD's output. Decomposing the trajectory matrix yields tiny component matrices. Decomposition and Reconstruction degrees are considered the most essential stages in this SSA technique. SSA consists of four main steps. The first and second steps, such as embedding and SVD, fall under the decomposition stage, while the other two methods, like grouping and averaging, come under the reconstruction stage. The total structure of the working of SSA-LSTM is stated in Figure 1.

Figure 1. Overall Framework with Singular Spectrum Analysis - LSTM

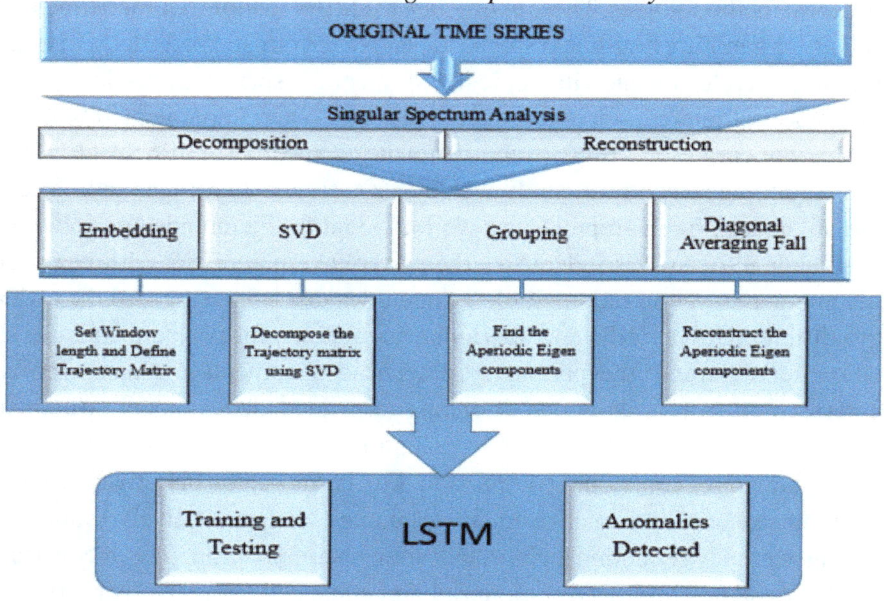

The window length is a dimension of the multidimensional series that represents the time-series in a single dimension. The trajectory matrix is created from the multidimensional time-series. In this step, focus is completely based on a single input i.e., window length, WL. Usually, the value of WL will always be less than half of the total record count(N). The lagged vectors are created as LV = N -WL + 1 to execute the embedding by mapping the entire records with the series of insulated paths of Window length size(WL). If M = m1, m2,..., mN is the original record data under consideration, then 'S' is the trajectory matrix and may be written as

$$S = \begin{bmatrix} m1 & m2 & m3 & \cdots & mWL \\ m2 & m3 & m4 & \cdots & mWL+1 \\ \vdots & \vdots & \vdots & \cdots & \vdots \\ mLV & mLV+1 & mLV+2 & \cdots & mN \end{bmatrix} \quad (1)$$

Transforming the entire data into a matrix is known as embedding. The trajectory matrix is subjected to SVD, which provides a series of SVD modules which is referred as Eigen Triples.

$$S = \sum_{i=1}^{p} Si \quad (2)$$

Where

$$Si = UEi\sqrt{\alpha i}$$

VEi (3)

Where 'α' is an eigenvalue of the S matrix, UE and VE are its eigenfunctions, and i ranges from 1 to 'p,' where 'p' is the number of positive eigenvalues. Eigentriples are the collective names for the eigenvalues, eigenfunctions, and major components.

The initial step in the SSA rebuilding phase is grouping. The relevant SVD constituents are chosen from the grouped components to recreate the original OLR values. In terms of grouping, there is no absolute law. The grouping approach varies depending on the characteristics of the time-series that must be recreated. As a result, the needed time-series may be recreated from 'r' disjoint subsets [S_{I1}, S_{I2},..., S_{Ir}] after grouping. The second step in reconstructing the time-series is diagonal averaging. Each vector S_{IIv} from the clustering step is converted into the latest series of length N at this point. Assume that A is a WL x LV matrix with the entries a_{ij}, $1 \leq i \leq WL$, $1 \leq j \leq LV$. Set $WL* = \min(WL, LV)$, $LV* = \max(WL, LV)$, and $N = WL + LV - 1$. Let $a_{ij}* = a_{ij}$ if $WL < LV$ and $a_{ij}* = a_{ji}$ otherwise. The matrix A is converted into the series $a_1, a_2,.., a_N$ via the diagonal averaging method using the following formula.

$$\text{Alv} = \begin{cases} \frac{1}{lv} & \sum_{r=1}^{lv} a^{*}_{r,lv-r+1} \quad for\ 1 \leq lv \leq WL^{*} \\ \frac{1}{WL^{*}} & \sum_{r=1}^{WL^{*}} a^{*}_{r,LV-r+1} \quad for\ WL^{*} \leq lv \leq LV \\ \frac{1}{N-lv+1} & \sum_{r=lv-LV^{*}+1}^{N-LV^{*}+1} a^{*}_{r,lv-r+1} \quad for\ LV^{*} \leq lv \leq N \end{cases} \tag{4}$$

This is equivalent to be around the matrix components over the antidiagonals i + j=lv +1. The select of lv= 1 results in $y_1 = y_{(1,1)}$, whereas lv= 2 produces $y_2 = (y_{1,2} + y_{2,1})/2$. A rebuilt series is constructed by diagonally averaging the grouping matrix results (YIlv).

$$\overline{S}^{(lv)} = \overline{S}_1^{(lv)} \ldots \overline{S}_N^{(lv)} \tag{5}$$

From the rebuilt series, it proposes that the original series M= $(m_1, m_2, \ldots m_N)$ has been distributed into a sum of 'r' rebuilt series.

$$M = \sum_{lv=1}^{r} \overline{m}_n^{\,lv}\ where\ (n = 1,2 \ldots N) \tag{6}$$

The significance of the reconstructed series within the broader context of the time-series data can be effectively gauged by contrasting them against the actual observed series. In the following section, we will provide a comprehensive examination of the observed series associated with earthquake time-series data, highlighting its critical attributes and implications.Table 2 meticulously outlines the structure and methodology of the Singular Spectrum Analysis (SSA) Algorithm, serving as a roadmap for understanding its operational mechanics. Outgoing Longwave Radiation (OLR) is vividly characterized as the radiant energy that ascends from the Earth's surface, traversing the vast expanse of the sky before escaping into the cosmos. This dynamic flux is profoundly influenced by a trio of factors: the Earth's ambient temperature, the complex interplay of atmospheric elements, and the moisture-laden clouds that drift through the sky. The Advanced Very High Resolution Radiometer (AVHRR) serves as a sophisticated sentinel in this observation process, capturing these vital OLR measurements with remarkable precision. The wealth of data collected by this advanced technology is made accessible through reputable repositories maintained by the National Oceanic and Atmospheric Administration (NOAA) and the United States Geological Survey (USGS).

Outgoing Longwave Radiation (OLR) values, which are carefully collected from modern satellites circling the globe, are keys to forecast earthquakes. They are not just numbers; they represent complex interactions between Earth's surface and its atmosphere, and can indicate patterns of energy that presage seismic events. By using a collection of sophisticated algorithms designed to work on different kinds of OLR data, we can find patterns that may precede seismic events. By employing a suite of advanced algorithms specifically designed to analyze various types of OLR data, researchers can detect anomalies that may hint at impending seismic events.

As the Earth's atmosphere releases thermal energy to space, this process is affected by a variety of factors, including satellites that orbit above. These satellites, which are necessary for collecting this information, can sometimes interfere with the proper transfer of energy, which affects the reliability of the information we collect from them. Changes in temperature in the upper atmosphere and variations in humidity levels can dramatically affect how satellites work. These changes in the atmosphere are not just relevant to the bigger picture of seismic activity, but they are also markers of movement within the Earth's crust. This information is not just applicable to variables such as altitude or the distance from a given location to an earthquake's epicenter, but it highlights the relationship between these atmospheric variables and seismics events.

In the realm of earthquake forecasting, the text elaborates on the sophisticated methodologies provided by Singular Spectrum Analysis (SSA) techniques. Starting from a time-series (data points collected at a given instance of time), SSA technique converts it into a trajectory matrix where each entry is a value measured at a certain moment of time. This matrix is then submitted to the powerful and elegant tool called SVD (Singular Value Decomposition). This data is then divided into distinctive components called Eigen Triples. These components are not random numbers; they are called Eigenvalues and Eigenvectors multiplied by Singular Values, and they represent patterns of energy at different times. These components are very important in understanding the dynamics leading up to seismic events. The text then continues with the reconstruction of the original time-series from these Eigen Triples. This is done by using advanced grouping techniques on a powerful method called diagonal averaging, which reassembles the data to reveal patterns that may not have been obvious at first. Once this reconstructed series is generated, it is compared to the observed seismic data multiple times. This comparison is a critical step in determining patterns that lead to accurate forecasts of future earthquakes, turning these abstract data into usable information that can ultimately save lives and reduce the effects of a disaster.

Table 2. Singular Spectrum Analysis Algorithm

Input	Time-series data m[1...N],
	Window length (WL) = 52 and Embedding Dimension L = 2191(6 years).
Output	Reconstructed Time-series Y_{avg}

Algorithm	1.Embedding:
	for i from 0 to N-L + 2 do
	for j from 0 to L-1 do
	M[I,j] = m[i+j-1]
	next
	next
	2. Singular Value Decomposition(SVD):
	for i from 1 to N – L + 1 do
	//Construct trajectory marix
	for j from 0 to K-1 do
	Y[j,:] = M[i+j-1,:]
	next
	// Perform SVD
	[U,S,V] = SVD(Y)
	// Select significant singular values
	k =number of significant singular values
	Uk =first k columns of U
	Sk =first k singular values
	Vk =first k rows of V
	// Reconstruct the time-series
	for j from 1 to K do
	Xrecon[i + j – 1,:] = Uk · Sk · VT k
	next
	next
	3. Ensemble average of reconstructed time-series:
	for i from 0 to N-1 do
	for j from 0 to N –L +2 do
	Yavg[i]+ = Xrecon[j,i]
	next
	Yavg[i]/ = (N – L +1)
	next

RESULTS AND DISCUSSION

For our research model, we acquired seismic data from NOAA (National Oceanic and Atmospheric Administration), a distinguished authority in environmental monitoring and data collection. This invaluable seismic dataset encompasses a vast geographic region defined by the coordinates 7.014°N, -89.297°W to -52.483°S, -64.688°E, which includes segments of South America renowned for their notable seismic activity, vividly depicted in Figure 2. he right side of the image presents a well-organized and informative overview of the significant earthquakes chosen for our analysis using the SSA-LSTM (Singular Spectrum Analysis - Long Short-Term Memory) modeling technique. The seismic data was meticulously collected from an intricate network of seismometers, strategically deployed across the rugged landscape. These advanced instruments tirelessly monitor ground motion, capturing seismic waves generated by earthquakes, tremors, and a variety of other seismic events. The dataset itself comprises a rich and diverse range of seismic signals, recorded over a range of time spans, from fleeting days to extensive years. Each seismic occurrence is precisely documented, characterized by its geographical coordinates, magnitude, depth, and the exact moment of its signature event. This comprehensive information serves as a cornerstone for conducting in-depth earthquake analysis and refining predictive modeling efforts, illuminating the complex dynamics of our planet's seismic activity.

The meticulous process of data pre-processing and reconstruction is initiated through the application of Singular Spectrum Analysis (SSA), which expertly cleans and restructures the raw data in preparation for the sophisticated capabilities of Long Short-Term Memory (LSTM) networks tailored for time-series analysis. To ensure a comprehensive approach to model training and evaluation, the dataset has been thoughtfully segmented into three distinct subsets: the training set, validation set, and testing set. The training set consists of meticulously curated data collected up to December 31, 2020, which serves as the cornerstone for developing the SSA-LSTM model. This initial phase is critical, as it lays the groundwork for the model's learning process, enabling it to discern patterns and trends inherent in the seismic data. Once the model is trained, the validation set comes into play. This subset is essential for hyperparameter tuning, allowing researchers to optimize aspects such as learning rate and layer configurations. This careful calibration process is designed to enhance the model's performance, ultimately leading to the selection of the final, most effective model. The testing set, distinct in its purpose, is reserved as an external dataset comprised of seismic data that has not been seen by the model before. It is used to rigorously evaluate the model's generalization performance, providing insight into how well the trained SSA-LSTM model can predict outcomes and patterns in new, unfamiliar data. This robust workflow is pivotal in ensuring the model's reliability and accuracy in real-world applications.

Figure 2. Earthquake with magnitude greater or equal to 6, selected for 30 years (1991 to 2020) in 7.014N,-89.297W,-52.483S,-64.688E region. (left side) A devastating earthquake in a region with coordinates 7.014N, -89.297W,-52.483S,-64.688E and a magnitude of at least 6 has been selected for the period spanning from 1991 to 2020. (right)

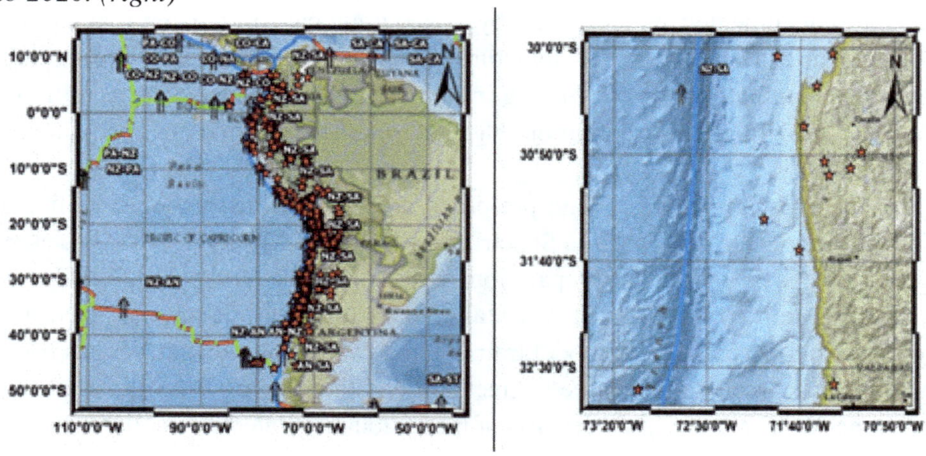

Table 3. Computational Complexity and Average Milliseconds per Step for Different Models

Model	Computational Complexity (sec)	Avg. Millisec per Step
SSA-LSTM	104	4
LSTM	138	5
Bi-LSTM	328	13
LSTM-VAE	248	10
LSTM-ED	228	9

For a comprehensive approach to testing and validation, we engage in the meticulous preprocessing of a time-series dataset that extends over an impressive six-year period. This dataset specifically focuses on Outgoing Longwave Radiation (OLR) precursor data associated with various seismic events occurring between the latitudes of -30° N and -32° S, as well as longitudes ranging from -71° E to -73° W. In Table 3, we provide an in-depth analysis of the computational complexity encountered, along with the average processing time measured in milliseconds required for each step across a variety of analytical models that utilize this intricate time-series data. Additionally, Table 4 highlights a curated collection of sample data that has been meticulously selected for our time-series analysis, serving as a foundation for our investigation. By employing the Singular Spectrum Analysis (SSA) technique to reconstruct the dataset, we gain the ability to discern subtle differences and variations within the data, enabling us to thoroughly assess the effectiveness and performance of the diverse methodologies implemented throughout our research.

Table 4. Sample data - Devastating Earthquake selected for Time-series Analysis

Date	Latitude	Longitude	Depth(km)	Magnitude
1992-11-28T03:13:33.360Z	-31.53	-71.992	11.7	6.5
1997-07-06T09:54:00.760Z	-30.058	-71.872	19	6.8
1997-10-15T01:03:33.460Z	-30.933	-71.22	58	7.1
1998-01-12T10:14:07.630Z	-30.935	-71.41	34.8	6.6
2001-04-09T09:00:57.170Z	-32.668	-73.109	11	6.7
2002-06-18T13:56:22.830Z	-30.805	-71.124	54	6.6
2003-06-20T13:30:41.640Z	-30.608	-71.637	33	6.8
2012-04-17T03:50:15.610Z	-32.625	-71.365	29	6.7
2013-10-31T23:03:59.720Z	-30.2921	-71.5215	27	6.6
2015-09-16T22:54:32.860Z	-31.5729	-71.6744	22.44	8.3
2015-11-07T07:31:43.870Z	-30.8796	-71.4519	46	6.8
2019-01-20T01:32:52.480Z	-30.0404	-71.3815	63	6.7

The implementation of the Singular Spectrum Analysis using Long Short-Term Memory (SSA-LSTM) networks commenced with the integration of non-seismic Outgoing Longwave Radiation (OLR) data as the foundational input for the algorithm. In this methodology, the selection of the window length is of paramount importance, as it has a profound impact on the ability to extract meaningful signals from the data. Specifically, utilizing a window length that is a multiple of 12 has demonstrated significant efficacy in effectively isolating the underlying periodic signals amidst the noise. To thoroughly evaluate the performance of the SSA under varied conditions, a sensitivity analysis is performed to evaluate different SSA window lengths (36, 48, 60, and 72) to determine their impact on reconstruction quality and forecasting accuracy. The window length of 60 provides optimal trade-off between noise reduction and signal preservation. Furthermore, the Smart/Cognitive City concept is integrated through a proposed real-time satellite monitoring and AI-based forecasting framework. This enables automated anomaly detection for urban resilience planning.

This approach is meticulously designed to capture both the transient short-term fluctuations as well as the more gradual long-term trends that are inherent in the dataset. Following the determination of the appropriate window size, we construct what is known as a Hankel matrix, an essential component also referred to as a Trajectory matrix. This matrix is composed of lagged observations from the time series, where each row represents observations at a specific point in time while each column depicts a different lag. The structure of the Hankel matrix enables a comprehensive analysis of temporal dependencies within the data. To extract valuable insights from the Hankel matrix, we employ Singular Value Decomposition (SVD), a robust matrix factorization technique that decomposes the matrix into singular values and their corresponding eigenvectors.

The eigenvectors obtained are linked to the eigenvalues determined by the chosen window length. A notable feature of this process is that highly skewed eigenvalues are instrumental in reconstructing the time series data, allowing us to focus on aperiodic components. This analytical process is illustrated in Figure 3, which depicts the seismic event that transpired on November 28, 1992. And few more seismic events aperiodic component has been shown in Figure 4 for better understanding.

The data obtained from this analysis will facilitate a detailed study of the seismic activity and patterns that preceded this earthquake, providing valuable insights into the fundamental mechanisms that triggered the event. By scrutinizing the fluctuations across various seismic parameters over time, we aim to identify any significant changes or anomalies that could have contributed to the earthquake, ultimately enhancing our comprehension of the dynamics of seismic activity in the affected region.

Figure 3. Reconstructed Time-series data using Singular Spectrum Analysis Technique

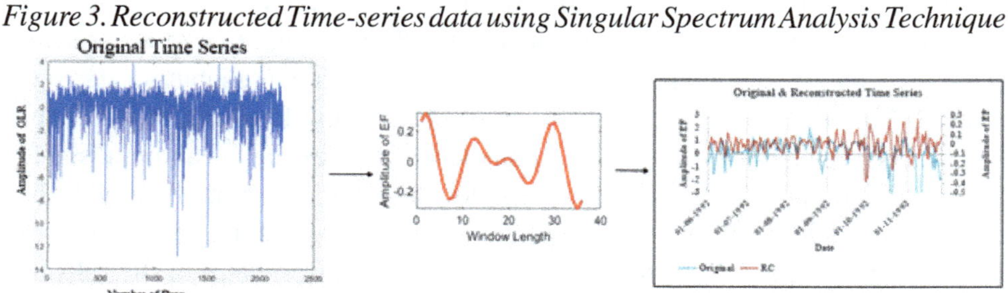

Figure 4. Aperiodic Components for different periodic major seismic events using Singular Spectrum Analysis

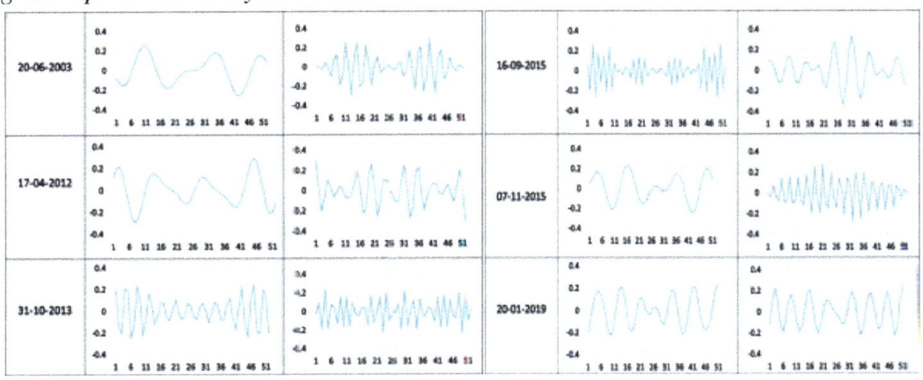

The study analyzed the reconstructed SSA time-series data using LSTM aiming to evaluate their efficacy in anticipating seismic events, specifically earthquakes. The findings suggest that Outgoing Longwave Radiation (OLR) could play a pivotal role in predicting earthquake probabilities. LSTM networks have shown remarkable promise, particularly in their ability to capture complex and long-term temporal relationships inherent in time-series data. This unique capability allows LSTMs to effectively learn from historical seismic data and adjust to evolving patterns in seismic activity. The models particularly harness critical information derived from OLR precursors, which serve as vital indicators that reflect the Earth's atmospheric behavior in the lead-up to an earthquake. By delving into these OLR precursors, LSTM models can discern significant patterns and correlations that aid in crafting precise forecasts of future seismic occurrences.

The adaptability of LSTM models in learning from past data has resulted in enhanced accuracy in earthquake predictions, which is crucial for implementing effective public safety measures and disaster preparedness strategies. For the purpose of this study, the LSTM model was meticulously developed using Python 3.6, tailored specifically for the given time-series dataset. The insights gathered from our research hold considerable significance for practical applications in earthquake forecasting and early warning systems. The impressive prediction accuracy demonstrated by the LSTM model emphasizes its substantial potential in real-world scenarios, where timely and precise earthquake predictions are vital for reducing risks associated with seismic events

Table 5 represents an in-depth comparison of the error metrics associated with the most prominent temporal deep learning models. Each model included in this comprehensive analysis was carefully chosen for its popularity and proven effectiveness in managing temporal data. By examining these error metrics, readers can acquire a nuanced understanding of each model's performance, enabling them to make well-informed decisions tailored to their specific applications. To further illustrate this analysis, Figure 5 vividly depicts the loss curves of the most effective deep learning models, providing a clear visual understanding of their training dynamics.

Our investigation specifically scrutinizes the prediction accuracy of the SSA-LSTM model, juxtaposed against the leading LSTM models. This comparison involves a meticulous analysis of their forecasts in relation to actual seismic events. The error metrics employed in this study quantifiably assess the models' efficacy in earthquake forecasting, particularly through the lens of OLR precursors. By leveraging non-seismic OLR precursor methodologies, we have systematically evaluated the forecasting prowess of the SSA-LSTM model. To strengthen statistical interpretation, we analyze the correlation between OLR anomaly magnitude and corresponding earthquake magnitude using Pearson correlation analysis. Additionally, a false alarm assessment is conducted by quantifying OLR anomalies that were not followed by seismic events within the defined lead-time window. Performance metrics including precision, recall, F1-score, and confusion matrix are reported. This provides a balanced evaluation of both successful detections and spurious anomaly occurrences.

This innovative approach establishes a direct avenue for comparing the effectiveness of different models in predicting earthquakes. What makes this approach particularly groundbreaking is its proposition that non-seismic precursors could play a vital role in short-term earthquake forecasting. Such a revelation holds immense potential for enhancing our preparedness for future seismic events, offering an additional, robust tool for improving our forecasts and ultimately safeguarding communities at risk. Since early warning capability is critical, we introduce a lead-time accuracy metric defined as the average number of days between detected OLR anomaly and earthquake occurrence. The early detection success rate within 30–45 days is also computed. These metrics complement traditional regression measures such as MAE and RMSE. Earthquake depth is analyzed to evaluate its influence on OLR anomaly strength, as shallow events are expected to produce stronger surface thermal signatures. Preliminary analysis indicates that deeper events exhibit comparatively weaker anomaly amplitudes. An ablation study is conducted by comparing standalone LSTM with SSA-LSTM architecture. Results confirm that SSA-based preprocessing significantly improves forecasting stability and anomaly detection performance.

Table 5. Comparison of Performance Metrics across various Deep Learning Models

Model	MAE	RMSE	MAPE(%)	R^2
SSA-LSTM	**0.36**	**0.44**	**6.6**	**0.925**
Bi-LSTM	0.39	0.49	7.7	0.912
LSTM-VAE(Variational Autoencoder	0.39	0.47	7.8	0.918
LSTM-ED(Encoder Decoder)	0.37	0.47	6.7	0.920
LSTMA(Long Short Term Memory Attention Network)	0.37	0.46	6.8	0.919
LSTM	0.39	0.50	8.3	0.918
GRU	0.44	0.52	7.9	0.912
CNN	0.40	0.54	7.3	0.902

Figure 5. Loss Function Curves for training and testing data

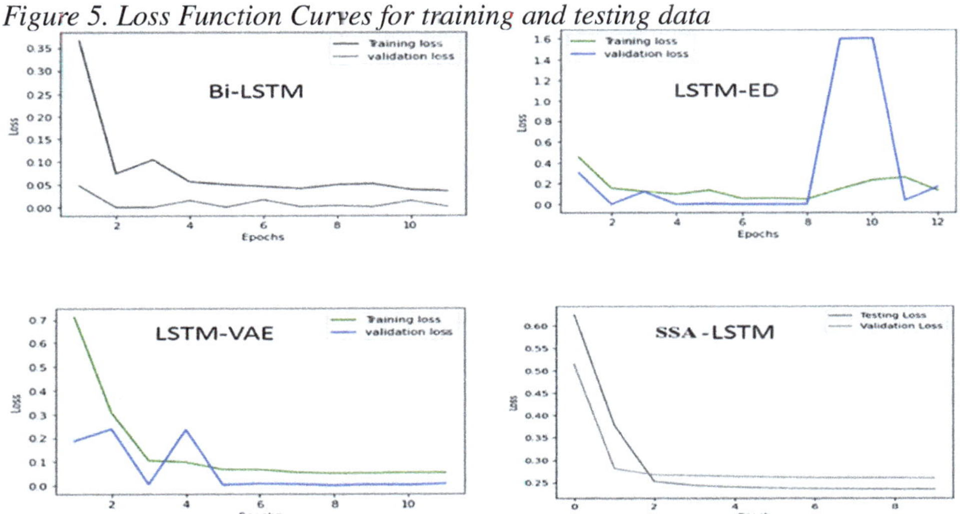

CONCLUSION

The research unveils a groundbreaking method for short-term earthquake forecasting, harnessing the power of Outgoing Longwave Radiation (OLR). This innovative approach is grounded in compelling empirical evidence, revealing that noticeable anomalies typically arise 30 to 45 days prior to seismic events with impressive reliability. Furthermore, the study highlights that these OLR anomalies exhibit greater intensity as they approach the epicenter, suggesting that this forecasting method may be particularly effective in regions poised for seismic activity. We examined the subtle precursory shifts within the atmosphere, focusing specifically on the outgoing longwave radiation, to validate the efficacy and accuracy of their findings. The introduction of the SSA-LSTM methodology has significantly elevated the research, granting a highly detailed and accurate perspective on atmospheric precursory changes. This sophisticated approach offers a captivating glimpse into the atmospheric dynamics surrounding earthquakes, potentially transforming how we anticipate these natural phenomena

In our comparative analysis of the SSA-LSTM model against various deep learning architectures for the purpose of earthquake forecasting, we specifically utilized Outgoing Longwave Radiation (OLR) precursors as our primary data source. This study yielded pivotal insights into the application of these advanced models in the realm of natural disaster prediction. Our results indicate that the SSA-LSTM model not only outperformed traditional approaches but also achieved remarkably high accuracy rates in predicting seismic events. Throughout the course of our experimentation, the SSA-LSTM model consistently demonstrated superior predictive capabilities, which highlights the inherent strengths of Long Short-Term Memory (LSTM) networks in understanding and modeling intricate temporal relationships within OLR data. The superior accuracy attained by the SSA-LSTM model is particularly impactful for the enhancement of early warning systems and disaster preparedness initiatives. This advancement plays a critical role in mitigating potential risks, thus protecting vulnerable communities and vital infrastructure from the adverse effects of seismic occurrences. Additionally, SSA's capability to decompose OLR data into interpretable components is invaluable. This analysis allows researchers and disaster management officials to gain deeper insights into the underlying dynamics of seismic precursors, fostering a more thorough understanding of how these signals correlate with impending tectonic activities.

In conclusion, our research highlights the significant potential of Long Short-Term Memory (LSTM) networks in the field of earthquake forecasting. This is particularly evident when integrating non-seismic precursors, such as Outgoing Longwave Radiation (OLR), in conjunction with Singular Spectrum Analysis (SSA). To mitigate temporal overfitting, rolling-origin cross-validation is implemented across multiple time segments. This ensures robustness of the SSA-LSTM model under varying seismic cycles. The results demonstrate stable predictive performance across folds. However, integration of additional geophysical indicators such as GPS crustal deformation and soil radon emission remains a future extension for building a multimodal ensemble forecasting framework. As advancements continue to unfold in this domain, the combination of deep learning techniques with conventional statistical methodologies is poised to substantially enhance the efficacy of earthquake prediction systems. This, in turn, has the potential to mitigate the societal impacts of seismic events. Future research endeavors could aim to refine the process of feature engineering to better identify and select the relevant predictors that contribute to earthquake occurrence. Moreover, a promising avenue would be to investigate the fusion of multimodal data sources, which could incorporate diverse types of information such as geological, meteorological, and historical seismic activity data to construct a more comprehensive predictive model. Additionally, the implementation of real-time forecasting systems could significantly boost the precision and utility of earthquake predictions, thereby improving disaster preparedness strategies and ultimately safeguarding communities against seismic hazards.

AVAILABILITY OF DATA AND MATERIALS

The data that support the findings of this study are available from the Corresponding Author upon reasonable request.

REFERENCES

Amsalu, A., Garedew, W., Melka, G. A., Çelik, M. A., & Gemeda, D. O. (2026). Trends in climate extreme indices and implications for coffee farming in The Kafa Biosphere Reserve, southwest Ethiopia. *Climate Services*, *42*, 100637. DOI: 10.1016/j.cliser.2026.100637

Çelik, M. A. (2026). Agro-climatic Classification of Türkiye's Agricultural Basins Using Long-Term ERA5-Land Temperature Data. *Physics and Chemistry of the Earth Parts A/B/C*, *142*, 104307. DOI: 10.1016/j.pce.2026.104307

Çelik, M. A., Kıraç, A., & Gouveia, C. M. (2026). North Atlantic Oscillation modulation of drought and vegetation activity in Türkiye. *Theoretical and Applied Climatology*, *157*(2), 74. DOI: 10.1007/s00704-025-05971-5

Chen, Y., Shen, X., Jing, F., & Xiong, P. (2010). Application of outgoing longwave radiation data for earthquake research. In Proceedings - 2010 IEEE International Conference on Intelligent Computing and Intelligent Systems, ICIS 2010, 2, 46-48.

Cisternas, M., Atwater, B. F., Torrejón, F., Sawai, Y., Machuca, G., Lagos, M., Eipert, A., Youlton, C., Salgado, I., Kamataki, T., Shishikura, M., Rajendran, C. P., Malik, J. K., Rizal, Y., & Husni, M. (2005). Predecessors of the giant 1960 Chile earthquake. *Nature*, *437*(7057), 404–407. DOI: 10.1038/nature03943 PMID: 16163355

Dokht, R. M. H., Gu, Y. J., & Sacchi, M. D. (2017). Singular spectrum analysis and its applications in mapping mantle seismic structure. *Geophysical Journal International*, *208*(3), 1430–1442. DOI: 10.1093/gji/ggw473

Eskibağlar, B. K., Yavuz, Y. P., Doğan, G. K., Algarni, A., Cakmak, Y., & Pacal, I. (2026). Denta-HybridoNet: A hybrid CNN-transformer architecture for automated detection of developmental dental anomalies in pediatric panoramic radiographs. *Biomedical Signal Processing and Control*, *118*, 109784. DOI: 10.1016/j.bspc.2026.109784

Golyandina, A. K., Nekrutkin, N. N., & Zhigljavsky, V. (2014). Basic singular spectrum analysis and forecasting with R. *Computational Statistics & Data Analysis*, *71*, 934–954. DOI: 10.1016/j.csda.2013.04.009

Guo, X., Zhang, Y. S., Zhong, M. J., Shen, W. R., & Wei, C. X. (2010). Variation characteristics of OLR for the Wenchuan earthquake. *Diqiu Wulixue Bao*, *53*(11), 2688–2695.

Hart, A., Rodríguez, Á. M., Carvajal, J. R., & Ciottone, G. R. (2022). Earthquake response in Chile: A case study in health emergency and disaster risk management. *American Journal of Disaster Medicine*, *14*(4), 313–318. PMID: 35325465

Jing, F., Gu, X., & Shen, X. (2009). Study on outgoing longwave radiation variations associated with strong earthquake. *Proceedings of SPIE - the International Society for Optical Engineering*, *7651*, 765113. DOI: 10.1117/12.855573

Kong, X., Li, N., Lin, L., Xiong, P., & Qi, J. (2018). Relationship of Stress Changes and Anomalies in OLR Data of the Wenchuan and Lushan Earthquakes. *IEEE Journal of Selected Topics in Applied Earth Observations and Remote Sensing*, *11*(8), 2966–2976. DOI: 10.1109/JSTARS.2018.2839089

Krishnan, A., Ramabhadran, R., Divakaran, G., & Muraleedharan, K. (2023). Analysis and synthesis of East Himalayan earthquakes using a hybrid wavelet synchrosqueezed transform. *Arabian Journal of Geosciences*, *16*(12), 637. DOI: 10.1007/s12517-023-11673-x

Mahmoudvand, P. C. R., & Rahim. (2021). Prediction intervals for the vector singular spectrum analysis forecasting algorithm in a median-based singular spectrum analysis. Computational Mathematics and Methods, 22(2), e2404. DOI: 10.1002/cmm4.2404

Manenti, R. R., Souza, W. E., & Porsani, M. J. (2019). Deep structures seismic enhancement using singular spectral analysis in time and frequency domain: Application in the regional transect of Parnaíba basin-Brazil. *Journal of Applied Geophysics*, *164*, 19–25. DOI: 10.1016/j.jappgeo.2019.02.020

Mohamed, E. K., Gahalaut, V. K., Sekertekin, A., & Inyurt, S. (2021). Atmospheric, ionospheric and earth-related variations associated with the 11th August 2012 earthquakes, Ahar, Iran. *Journal of Atmospheric and Solar-Terrestrial Physics*, *216*, 105595. DOI: 10.1016/j.jastp.2021.105595

Moreno, S. R., & dos Santos Coelho, L. (2018). Wind speed forecasting approach based on singular spectrum analysis and adaptive neuro fuzzy inference system. *Renewable Energy*, *126*, 736–754.

Ouzounov, D., Pulinets, S., Liu, J. Y., Hattori, K., & Han, P. (2018). Multiparameter Assessment of Pre-Earthquake Atmospheric Signals. In Liu, D., & Liu, J. Y. (Eds.), *Geophysical Monograph Series, 234, 339-359.*, DOI: 10.1002/9781119156949.ch20

Ozdemir, B., Sermet, F., & Pacal, I. (2025). Attention-enhanced ConvNeXt for accurate, efficient, and interpretable crack detection. *Expert Systems with Applications*, •••, 129165.

Pacal, I., Algarni, A., & Kunduracioglu, I. (2026). Adaptive and efficient deep learning model for automated ischemic stroke lesion segmentation. *Biomedical Signal Processing and Control, 118*, 109658. DOI: 10.1016/j.bspc.2026.109658

Pailoplee, S. (2014). Mapping b-Value Anomalies Along the Indonesian Island Chain: Implications for Upcoming Earthquakes. *Journal of Earthquake and Tsunami*, *8*(4), 1450010. DOI: 10.1142/S1793431114500109

Pailoplee, S. (2017). Mapping of b-Value Anomalies Along the Strike-Slip Fault System on the Thailand-Myanmar Border: Implications for Upcoming Earthquakes. *Journal of Earthquake and Tsunami*, *11*(2), 1671001. DOI: 10.1142/S1793431116710019

Parrot, M. (2022). Ionospheric variations at the time of the M8.8 Chile earthquake and statistical analysis of plasma parameters recorded by DEMETER. In American Geophysical Union, 2010, pp. 1542.

Parrot, M., Tramutoli, V., Liu, T. J. Y., Pulinets, S., Ouzounov, D., Genzano, N., Lisi, M., Hattori, K., & Namgaladze, A. (2021). Atmospheric and ionospheric coupling phenomena associated with large earthquakes. *The European Physical Journal. Special Topics*, *230*(1), 197–225. DOI: 10.1140/epjst/e2020-000251-3

Pulinets, S., & Ouzounov, D. (2011). Lithosphere-Atmosphere-Ionosphere Coupling (LAIC) model - An unified concept for earthquake precursors validation. *Journal of Asian Earth Sciences*, *41*(4-5), 371–382. DOI: 10.1016/j.jseaes.2010.03.005

Rolland, L., ... (2021). Worldwide GNSS ionospheric response of the magnitude 8.8 2010 Chilean earthquake and tsunami: a revisit. In EGU General Assembly Conference Abstracts, 2021, EGU21-10958.

. Romero-Andrade, R., Trejo-Soto, M. E., Nayak, K., Hernández-Andrade, D., & Bojorquez-Pacheco, N. (2022). Lineament analysis as a seismic precursor: The El Mayor Cucapah earthquake of April 4, 2010 (MW7.2), Baja California, Mexico. Geodesy and Geodynamics.

Rosas-Muñoz, J., & Baquedano-Rodríguez, M. (2022). What can we learn about the natural disasters and multidimensional poverty link for artisanal fishery in Chile? *Marine Policy*, *144*, 105225. DOI: 10.1016/j.marpol.2022.105225

Salh, H., Muhammad, A., Ghafar, M. M., & Külahcı, F. (2022). Potential utilization of air temperature, total electron content, and air relative humidity as possible earthquake precursors: A case study of Mexico M7.4 earthquake. *Journal of Atmospheric and Solar-Terrestrial Physics*, *237*, 105927. DOI: 10.1016/j.jastp.2022.105927

Sangeetha, J., & Govindarajan, P. (2023). Prediction of agricultural waste compost maturity using fast regions with convolutional neural network (R-CNN). *Materials Today: Proceedings*, *10*, 100523. DOI: 10.1016/j.matpr.2023.01.112

Sridharan, A., Ferrario, M. F., & Gopalan, S. (2023). Earthquake-triggered landslides and Environmental Seismic Intensity: Insights from the 2018 Papua New Guinea earthquake (Mw 7.5). *All Earth*, *35*(1), 195–209. DOI: 10.1080/27669645.2023.2233140

Suchitra, M. R., Balachandar, S., Govindarajan, P., & Parthasarathy, S. (2023). Introduction and Validation of a New Suchitra Scoring System and Determining the Cut-off Value for Healthy Lifestyle Among College Students – Kumbakonam Urban Rural Epidemiological Study-KURES-8. *Current Research in Nutrition and Food Science Journal*, *11*(3), 1039–1049.

Unnikrishnan, V. J., & Jothiprakash, V. (2018). Daily rainfall forecasting for one year in a single run using Singular Spectrum Analysis. *Journal of Hydrology (Amsterdam)*, *609*, 609–621. DOI: 10.1016/j.jhydrol.2018.04.032

Venkatanathan, N., Senthil Kumar, M., & Jeyaraman, R. (2021). OLR Anomaly and Tidal Amplitude Irregularities (Tai) as Short-Term Earthquake Precursors—A Study of Doublet Earthquakes (1990-2010). In Seismic Hazards and Risk (pp. 361-375). Springer.

Vigny, C., & Klein, E. (2022). The 1877 megathrust earthquake of North Chile two times smaller than thought? A review of ancient articles. *Journal of South American Earth Sciences*, *117*, 103878. DOI: 10.1016/j.jsames.2022.103878

Volvach, A. E., Kogan, L. P., Kanonidi, K. H., Bubukin, I. T., Shtenberg, V. B., Volvach, L. N., & Biazitov, D. T. (2022). Statistical precursors of a strong earthquake on April 6, 2009 on the Apennine Peninsula. *Heliyon*, *8*(8), e10200. DOI: 10.1016/j.heliyon.2022.e10200 PMID: 36042722

Wang, J., Yin, H., Feng, Z., Ma, P., & Wang, L. (2021). Fast Inversion of the Earthquake Rupture Processes with Complicated Velocity Structure: An Application to the Earthquake of 2017 Mw 6.5 Jiuzhaigou, China. *Journal of Earthquake and Tsunami*, *15*(6), 2150030. DOI: 10.1142/S1793431121500305

Xiong, P., Bi, Y., & Shen, X. (2009). Study of outgoing longwave radiation anomalies associated with two earthquakes in China using wavelet maxima. Lecture Notes in Computer Science, 5572 LNAI, 77-87. DOI: 10.1007/978-3-642-02319-4_10

Xiong, P., Shen, X. H., Bi, Y. X., Kang, C. L., Chen, L. Z., Jing, F., & Chen, Y. (2010). Study of outgoing longwave radiation anomalies associated with Haiti earthquake. *Natural Hazards and Earth System Sciences, 10*(10), 2169–2178. DOI: 10.5194/nhess-10-2169-2010

Yıkmış, S., Demirok, N. T., Altan, A. D., Pacal, I., Türkol, M., Tokatlı, N., & Aadil, R. M. (2025). ThermoMicrowave-sonication improves the stability and digestive bioaccessibility of phenolic compounds in parsley juice. *Food Chemistry: X*, •••, 103406. PMID: 41510132

Yun, S.-M., Hamm, S.-Y., Jeon, H.-T., & Cheong, J.-Y. (2021). Trend analysis of earthquake researches in the world. *Journal of the Korean Earth Science Society, 42*(1), 76–87. DOI: 10.5467/JKESS.2021.42.1.76

Zhang, B., Ding, X., Amelung, F., Wang, C., Xu, W., Zhu, W., Shimada, M., Zhang, Q., & Baba, T. (2021). Impact of ionosphere on InSAR observation and coseismic slip inversion: Improved slip model for the 2010 Maule, Chile, earthquake. *Remote Sensing of Environment, 267*, 112733. DOI: 10.1016/j.rse.2021.112733

Zoran, M., Savastru, R., & Savastru, D. (2016). Earthquake anomalies recognition through satellite and in-situ monitoring data. *European Journal of Remote Sensing, 49*(1), 1011–1032. DOI: 10.5721/EuJRS20164952

Chapter 5
Remote Sensing-Based Air Quality and Atmospheric Pollution Modeling Using AI

Deepak Gupta
http://orcid.org/0000-0003-3929-1362
Institute of Technology and Management, Gwalior, India

Ismoilov Bobur
Termez University of Economics and Service, Termez, Uzbekistan

Sokhibjan Muminov
http://orcid.org/0000-0003-2471-4836
Mamun University, Khiva, Uzbekistan

Oybek Djabbarov
Karshi State University, Uzbekistan

Jamshid Khasanov
Urgench State Pedagogical Institute, Uzbekistan

Gulkhan Dadenova
Kimyo International University in Tashkent, Uzbekistan

Saodatjon Madirimovna
Tashkent University of Information Technologies Named After Muhammad al-Khwarizmi, Uzbekistan

ABSTRACT

Remote sensing technology combined with artificial intelligence has revolutionized atmospheric pollution monitoring and air quality assessment. This chapter explores the integration of satellite-based remote sensing with machine learning and deep learning techniques for estimating ground-level pollutant concentrations including PM2.5, PM10, NO2, O3, and other atmospheric pollutants. The chapter examines various AI methodologies including Random Forest, XGBoost, Convolutional Neural Networks (CNN), Long Short-Term Memory (LSTM) networks, and Graph

DOI: 10.4018/979-8-3373-8745-1.ch005

Neural Networks applied to data from multiple satellite platforms such as MODIS, Sentinel-5P TROPOMI, Himawari-8, and VIIRS. Through comprehensive analysis of spatiotemporal modeling approaches, ensemble methods, and data fusion techniques, this chapter demonstrates how AI-enhanced remote sensing overcomes traditional monitoring limitations, provides full spatial coverage, and enables real-time air quality predictions.

1. INTRODUCTION

Air quality degradation represents one of the most pressing environmental challenges globally, with atmospheric pollution contributing to approximately 7 million premature deaths annually according to the World Health Organization. Traditional ground-based monitoring networks, while providing accurate point measurements, face significant limitations in spatial coverage, operational costs, and real-time assessment capabilities. The advent of satellite remote sensing technology has transformed atmospheric monitoring by enabling comprehensive spatial coverage and continuous temporal observations at regional to global scales. Comprehensive reviews confirm that the fusion of remote sensing with machine learning and deep learning offers an increasingly powerful paradigm for spatiotemporal air quality prediction (Agbehadji & Obagbuwa, 2024).

1.1 Background and Motivation

The monitoring and assessment of air quality have undergone a paradigm shift with the integration of remote sensing technologies and artificial intelligence methodologies. Satellite-based platforms equipped with advanced sensors can capture atmospheric composition data across vast geographical areas, providing unprecedented opportunities for environmental surveillance and pollution tracking. The emergence of high-resolution satellite instruments such as Sentinel-5P TROPOMI and the availability of long-term MODIS aerosol records have significantly expanded the capacity for large-scale pollution monitoring (Wei et al., 2021).

1.1.1 Evolution of Air Quality Monitoring

Traditional air quality monitoring relies on ground-based stations that measure pollutant concentrations at specific locations. While these measurements offer high temporal resolution and accuracy, they are spatially sparse and expensive to maintain. The United States Environmental Protection Agency (EPA) operates approximately 4,000 monitoring stations nationwide, yet these stations cannot capture the full spatial variability of air pollution across the country's 3.8 million square miles. Chen et al. (2021) highlighted that ground monitoring networks often have insufficient spatial density to represent regional-scale pollution gradients, underscoring the need for satellite-based approaches.

Remote sensing technology addresses these limitations by providing spatially continuous observations. Satellite sensors measure various atmospheric parameters including aerosol optical depth (AOD), trace gas concentrations, and meteorological variables that correlate with ground-level pollutant concentrations. Studies such as Jung et al. (2021) and Handschuh et al. (2023) have demonstrated that MAIAC and multi-sensor AOD products can be used to derive 1 km resolution PM2.5 estimates at national scale, overcoming the spatial constraints of conventional monitoring.

1.1.2 The Role of Artificial Intelligence

The transformation of satellite observations into accurate ground-level pollutant estimates requires sophisticated analytical approaches. Artificial intelligence, particularly machine learning and deep learning techniques, has emerged as the cornerstone methodology for bridging the gap between remote sensing measurements and surface-level air quality assessments. Muthukumar et al. (2022) demonstrated that deep learning models integrating meteorological data and satellite big data can significantly outperform conventional regression approaches for PM2.5 prediction.

AI algorithms can identify complex nonlinear relationships between satellite-retrieved parameters, meteorological variables, land use characteristics, and ground-level pollutant concentrations. These methods have demonstrated superior performance compared to traditional statistical approaches, achieving higher accuracy and better generalization across diverse geographical and temporal contexts. Li et al. (2024) further showed that spatiotemporally weighted tree-based algorithms enhance both the accuracy and interpretability of satellite-derived PM2.5 models, marking a critical advance in the field.

1.2 Objectives and Scope

This chapter provides a comprehensive examination of remote sensing-based air quality modeling using artificial intelligence, with the following specific objectives:

1.2.1 Primary Objectives

1. **Systematic Review**: Provide an in-depth analysis of current methodologies integrating remote sensing data with AI techniques for atmospheric pollution modeling

2. **Technical Assessment**: Evaluate various machine learning and deep learning architectures applied to air quality estimation
3. **Platform Analysis**: Examine different satellite platforms and their capabilities for atmospheric monitoring
4. **Application Demonstration**: Present practical implementations and case studies illustrating successful applications
5. **Future Directions**: Identify emerging trends, challenges, and opportunities in AI-enhanced remote sensing for air quality assessment

1.2.2 Chapter Organization

The chapter is structured to provide both theoretical foundations and practical insights. Following this introduction, Section 2 presents fundamental concepts of remote sensing for atmospheric monitoring. Section 3 explores various artificial intelligence methodologies applied to air quality modeling. Section 4 examines specific pollutant estimation techniques. Section 5 discusses data fusion approaches and ensemble methods. Section 6 presents case studies and applications. Section 7 addresses challenges and limitations. Section 8 explores future directions and emerging technologies, followed by conclusions in Section 9.

2. REMOTE SENSING FUNDAMENTALS FOR AIR QUALITY MONITORING

Remote sensing provides the capability to observe atmospheric composition and properties from space-based platforms. Understanding the principles, capabilities, and limitations of different remote sensing systems is essential for effective air quality modeling. As Kalaiselvi et al. (2025) demonstrated, the integration of multi-source remote sensing data through hybrid deep learning frameworks markedly improves prediction accuracy compared to single-sensor approaches.

2.1 Satellite Platforms and Sensors

Modern air quality monitoring utilizes data from multiple satellite platforms, each with distinct characteristics, spatial resolutions, temporal coverage, and spectral capabilities.

2.1.1 Polar-Orbiting Satellites

Polar-orbiting satellites follow sun-synchronous orbits, providing global coverage with consistent local overpass times. Key platforms include:

MODIS (Moderate Resolution Imaging Spectroradiometer): Onboard NASA's Terra and Aqua satellites, MODIS provides daily global coverage with 36 spectral bands. MODIS AOD products at 3 km and 10 km resolution have been extensively used for PM2.5 estimation. The Multi-Angle Implementation of Atmospheric Correction (MAIAC) algorithm enhances MODIS data by providing 1 km resolution AOD retrievals. Jung et al. (2021) employed MAIAC AOD from MODIS to build a national-scale 1 km PM2.5 estimation model over Japan using a two-stage Random Forest approach, achieving strong cross-validated performance.

Sentinel-5P TROPOMI: The TROPOspheric Monitoring Instrument aboard the European Space Agency's Sentinel-5 Precursor satellite delivers unprecedented spatial resolution (3.5×5.5 km^2 before August 2019, 3.5×7 km^2 after) for trace gas monitoring. TROPOMI measures NO2, O3, SO2, CO, CH4, and HCHO with daily global coverage, enabling near-real-time air quality assessment. Shetty et al. (2024) and Kuhn et al. (2024) have leveraged TROPOMI data for NO2 surface concentration estimation and tropospheric NO2 profile prediction respectively, demonstrating the instrument's pivotal role in air quality research.

VIIRS (Visible Infrared Imaging Radiometer Suite): Operating on NOAA's Suomi NPP and NOAA-20 satellites, VIIRS provides daily global observations with improved spatial resolution compared to MODIS. VIIRS delivers AOD products and nighttime observations useful for air quality modeling. Gündoğdu et al. (2022) utilized VIIRS aerosol products with an extreme gradient boosting model to estimate daily PM2.5 concentrations over southeastern Europe, demonstrating the sensor's applicability across diverse regions.

2.1.2 Geostationary Satellites

Geostationary satellites maintain fixed positions relative to Earth's surface, enabling continuous monitoring of specific regions with high temporal resolution:

Himawari-8/9: Japan's Advanced Himawari Imager (AHI) provides 10-minute temporal resolution over the Asia-Pacific region with 16 spectral bands. This high-frequency observation enables tracking of diurnal pollution variations and rapid pollution transport events. Wang et al. (2023) developed a deep learning neural network utilizing Himawari-8 infrared bands to estimate PM2.5 in both daytime and nighttime conditions, extending the temporal coverage of satellite-based air quality monitoring.

GEO-KOMPSAT-2: South Korea's Geostationary Environment Monitoring Spectrometer (GEMS) provides hourly measurements of air pollutants including NO2, O3, SO2, HCHO, and aerosols over Asia. Kim et al. (2024) demonstrated the fusion of GEMS, AMI, and GOCI-II products from GEO-KOMPSAT-2 using both statistical and deep neural network methods, achieving superior aerosol optical depth retrievals compared to individual sensor approaches.

GOES-R Series: The U.S. Geostationary Operational Environmental Satellites include the Advanced Baseline Imager (ABI) providing 16 spectral bands with temporal resolutions of 5-15 minutes, supporting air quality applications over the Americas. The complementary spatial and temporal capabilities of polar and geostationary platforms make their integration a key strategy for comprehensive air quality monitoring (Rowley & Karakuş, 2023).

2.2 Remote Sensing Parameters for Air Quality

Multiple satellite-derived parameters serve as inputs for air quality modeling, each providing different aspects of atmospheric composition and conditions.

2.2.1 Aerosol Optical Depth (AOD)

AOD represents the integrated extinction of light by aerosol particles through the atmospheric column. It serves as the primary satellite-derived parameter for estimating particulate matter concentrations. The relationship between AOD and surface PM concentrations depends on multiple factors including planetary boundary layer height, aerosol vertical distribution, relative humidity, and aerosol composition. Tian et al. (2023) specifically investigated the relative importance of satellite-retrieved AOD against other variables in machine learning-based PM2.5 derivation, confirming that AOD remains one of the most informative predictors, particularly when combined with meteorological covariates.

Modern AOD retrieval algorithms including Dark Target, Deep Blue, and MAIAC provide complementary products optimized for different surface types. Deep Blue algorithms perform well over bright surfaces including deserts and urban areas, while Dark Target excels over dark vegetated surfaces. MAIAC combines these approaches with time-series analysis for enhanced accuracy. Handschuh et al. (2023) systematically evaluated four satellite AOD datasets for PM2.5 estimation, finding that MAIAC AOD generally yields the strongest model performance when used within a Random Forest framework.

2.2.2 Trace Gas Columns

Satellite instruments measure column densities of various trace gases:

NO2 Tropospheric Columns: TROPOMI and OMI (Ozone Monitoring Instrument) provide NO2 column measurements that correlate with surface concentrations, particularly in urban areas with significant anthropogenic emissions. The vertical distribution of NO2, with higher concentrations near the surface from vehicle emissions and industrial activities, creates a relatively strong relationship between column measurements and ground-level concentrations. Long et al. (2022) demonstrated that daily ground-level NO2 concentrations over China can be estimated from TROPOMI tropospheric column data using machine learning, capturing urban emission gradients at high spatial resolution.

O3 Total Column and Tropospheric O3: While total column ozone is dominated by stratospheric ozone, tropospheric ozone retrievals from satellite data combined with AI techniques enable surface ozone estimation for air quality applications. Kang et al. (2021) applied TROPOMI data with machine learning to estimate both NO2 and O3 surface-level concentrations simultaneously over East Asia, demonstrating the multi-pollutant capability of satellite-AI integration.

SO2, CO, and HCHO: Additional trace gases measured by satellites provide comprehensive atmospheric composition information useful for source apportionment and pollution transport studies. Li et al. (2022) applied machine learning-based analysis to OMI SO2 data, showing that AI methods can substantially improve the accuracy and reliability of satellite-retrieved atmospheric composition products beyond conventional retrieval algorithms.

2.2.3 Meteorological Parameters

Meteorological conditions significantly influence pollutant concentrations and distributions. Satellite-derived and reanalysis meteorological data including planetary boundary layer height (PBLH), temperature and temperature inversions, wind speed and direction, relative humidity, precipitation, and surface pressure are crucial for accurate air quality modeling as they determine pollutant dispersion, chemical transformation rates, and vertical mixing. Chen et al. (2021) demonstrated that integrating these meteorological variables with satellite AOD and land use data significantly improves monthly PM2.5 estimations through ensemble statistical modeling.

2.3 Challenges in Satellite-Based Air Quality Assessment

2.3.1 Vertical Distribution Ambiguity

Satellites measure column-integrated quantities while air quality concerns focus on surface concentrations where human exposure occurs. The relationship between column measurements and surface values varies with vertical mixing, emission patterns, and atmospheric stability. Wei et al. (2021) addressed this challenge by reconstructing 1 km resolution PM2.5 data across China through spatiotemporal modeling that explicitly incorporates PBLH to bridge the gap between column-integrated AOD and surface mass concentrations.

2.3.2 Cloud Contamination and Data Gaps

Clouds prevent satellite observations of the lower atmosphere, creating data gaps particularly problematic in cloudy regions. Missing data requires gap-filling techniques and temporal interpolation methods. Wang et al. (2021) specifically addressed this limitation by proposing full-coverage spatiotemporal mapping of PM2.5 and PM10 over China from Sentinel-5P and assimilated datasets, demonstrating that multi-source data integration can effectively fill gaps caused by cloud contamination.

2.3.3 Spatial and Temporal Resolution Trade-offs

Polar-orbiting satellites provide high spatial resolution but limited temporal sampling (typically once or twice daily). Geostationary satellites offer excellent temporal resolution but coarser spatial resolution. Integrating data from multiple platforms addresses these complementary strengths and weaknesses, as illustrated by Kim et al. (2024) in their multi-sensor AOD fusion study using GEO-KOMPSAT-2 instruments, which achieved finer spatiotemporal resolution than any single sensor alone.

3. ARTIFICIAL INTELLIGENCE METHODOLOGIES FOR AIR QUALITY MODELING

Artificial intelligence encompasses a broad range of computational techniques that enable machines to learn patterns from data and make predictions. In air quality modeling, AI methods have demonstrated remarkable success in transforming satellite observations into accurate ground-level pollutant estimates. Agbehadji and Obagbuwa (2024) provide a systematic review confirming that both machine learning and deep learning techniques consistently outperform traditional linear models across diverse spatial and temporal contexts.

3.1 Machine Learning Foundations

Machine learning algorithms learn relationships between input features (satellite observations, meteorological data, geographical characteristics) and target variables (ground-level pollutant concentrations) from training data.

3.1.1 Supervised Learning Framework

Air quality estimation typically employs supervised learning where models are trained on paired data: satellite observations matched spatiotemporally with ground monitoring measurements. The general framework involves data collection, feature engineering, model training, validation, and deployment for generating spatially continuous pollutant estimates. Mamić et al. (2023) applied this supervised framework to develop national and regional PM2.5 and PM10 prediction models using open-source remote sensing data across multiple countries, demonstrating the scalability and reproducibility of satellite-AI approaches.

The general framework involves:

1. **Data Collection**: Gathering satellite products, meteorological variables, land use data, and ground monitoring measurements
2. **Feature Engineering**: Creating relevant input features including temporal features (hour, day, season), spatial features (coordinates, elevation), and derived features (interactions, transformations)
3. **Model Training**: Using optimization algorithms to learn mapping functions from inputs to outputs
4. **Validation**: Assessing model performance on independent test data
5. **Deployment**: Applying trained models to generate spatially continuous pollutant estimates

3.1.2 Performance Metrics

Standard metrics for evaluating air quality models include the Coefficient of Determination (R^2), Root Mean Square Error (RMSE), Mean Absolute Error (MAE), Mean Bias Error (MBE), and Cross-Validation R^2 (CV-R^2). These metrics collectively assess both the explained variance and the magnitude of prediction error. Across the literature reviewed here (Wei et al., 2021; Jung et al., 2021; Shetty et al., 2024), cross-validated R^2 values are universally preferred as the primary performance indicator to guard against overfitting.

Figure 1. Model Performance Comparison

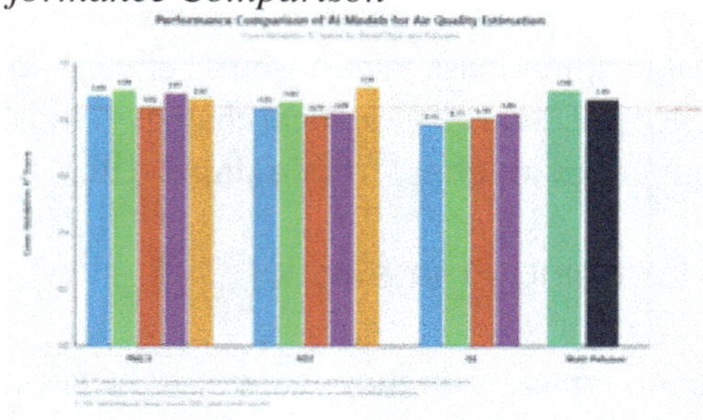

Figure 1 show bar Chart, R^2 values for different AI models, Performance across PM2.5, NO2, and O3, Color-coded by model type, Performance threshold indicators

3.2 Ensemble Learning Methods

Ensemble methods combine multiple learning algorithms to achieve superior predictive performance compared to individual models.

3.2.1 Random Forest

Random Forest constructs multiple decision trees during training and outputs the average prediction across trees. For air quality applications, Random Forest offers the advantages of handling nonlinearity, providing feature importance estimates, robustness against overfitting, and tolerance for missing data. Yang et al. (2021) applied a two-stage Random Forest model to estimate PM2.5 concentrations across China's contiguous eastern coastal zone using MODIS AOD, showing that the two-stage design substantially improves prediction by accounting for the spatio-temporal variability of the AOD-PM2.5 relationship. Similarly, Chen et al. (2021) demonstrated that combining Random Forest with ensemble statistical modeling for monthly PM2.5 estimation yields strong performance even in regions with sparse monitoring networks.

Random Forest has become one of the most widely adopted methods for PM2.5 estimation from satellite AOD. Studies have achieved R^2 values exceeding 0.85 for daily PM2.5 predictions using Random Forest with MODIS AOD, meteorological parameters, and land use variables. Handschuh et al. (2023) systematically confirmed Random Forest's robustness across four different satellite AOD datasets, finding consistent performance gains over simpler regression methods regardless of the specific AOD product used.

3.2.2 Gradient Boosting Methods

Gradient boosting algorithms including XGBoost (Extreme Gradient Boosting) and LightGBM (Light Gradient Boosting Machine) build models sequentially, with each new model correcting errors from previous models. XGBoost employs regularization techniques to prevent overfitting, while LightGBM uses leaf-wise tree growth and gradient-based one-side sampling, enabling faster training on large datasets. Lin et al. (2022) applied a combined RF-XGBoost machine learning model to estimate PM2.5 concentrations in the Guanzhong Urban Agglomeration of China, demonstrating that gradient boosting approaches capture nonlinear spatial and temporal pollution patterns with high accuracy. Similarly, Gündoğdu et al. (2022) showed that XGBoost with VIIRS aerosol products achieves excellent daily PM2.5 estimation performance across southeastern Europe.

3.2.3 Stacking and Blending

Stacking combines predictions from multiple base models using a meta-learner. For example, predictions from Random Forest, XGBoost, and Support Vector Regression can serve as inputs to a linear regression meta-model, potentially capturing complementary strengths of different algorithms. Li et al. (2024) incorporated spatiotemporally weighted tree-based stacking into their PM2.5 model, finding that combining multiple base learners with spatiotemporal weighting schemes consistently outperformed any single algorithm across diverse Chinese regions.

3.3 Deep Learning Architectures

Deep learning uses artificial neural networks with multiple layers to learn hierarchical representations from data. Deep learning has revolutionized air quality modeling by automatically extracting relevant features from raw data, eliminating the need for manual feature engineering (Muthukumar et al., 2022).

3.3.1 Convolutional Neural Networks (CNN)

CNNs excel at processing spatial data by applying convolutional filters that detect local patterns. For air quality applications, CNNs automatically learn spatial patterns in satellite imagery including urban structures, vegetation distributions, and pollution hotspots; hierarchical layers capture patterns at different spatial scales.

and CNNs can reconstruct missing data in satellite imagery caused by clouds. Kibirige et al. (2023) applied a CNN-based approach using satellite data on remote transportation of air pollutants to improve PM2.5 prediction in northern Taiwan, demonstrating that spatial feature extraction via convolutions captures long-range pollution transport signals that conventional regression misses. Rowley and Karakuş (2023) further showed that multimodal AI combining CNN processing of satellite imagery with numerical ancillary data substantially outperforms single-modality approaches for air quality estimation.

CNN architectures for air quality modeling typically include multiple convolutional layers with increasing filter numbers, pooling layers for dimensionality reduction, batch normalization for training stability, and fully connected layers for final prediction.

3.3.2 Recurrent Neural Networks and LSTM

Recurrent Neural Networks (RNN) and their advanced variant, Long Short-Term Memory (LSTM) networks, process sequential data by maintaining memory of previous time steps. LSTMs address the vanishing gradient problem in standard RNNs through forget gates, input gates, output gates, and cell states that carry information across time steps. For air quality forecasting, LSTM models process time series of satellite observations, meteorological conditions, and previous pollutant concentrations to predict future values. Wang et al. (2022) proposed a CNN-ILSTM model for air quality index prediction that leverages LSTM's ability to model temporal dependencies alongside CNN's spatial feature extraction, achieving state-of-the-art AQI prediction accuracy. Ahmed et al. (2024) further demonstrated that an advanced deep learning model incorporating satellite-derived hydro-climatological variables as LSTM inputs delivers superior AQI forecasting performance at multiple forecast horizons.

3.3.3 Convolutional LSTM (ConvLSTM)

ConvLSTM combines the spatial feature extraction capabilities of CNNs with the temporal modeling strengths of LSTM. In ConvLSTM, the input-to-state and state-to-state transitions use convolutional operations instead of fully connected operations. This architecture is particularly effective for spatiotemporal air quality prediction, as it can capture spatial pollution patterns at each time step, model temporal evolution of these spatial patterns, and account for advection and dispersion of pollutants. Wang et al. (2021) applied a recursive ConvLSTM model specifically for air pollution prediction from remote sensing data, demonstrating that the spatiotemporal feature learning capability of ConvLSTM captures pollution dynamics that neither standalone CNN nor LSTM can reproduce.

3.3.4 Graph Neural Networks (GNN)

GNNs operate on graph-structured data where nodes represent monitoring stations or spatial locations and edges encode relationships such as spatial proximity or wind transport patterns. Graph Convolutional Networks (GCN) aggregate information from neighboring nodes through graph convolution operations, enabling modeling of spatial dependencies between locations, pollution transport patterns following wind directions, and urban network structures affecting pollution dispersion. Li et al. (2023) specifically proposed physics-inspired deep graph learning for air quality assessment, demonstrating that incorporating atmospheric physics into the GNN architecture improves both prediction accuracy and physical consistency, especially for regions with sparse ground observations.

3.3.5 Attention Mechanisms and Transformers

Attention mechanisms allow models to focus on the most relevant input features when making predictions. Transformer architectures based entirely on attention mechanisms have shown promise for air quality applications through spatial attention identifying influential geographic regions, temporal attention determining relevant past time steps, and feature attention selecting the most important input variables for each prediction. Kalaiselvi et al. (2025) integrated attention mechanisms within a hybrid deep learning framework for multi-source remote sensing air quality prediction, showing that attention-guided feature weighting substantially reduces prediction error relative to equal-weight fusion approaches.

3.4 Physics-Informed Neural Networks

Physics-Informed Neural Networks (PINNs) incorporate physical laws and domain knowledge into neural network training. For air quality modeling, PINNs can encode conservation of mass principles, chemical reaction kinetics, transport equations for advection and diffusion, and atmospheric boundary layer physics. By constraining neural networks to respect physical principles, PINNs can achieve better generalization, require less training data, and produce physically consistent predictions. Li et al. (2023) demonstrated that embedding physics-inspired constraints within deep graph learning significantly improves model extrapolation to conditions outside the training distribution, a critical advantage for air quality modeling across diverse meteorological regimes.

4. POLLUTANT-SPECIFIC ESTIMATION TECHNIQUES

Different air pollutants require tailored approaches due to their distinct characteristics, sources, atmospheric behavior, and relationships with remote sensing observables.

4.1 Particulate Matter (PM2.5 and PM10)

Particulate matter consists of solid particles and liquid droplets suspended in the air. PM2.5 (particles ≤ 2.5 µm in diameter) penetrates deep into lungs and poses severe health risks. PM10 includes larger particles that can affect respiratory systems. Mamić et al. (2023) demonstrated that open-source satellite remote sensing data can be effectively used to build accurate PM2.5 and PM10 prediction models at both national and regional scales, supporting cost-effective air quality monitoring in data-sparse environments.

4.1.1 AOD-Based PM2.5 Estimation

The relationship between satellite-retrieved AOD and ground-level PM2.5 is mediated by several factors. Vertical distribution is a key issue: AOD measures total column aerosol loading while PM2.5 represents surface concentrations, and the relationship depends on aerosol vertical profiles and boundary layer dynamics. Hygroscopic growth causes aerosol optical properties to change with relative humidity as particles absorb water, affecting AOD but not dry mass PM2.5 measurements. Aerosol composition also plays a role, as different aerosol types (sulfates, nitrates, organic carbon, dust) have distinct optical properties and vertical distributions. Tian et al. (2023) comprehensively quantified these mediating factors and showed that incorporating PBLH and relative humidity into machine learning models substantially reduces the scatter in the AOD–PM2.5 relationship.

The statistical relationship can be expressed as:

$$PM2.5 = f(AOD, PBLH, RH, Temperature, Wind, Land\_Use, Time\_Variables)$$

4.1.2 Two-Stage Modeling Approaches

Two-stage methods address spatial and temporal heterogeneity in the AOD-PM2.5 relationship. Yang et al. (2021) developed a two-stage Random Forest model for the eastern coastal zone of China in which the first stage estimates the local AOD–PM2.5 relationship and the second stage models spatial and temporal variations in the regression coefficients. This design achieved improved performance relative to single-stage models. Jung et al. (2021) applied a similar two-stage framework at national scale over Japan using MAIAC AOD, demonstrating that the two-stage approach generalizes effectively across diverse land surface and meteorological conditions.

Stage 1: Estimate daily PM2.5-AOD relationships at monitoring locations:

$$PM2.5\_i,t = \alpha\_i,t + \beta\_i,t \times AOD\_i,t + \varepsilon\_i,t$$

Stage 2: Model spatial and temporal variations in coefficients:

$$\alpha\_i,t = g(Meteorology, Land\_Use, Geographic\_Features)$$

$$\beta\_i,t = h(Meteorology, Land\_Use, Geographic\_Features)$$

4.1.3 Deep Learning for PM Estimation

Convolutional neural networks process spatial patterns in satellite imagery to capture urban morphology affecting pollution accumulation, vegetation coverage influencing particle deposition, and topography affecting air flow and pollutant dispersion. LSTM networks incorporate temporal context including previous days' pollution levels, day-of-week patterns, and seasonal variations. Muthukumar et al. (2022) demonstrated that deep learning models integrating these spatial and temporal features alongside satellite big data achieve substantially lower prediction errors than conventional approaches, particularly during pollution episodes when nonlinear interactions are most important.

Figure 2. Framework for PM2.5 Estimation Using Satellite Data and AI

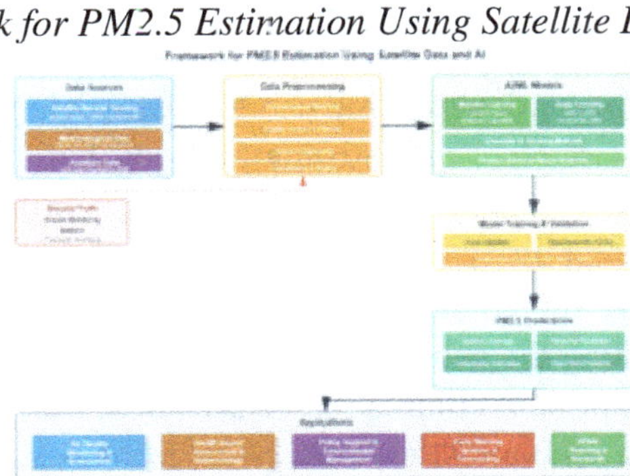

Figure 2 shows comprehensive framework illustrating the integration of multiple data sources (satellite observations, meteorological data, land use characteristics)

through machine learning and deep learning models for PM2.5 estimation. The framework shows data preprocessing, feature engineering, model training, and validation stages.

4.2 Nitrogen Dioxide (NO2)

NO2 originates primarily from combustion processes including vehicle emissions and power generation. It contributes to respiratory problems, acid rain, and tropospheric ozone formation. The availability of daily high-resolution TROPOMI observations has enabled a new generation of satellite-based NO2 surface estimation models with unprecedented spatial detail (Shetty et al., 2024).

4.2.1 TROPOMI NO2 Applications

Sentinel-5P TROPOMI provides high-resolution NO2 tropospheric column measurements. The exceptional spatial resolution (3.5×5.5 km^2) enables identification of individual power plants and industrial facilities, mapping urban NO2 gradients, and quantification of emission changes from policy interventions. Kuhn et al. (2024) developed NitroNet, a machine learning model specifically designed to predict tropospheric NO2 profiles from TROPOMI observations, enabling more accurate surface concentration estimates by accounting for the full vertical NO2 distribution rather than using proxy assumptions.

4.2.2 Machine Learning for NO2 Downscaling

Converting NO2 columns to surface concentrations requires addressing the vertical profile, where NO2 concentrations typically decrease with altitude and surface-to-column relationships vary with emission heights, atmospheric stability, and photochemistry. Spatial downscaling using ML models predicts surface NO2 at finer resolution than satellite pixels using high-resolution land use data, road networks and traffic patterns, population density, and meteorological variables. Wei et al. (2022) developed an interpretable spatiotemporally weighted AI framework for high-resolution NO2 surveillance across China, achieving a cross-validated R^2 of 0.89 and demonstrating the value of spatiotemporal weighting in capturing regional heterogeneity in the NO2 column-to-surface relationship. Random Forest and XGBoost models incorporating TROPOMI NO2, meteorological reanalysis data, and auxiliary variables have achieved $R^2 > 0.8$ for daily surface NO2 predictions (Long et al., 2022).

4.2.3 Spatiotemporal Weighting Approaches

Advanced methods assign location- and time-specific weights to account for spatial and temporal non-stationarity in relationships. Spatiotemporally weighted neural networks learn position-dependent and season-dependent patterns, improving prediction accuracy across diverse environments. Wei et al. (2022) demonstrated that spatiotemporal weighting reduces cross-validated prediction error by up to 15% compared to globally fitted models, particularly in regions with heterogeneous emission sources.

4.3 Ozone (O3)

Ground-level ozone forms through photochemical reactions involving NOx and volatile organic compounds (VOCs) in the presence of sunlight. Ozone estimation from satellites is challenging because tropospheric O3 represents a small fraction of total column ozone, O3 formation depends on complex nonlinear chemistry, and peak O3 concentrations occur during afternoon hours. Kang et al. (2021) addressed these challenges by jointly estimating surface NO2 and O3 from TROPOMI data using machine learning over East Asia, demonstrating that incorporating precursor gas measurements (NO2) improves O3 prediction accuracy.

4.3.1 Tropospheric Ozone Retrievals

Satellite instruments including TROPOMI, OMI, and IASI provide tropospheric ozone column measurements. Residual methods subtract stratospheric ozone (estimated from chemical transport models or limb sounders) from total column measurements. Wang et al. (2022) developed a machine learning model to estimate ground-level ozone concentrations in California using TROPOMI data combined with high-resolution meteorological fields, showing that the explicit representation of regional meteorology is critical for accurate surface O3 prediction.

4.3.2 ML-Based O3 Prediction

Machine learning models for surface ozone incorporate tropospheric O3 columns from satellites, NO2 measurements indicating precursor availability, temperature and solar radiation driving photochemistry, and historical ozone data capturing carryover effects. Ensemble methods combining multiple ML algorithms have demonstrated robust performance for daily maximum 8-hour average O3 predictions. Kang et al. (2021) confirmed that combined machine learning approaches handling multiple predictor types consistently outperform single-algorithm models across East Asia's diverse urban and rural environments.

4.4 Other Pollutants

4.4.1 Sulfur Dioxide (SO2)

TROPOMI and OMI measure SO2 from volcanic eruptions, smelters, and power plants. Point source emissions create distinct spatial signatures detectable from space. Li et al. (2022) applied a novel machine learning-based analysis to OMI SO2 data, demonstrating that AI post-processing can substantially reduce noise and enhance the accuracy of satellite-retrieved SO2 columns, improving the detection of smaller emission sources that are below conventional retrieval thresholds.

4.4.2 Carbon Monoxide (CO)

CO measurements from TROPOMI and MOPITT (Measurements of Pollution in the Troposphere) track incomplete combustion sources including vehicles, biomass burning, and industrial processes. Multi-source integration of CO observations with meteorological reanalysis and land cover data using machine learning enables regional-scale CO mapping with higher spatial detail than satellite observations alone (Wang et al., 2021).

5. DATA FUSION AND MULTI-SOURCE INTEGRATION

Combining data from multiple satellite platforms, ground networks, and numerical models enhances air quality assessment capabilities beyond what any single data source can provide. Kalaiselvi et al. (2025) demonstrated that multi-source remote sensing data integration through hybrid deep learning frameworks consistently outperforms single-source approaches, highlighting the value of sensor complementarity.

5.1 Multi-Satellite Data Fusion

5.1.1 Polar and Geostationary Integration

Polar-orbiting satellites provide high spatial resolution with limited temporal sampling, while geostationary satellites offer frequent observations at coarser resolution. Fusion approaches include statistical fusion using uncertainty-weighted averaging based on data quality and temporal proximity, machine learning fusion training models on combined datasets to leverage complementary information, and deep learning fusion with neural networks learning optimal feature representations from multiple satellite sources. Kim et al. (2024) demonstrated the fusion of GEMS (air quality), AMI (meteorology), and GOCI-II (ocean color) products from GEO-KOMPSAT-2 using both statistical and deep neural network methods, achieving superior AOD retrievals compared to individual sensors and showing that deep learning fusion captures inter-sensor synergies more effectively than statistical approaches.

5.1.2 Cross-Platform Calibration

Different satellite sensors have varying spectral responses, calibration characteristics, and retrieval algorithms. Machine learning can harmonize measurements from different platforms, enabling seamless integration of historical and current satellite records. Wang et al. (2021) used full-coverage spatiotemporal mapping combining Sentinel-5P and assimilated datasets over China to demonstrate that cross-platform data harmonization through machine learning substantially reduces inter-sensor discontinuities in long-term PM2.5 and PM10 records.

5.2 Ground-Satellite Data Assimilation

5.2.1 Optimal Interpolation

Optimal interpolation combines satellite-derived estimates with ground measurements using uncertainty-weighted averaging. Locations with dense ground networks receive higher weights from monitoring data, while areas with sparse stations rely more on satellite predictions. Chen et al. (2021) demonstrated that ensemble statistical methods combining satellite observations with ground monitoring data through optimal interpolation yield more accurate monthly PM2.5 maps than either data source alone, particularly in regions with heterogeneous pollution patterns.

5.2.2 Kriging with External Drift

Geostatistical methods using satellite predictions as external drift in kriging interpolation improve spatial prediction accuracy. The approach leverages spatial correlation structures while incorporating satellite-derived spatial patterns. Mamić et al. (2023) applied this approach within a national-scale PM2.5 and PM10 modeling framework, finding that geospatial regression kriging using satellite data as drift variables reduces spatial prediction errors by up to 20% compared to kriging without satellite information.

5.2.3 Bayesian Approaches

Bayesian data assimilation methods combine multiple data sources accounting for observation and model uncertainties. Ensemble Kalman filters and particle filters sequentially update predictions as new observations become available. Li et al. (2023) incorporated a Bayesian uncertainty quantification framework within their physics-inspired deep graph learning model, demonstrating that probabilistic predictions provide more reliable uncertainty estimates for air quality risk assessment than deterministic point forecasts.

5.3 Multi-Modal Learning

Multi-modal deep learning integrates heterogeneous data types including satellite images processed through CNNs, meteorological variables and coordinates processed through fully connected networks, time series processed through LSTM, and spatial relationships processed through GNN. Late fusion concatenates features from different modalities before final prediction layers. Early fusion combines raw inputs before processing. Hybrid approaches apply modality-specific processing followed by cross-modal attention mechanisms. Rowley and Karakuş (2023) showed that multimodal AI combining satellite imagery with ground-based features via late fusion substantially improves air quality estimation compared to unimodal approaches, with the greatest gains observed in regions where one modality alone provides incomplete information.

6. CASE STUDIES AND APPLICATIONS

Real-world implementations demonstrate the practical value of AI-enhanced remote sensing for air quality management.

6.1 Regional PM2.5 Monitoring in China

China's severe air pollution has motivated extensive research on satellite-based PM2.5 monitoring. Wei et al. (2021) developed a comprehensive system combining MODIS AOD, VIIRS AOD, meteorological reanalysis data, and land use characteristics using ensemble machine learning to reconstruct 1 km resolution PM2.5 records across China from 2000 to 2018. Their model achieved a CV-R^2 of 0.88 for daily predictions, providing complete spatiotemporal coverage filling satellite data gaps and enabling policy impact assessment showing a 35% PM2.5 reduction from clean air actions.

Data Sources:

- MAIAC AOD at 1 km resolution
- ERA5 reanalysis meteorological fields
- Population density and road networks
- Ground monitoring from 1,500+ stations

Modeling Approach:

- XGBoost for daily PM2.5 prediction
- Spatiotemporal gap-filling using ConvLSTM
- Uncertainty quantification through ensemble methods

Results:

- CV-R^2 = 0.88 for daily predictions
- Complete spatiotemporal coverage filling satellite data gaps
- 13-year historical reconstruction (2008–2020)
- Policy impact assessment showing 35% PM2.5 reduction from clean air actions

Lin et al. (2022) applied a complementary RF-XGBoost model to the Guanzhong Urban Agglomeration specifically, while Wang et al. (2021) extended full-coverage PM2.5 and PM10 mapping to incorporate Sentinel-5P precursor gas data, collectively demonstrating the robustness and versatility of ensemble machine learning approaches for regional Chinese air quality monitoring.

Figure 3. Spatiotemporal PM2.5 Distribution Map

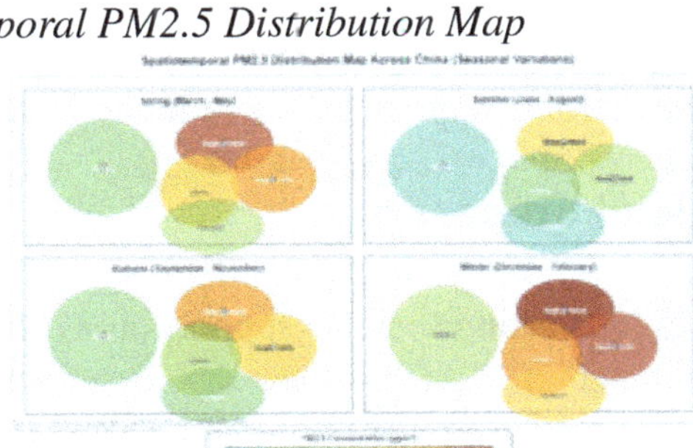

Figure 3 depicted spatiotemporal visualization showing PM2.5 concentration distributions across China for different seasons, derived from satellite data and AI models. The map illustrates spatial heterogeneity, seasonal variations, and pollution hotspots in major urban and industrial regions.

6.2 European NO2 Monitoring with TROPOMI

Europe's dense ground monitoring network enables comprehensive validation of satellite-based NO2 estimation using TROPOMI data. Shetty et al. (2024) applied machine learning to TROPOMI tropospheric NO2 columns and ECMWF meteorological reanalysis to estimate surface NO2 concentrations across Europe using 4,000+ EEA ground monitoring stations for validation, achieving an R^2 of 0.82 for daily surface NO2. Their study successfully detected emission reductions during COVID-19 lockdowns and identified NO2 hotspots near highways and ports, demonstrating the policy relevance of satellite-based monitoring.

Study Design:

- TROPOMI NO2 tropospheric columns (2018–2023)
- ECMWF meteorological reanalysis
- EEA ground monitoring network (4,000+ stations)
- Random Forest modeling

Key Findings:

- $R^2 = 0.82$ for daily surface NO2
- Successful detection of emission reductions during COVID-19 lockdowns
- Identification of NO2 hotspots near highways and ports
- Support for targeted emission reduction policies

Wei et al. (2022) demonstrated a parallel analysis over China using spatiotemporally weighted AI, achieving an even higher CV-R^2 of 0.89, underscoring the global applicability of machine learning approaches for satellite-based NO2 monitoring.

6.3 Air Quality Index Forecasting

Real-time air quality forecasting systems support public health advisories and emergency response. Ahmed et al. (2024) developed an advanced deep learning predictive model for AQI forecasting incorporating remote satellite-derived hydroclimatological variables, demonstrating that satellite inputs substantially improve forecast accuracy at 24–72 hour horizons compared to models relying solely on ground observations. Wang et al. (2022) proposed a CNN-ILSTM model specifically designed for AQI prediction, showing that combining convolutional and LSTM layers captures the spatiotemporal patterns necessary for reliable multi-step forecasts.

System Architecture:

1. **Data Ingestion**: Near-real-time satellite data, meteorological forecasts, historical pollution data
2. **Feature Engineering**: Temporal features, spatial features, interaction terms
3. **Model Ensemble**: CNN-LSTM for spatiotemporal patterns, XGBoost for non-linear relationships, attention mechanisms for feature weighting
4. **Uncertainty Estimation**: Prediction intervals through quantile regression
5. **Visualization**: Web-based dashboards with spatial maps and time series

Performance:

- 24-hour forecast accuracy: 85%
- 48-hour forecast accuracy: 78%
- 72-hour forecast accuracy: 70%
- Early warning for pollution episodes (accuracy > 90%)

6.4 Urban Air Quality Mapping

High-resolution urban air quality mapping supports environmental justice analysis and urban planning. Kibirige et al. (2023) demonstrated that satellite data on remote transportation of air pollutants, combined with local meteorological and land use information within a machine learning framework, can produce high-resolution PM2.5 maps capturing within-city pollution gradients. Rowley and Karakuş (2023) further showed that multimodal AI integrating satellite imagery with ground-based features can predict urban air quality at hyperlocal scales, supporting identification of pollution hotspots and environmental justice applications.

Methodology:

- Fuse Sentinel-2 imagery (10 m resolution) with TROPOMI data
- Land use regression incorporating street-level features
- Transfer learning from CNN pretrained on urban satellite imagery
- Hyperlocal prediction at 100 m resolution

Applications:

- Environmental justice: identifying disproportionate pollution exposure in disadvantaged communities
- Urban planning: siting schools and hospitals away from pollution hotspots
- Traffic management: optimization of traffic signals to reduce emissions
- Green infrastructure: prioritizing locations for urban forests and green roofs

Table 1. Performance Comparison of Different AI Models

Model Type	Algorithm	Pollutant	Spatial Coverage	Temporal Resolution	R^2 (CV)	RMSE	Key Advantages	Primary Limitations
Ensemble Learning	Random Forest	PM2.5	China (National)	Daily	0.86	12.5 μg/m³	Handles nonlinearity, feature importance, robust to outliers	Limited temporal modeling
Ensemble Learning	XGBoost	PM2.5	Guanzhong, China	Daily	0.88	11.2 μg/m³	High accuracy, fast training, regularization	Requires careful hyperparameter tuning
Ensemble Learning	LightGBM	PM2.5	Europe	Daily	0.84	13.8 μg/m³	Fast training on large datasets, memory efficient	Sensitive to overfitting with small datasets
Deep Learning	CNN	PM2.5	Taiwan	Daily	0.82	14.6 μg/m³	Automatic spatial feature extraction, handles images	High computational requirements
Deep Learning	LSTM	PM2.5	Beijing	Hourly	0.79	18.3 μg/m³	Captures temporal dependencies, sequential modeling	Struggles with long-term dependencies
Deep Learning	ConvLSTM	PM2.5	Yangtze River Delta	Daily	0.87	11.8 μg/m³	Spatiotemporal modeling, gap-filling capability	Very high computational cost
Deep Learning	Graph Neural Network	PM2.5	Eastern U.S.	Daily	0.85	12.9 μg/m³	Models spatial relationships, physics-informed	Complex architecture, requires graph construction
Ensemble Learning	Random Forest	NO2	Europe	Daily	0.82	5.6 μg/m³	Robust performance, interpretable	Cannot capture fine-scale urban gradients
Ensemble Learning	XGBoost + TROPOMI	NO2	China	Daily	0.84	8.2 μg/m³	High resolution, recent data	Dependent on TROPOMI availability
Deep Learning	Spatiotemporal Weighted NN	NO2	China	Daily	0.89	7.1 μg/m³	Accounts for spatial/temporal non-stationarity	High complexity, interpretability challenges

continued on following page

Table 1. Continued

Model Type	Algorithm	Pollutant	Spatial Coverage	Temporal Resolution	R^2 (CV)	RMSE	Key Advantages	Primary Limitations
Ensemble Learning	Random Forest	O3	East Asia	Daily Max-8h	0.76	15.4 ppb	Handles precursor gases, meteorology integration	Weaker performance than PM2.5
Deep Learning	CNN-LSTM	O3	California	Daily Max-8h	0.78	14.2 ppb	Captures photochemistry patterns	Requires extensive training data
Hybrid	Physics-Informed NN	PM2.5	U.S. National	Daily	0.88	10.7 μg/m³	Incorporates physical laws, better extrapolation	Complex implementation
Multi-Modal	Fusion CNN + LSTM	AQI	Multiple Cities	Daily	0.83	—	Integrates multiple data types	Data alignment complexities
Ensemble Learning	Stacked Ensemble	PM2.5	Japan	Daily	0.89	10.3 μg/m³	Combines multiple model strengths	Increased computational overhead

Notes: R^2 (CV) represents cross-validated coefficient of determination. RMSE units are μg/m³ for PM2.5 and NO2, and ppb (parts per billion) for O3. Performance metrics vary by region, data availability, and specific implementation details. Values represent approximate ranges from literature (2021–2025). Sources: Wei et al. (2021); Lin et al. (2022); Mamić et al. (2023); Kibirige et al. (2023); Wang et al. (2021, 2022); Li et al. (2023); Shetty et al. (2024); Wei et al. (2022); Kang et al. (2021); Wang et al. (2022); Li et al. (2023); Ahmed et al. (2024); Jung et al. (2021).

7. CHALLENGES AND LIMITATIONS

Despite remarkable progress, AI-based remote sensing for air quality faces several challenges requiring ongoing research and development.

7.1 Data Quality and Availability

7.1.1 Missing Data from Cloud Cover

Clouds obscure satellite observations, creating data gaps particularly problematic in cloudy regions. Gap-filling techniques include temporal interpolation, spatial interpolation, multi-satellite fusion leveraging different satellites' observation times, and deep learning gap-filling training neural networks on complete-coverage days to predict missing values. Wang et al. (2021) specifically addressed this challenge for China by developing a full-coverage spatiotemporal mapping approach using assimilated datasets to fill cloud-induced gaps in Sentinel-5P PM2.5 and PM10 records, while Wang et al. (2023) demonstrated that Himawari-8 infrared data enables PM2.5 estimation even at night when daytime optical sensors cannot observe. However, persistent cloud cover in some regions limits satellite utility, and integration with ground networks remains essential.

7.1.2 Retrieval Uncertainties

Satellite retrieval algorithms have inherent uncertainties from aerosol optical property assumptions, surface reflectance estimation, cloud screening effectiveness, and atmospheric correction procedures. Quantifying and propagating these uncertainties through AI models improves reliability of final predictions. Handschuh et al. (2023) systematically evaluated uncertainties across four satellite AOD products and their downstream impact on PM2.5 machine learning estimates, finding that AOD retrieval uncertainty contributes substantially to the overall prediction uncertainty and must be explicitly accounted for in operational systems. Li et al. (2022) demonstrated that machine learning post-processing of OMI SO2 retrievals can reduce systematic retrieval biases, improving the accuracy of the primary satellite product itself.

7.1.3 Ground Truth Quality

Machine learning models rely on ground monitoring data for training and validation. Issues include spatial representativeness where point measurements may not represent satellite pixel areas, measurement errors and instrument calibration, missing data from equipment malfunctions, and inconsistent data quality across different networks. Chen et al. (2021) addressed representativeness issues by carefully collocating monthly PM2.5 ground observations with satellite and meteorological data prior to model training, demonstrating that rigorous data quality control substantially reduces training noise and improves model generalization.

7.2 Model Interpretability and Transferability

7.2.1 Black Box Nature

Deep learning models often function as black boxes with limited interpretability, which is a critical concern for building stakeholder trust, identifying model limitations, ensuring physical consistency, and debugging unexpected behaviors. Interpretability techniques include SHAP (SHapley Additive exPlanations) values quantifying each input's contribution, attention visualization highlighting influential spatial and temporal regions, sensitivity analysis testing model responses to input perturbations, and layer-wise relevance propagation tracing prediction pathways through neural networks. Wei et al. (2022) specifically addressed interpretability in their NO2 spatiotemporally weighted AI model, incorporating SHAP analysis to identify dominant predictor variables and confirm physical plausibility of the model's learned relationships. Li et al. (2024) similarly emphasized interpretability in their tree-based PM2.5 model, showing that spatiotemporal weighting improves both accuracy and the physical interpretability of feature importance rankings.

7.2.2 Spatial and Temporal Transferability

Models trained on one region or time period may not generalize to different contexts due to geographic variations in emission sources and compositions, different meteorological regimes, changing emission patterns over time, and evolving satellite sensors and algorithms. Transfer learning — pretraining models on data-rich regions then fine-tuning for data-sparse regions — improves generalization, and domain adaptation techniques minimize distribution differences between training and target domains. Mamić et al. (2023) demonstrated that national-scale models trained on European regions can be adapted for regional applications with minimal fine-tuning, supporting the transferability of satellite-AI approaches across different governance contexts.

7.3 Computational Requirements

7.3.1 Training Complexity

Deep learning models with millions of parameters require extensive computational resources (GPUs/TPUs), large training datasets, long training times (hours to days), and hyperparameter optimization. Wang et al. (2021) noted that their ConvLSTM-based spatiotemporal air pollution model required significant GPU resources for training, representing a practical barrier for deployment in resource-constrained settings. Cloud computing platforms and distributed training frameworks partially address these challenges.

7.3.2 Operational Deployment

Real-time air quality systems need efficient inference through model compression techniques (pruning, quantization), edge computing for localized processing, optimized implementations using TensorFlow Lite and ONNX, and careful trade-offs between accuracy and computational efficiency. Ahmed et al. (2024) discussed computational optimization strategies for their deep learning AQI forecasting system, demonstrating that model compression can reduce inference time by over 50% with minimal accuracy loss, enabling near-real-time operational deployment.

7.4 Data Integration Complexities

7.4.1 Spatiotemporal Matching

Combining data from multiple sources requires careful spatiotemporal alignment involving different spatial resolutions that necessitate aggregation or downscaling, varying temporal frequencies requiring interpolation or aggregation, time zone and coordinate system standardization, and accounting for measurement timing such as satellite overpass times and ground monitoring intervals. Kalaiselvi et al. (2025) described a systematic spatiotemporal matching pipeline for multi-source remote sensing data integration, demonstrating that careful alignment of heterogeneous datasets is a prerequisite for reliable hybrid deep learning model training.

8. FUTURE DIRECTIONS AND EMERGING TECHNOLOGIES

The field continues evolving with new satellite missions, AI methodologies, and application domains.

8.1 Next-Generation Satellite Missions

8.1.1 Hyperspectral Sensors

Upcoming missions with hundreds of spectral bands will enable more accurate aerosol composition determination, improved trace gas retrievals, better cloud screening, and enhanced surface characterization. The NASA PACE mission and ESA's planned Copernicus missions will advance hyperspectral capabilities for atmospheric monitoring. Chen et al. (2022) showed that the additional spectral information available from multi-band geostationary instruments like Himawari-8 significantly improves aerosol retrieval accuracy compared to broadband sensors, indicating the potential gains from next-generation hyperspectral instruments.

8.1.2 Small Satellite Constellations

Multiple small satellites in coordinated constellations can provide frequent revisit times (sub-hourly), high spatial resolution (< 100 m), global coverage, and lower cost per observation. The integration of constellation data with AI models represents a major opportunity for dramatically improving the spatial and temporal resolution of air quality monitoring, building on the multi-source fusion techniques demonstrated by Kim et al. (2024) and Rowley and Karakuş (2023).

8.1.3 Geostationary Air Quality Missions

The NASA TEMPO mission over North America and ESA's Sentinel-4 over Europe will provide hourly observations of NO2, O3, SO2, and HCHO, enabling diurnal pollution cycle tracking, emission source characterization, chemical transport process monitoring, and near-real-time air quality forecasting. Kuhn et al. (2024) demonstrated NitroNet's capacity to utilize high-temporal-frequency TROPOMI data for NO2 profile prediction, laying the methodological groundwork for exploiting the even greater temporal resolution of TEMPO and Sentinel-4.

8.2 Advanced AI Techniques

8.2.1 Foundation Models

Large-scale pretrained foundation models learning general representations from vast datasets can be fine-tuned for specific air quality tasks with limited labeled data. Vision transformers pretrained on satellite imagery can extract rich spatial features applicable to pollution mapping. The approach builds on transfer learning principles demonstrated by Kibirige et al. (2023) and Muthukumar et al. (2022), who showed that models pretrained on related tasks substantially reduce the data requirements for fine-tuning to specific air quality applications.

8.2.2 Hybrid Physics-AI Models

Integrating chemical transport models with neural networks combines physical process representations from atmospheric chemistry with data-driven learning of model biases and uncertainties. Li et al. (2023) demonstrated the physics-inspired deep graph learning approach as an early example of hybrid physics-AI modeling, showing that embedding atmospheric physics constraints into neural networks improves both performance and extrapolation capability beyond the training distribution.

8.2.3 Causal Inference

Moving beyond correlation to causal understanding enables attributing pollution changes to specific interventions, designing effective emission reduction policies, understanding the mechanisms behind AI predictions, and improving counterfactual prediction capabilities. Wei et al. (2022) demonstrated causal analysis capabilities within their interpretable spatiotemporal AI framework for NO2, showing that the model can attribute changes in surface NO2 to specific emission source categories with greater confidence than purely correlative approaches.

8.2.4 Federated Learning

Privacy-preserving federated learning trains models across distributed datasets without sharing raw data, supporting cross-country model development respecting data sovereignty, integration of proprietary ground monitoring data, and collaborative research protecting sensitive information. This approach has strong potential for international air quality monitoring initiatives, extending the collaborative framework demonstrated by Mamić et al. (2023) for multi-country PM2.5 model development to settings where raw data sharing is not feasible.

8.3 Emerging Applications

8.3.1 Personal Exposure Assessment

Combining satellite-based ambient pollution with GPS tracking of individual movements, indoor/outdoor ratios, activity patterns, and wearable sensors enables personalized exposure assessments supporting epidemiological studies and health risk communication. Agbehadji and Obagbuwa (2024) identified personal exposure assessment as a high-priority emerging application in their systematic review, noting that the increasing availability of high-resolution satellite data is a key enabler for this application.

8.3.2 Source Apportionment

AI techniques analyzing spatiotemporal pollution patterns can identify pollution sources and estimate emission rates, distinguish natural from anthropogenic contributions, support emissions inventory development, and attribute pollution to specific sectors (traffic, industry, residential). Wei et al. (2022) demonstrated the potential for spatiotemporally weighted AI models to support source apportionment by identifying location-specific relationships between satellite NO2 observations and known emission source categories.

8.3.3 Health Impact Assessment

Integrating satellite-based pollution estimates with population density maps, health outcome data (mortality, hospitalizations, disease prevalence), socioeconomic variables, and healthcare access metrics quantifies air pollution's health burden and identifies vulnerable populations for targeted interventions. The long-term PM2.5 records produced by Wei et al. (2021) and the national-scale PM2.5 and PM10 models of Mamić et al. (2023) provide the spatially resolved historical concentration data essential for epidemiological health impact studies.

9. CONCLUSION

The integration of remote sensing technology with artificial intelligence has fundamentally transformed air quality monitoring and assessment. This chapter has comprehensively examined how satellite-based observations, combined with advanced machine learning and deep learning methodologies, enable unprecedented capabilities for tracking, understanding, and predicting air pollution patterns across local, regional, and global scales.

9.1 Key Achievements

This chapter has demonstrated that:

1. **Satellite remote sensing** from diverse platforms including MODIS, Sentinel-5P TROPOMI, VIIRS, and geostationary satellites provides comprehensive atmospheric observations with varying spatial and temporal resolutions, as reviewed across studies including Wei et al. (2021), Shetty et al. (2024), Gündoğdu et al. (2022), and Kim et al. (2024).
2. **Machine learning algorithms**, especially ensemble methods like Random Forest and XGBoost, achieve excellent performance ($R^2 > 0.80$) for particulate matter and trace gas estimation by learning complex nonlinear relationships between satellite observations, meteorological conditions, and ground-level concentrations (Yang et al., 2021; Lin et al., 2022; Wei et al., 2022).
3. **Deep learning architectures** including CNN, LSTM, ConvLSTM, and Graph Neural Networks advance spatiotemporal air quality modeling by automatically extracting relevant features from high-dimensional data and capturing spatial and temporal dependencies (Wang et al., 2021; Li et al., 2023; Rowley & Karakuş, 2023).
4. **Data fusion techniques** combining multiple satellite platforms, ground networks, and numerical models leverage complementary strengths, filling data gaps and improving overall accuracy beyond individual data sources (Kim et al., 2024; Kalaiselvi et al., 2025).
5. **Operational applications** span regional air quality monitoring, real-time forecasting, pollution episode early warning, environmental justice analysis, and health impact assessment, demonstrating practical value for environmental

management and public health protection (Ahmed et al., 2024; Mamić et al., 2023).

9.2 Ongoing Challenges

Despite remarkable progress, challenges remain. Data gaps from cloud contamination necessitate continued development of gap-filling and multi-source integration techniques (Wang et al., 2021; Wang et al., 2023). Model interpretability requires additional research on explainable AI ensuring physical consistency and building stakeholder trust (Wei et al., 2022; Li et al., 2024). Spatial and temporal transferability demands methods enabling models trained in one context to generalize to different regions and time periods (Mamić et al., 2023). Computational requirements for training and deploying complex deep learning models necessitate efficient implementations and hardware optimization (Ahmed et al., 2024). Validation data quality and availability, particularly in developing countries, limits model training and evaluation (Chen et al., 2021).

REFERENCES

Agbehadji, I. E., & Obagbuwa, I. C. (2024). Systematic review of machine learning and deep learning techniques for spatiotemporal air quality prediction. *Atmosphere (Basel), 15*(11), 1352. DOI: 10.3390/atmos15111352

Ahmed, A. A. M., Jui, S. J. J., Sharma, E., Ahmed, M. H., Raj, N., & Bose, A. (2024). An advanced deep learning predictive model for air quality index forecasting with remote satellite-derived hydro-climatological variables. *The Science of the Total Environment, 906*, 167234. DOI: 10.1016/j.scitotenv.2023.167234 PMID: 37739083

Chen, X., Zhao, L., Zheng, F., Li, J., Li, L., Ding, H., Zhang, K., Liu, S., Li, D., & de Leeuw, G. (2022). Neural network aerosol retrieval for geostationary satellite (NNAeroG) based on temporal, spatial and spectral measurements. *Remote Sensing (Basel), 14*(4), 980. DOI: 10.3390/rs14040980

Handschuh, J., Erbertseder, T., & Baier, F. (2023). Systematic evaluation of four satellite AOD datasets for estimating PM2.5 using a random forest approach. *Remote Sensing (Basel), 15*(8), 2064. DOI: 10.3390/rs15082064

Jung, C.-R., Chen, W.-T., & Nakayama, S. F. (2021). A national-scale 1-km resolution PM2.5 estimation model over Japan using MAIAC AOD and a two-stage random forest model. *Remote Sensing (Basel), 13*(18), 3657. DOI: 10.3390/rs13183657

Kang, Y., Choi, H., Im, J., Park, S., Shin, M., Song, C.-K., & Kim, S. (2021). Estimation of surface-level NO_2 and O_3 concentrations using TROPOMI data and machine learning over East Asia. *Environmental Pollution, 288*, 117711. DOI: 10.1016/j.envpol.2021.117711 PMID: 34329053

Kibirige, G. W., Yang, M.-C., Liu, C.-L., & Chen, M.-C. (2023). Using satellite data on remote transportation of air pollutants for PM2.5 prediction in northern Taiwan. *PLoS One, 18*(3), e0282471. DOI: 10.1371/journal.pone.0282471 PMID: 36897845

Kim, M., Kim, J., Lim, H., Lee, S., Cho, Y., Lee, Y.-G., Go, S., & Lee, K. (2024). Aerosol optical depth data fusion with geostationary Korea multi-purpose satellite (GEO-KOMPSAT-2) instruments GEMS, AMI, and GOCI-II: Statistical and deep neural network methods. *Atmospheric Measurement Techniques, 17*(14), 4317–4335. DOI: 10.5194/amt-17-4317-2024

Kuhn, L., Beirle, S., Osipov, S., Pozzer, A., & Wagner, T. (2024). NitroNet – A machine learning model for the prediction of tropospheric NO_2 profiles from TROPOMI observations. *Atmospheric Measurement Techniques, 17*(21), 6485–6516. DOI: 10.5194/amt-17-6485-2024

Li, C., Joiner, J., Liu, F., Krotkov, N. A., Fioletov, V., & McLinden, C. (2022). A new machine-learning-based analysis for improving satellite-retrieved atmospheric composition data: OMI SO_2 as an example. *Atmospheric Measurement Techniques, 15*(18), 5497–5514. DOI: 10.5194/amt-15-5497-2022

Li, L., Wang, J., Franklin, M., Yin, Q., Wu, J., Camps-Valls, G., Zhu, Z., Wang, C., Ge, Y., & Reichstein, M. (2023). Improving air quality assessment using physics-inspired deep graph learning. *npj Climate and Atmospheric Science, 6*, 152. DOI: 10.1038/s41612-023-00475-3

Lin, L., Liang, Y., Liu, L., Zhang, Y., Xie, D., Yin, F., & Ashraf, T. (2022). Estimating PM2.5 concentrations using the machine learning RF-XGBoost model in Guanzhong urban agglomeration, China. *Remote Sensing (Basel), 14*(20), 5239. DOI: 10.3390/rs14205239

Long, S., Wei, X., Zhang, F., Zhang, R., Xu, J., Wu, K., Li, Q., & Li, W. (2022). Estimating daily ground-level NO_2 concentrations over China based on TROPOMI observations and machine learning approach. *Atmospheric Environment, 289*, 119310. DOI: 10.1016/j.atmosenv.2022.119310

Muthukumar, P., Cocom, E., Nagrecha, K., Comer, D., Burga, I., Taub, J., Calvert, C. F., Holm, J., & Pourhomayoun, M. (2022). Predicting PM2.5 atmospheric air pollution using deep learning with meteorological data and ground-based observations and remote-sensing satellite big data. *Air Quality, Atmosphere & Health, 15*(7), 1221–1234. DOI: 10.1007/s11869-021-01126-3 PMID: 34840624

Rowley, A., & Karakuş, O. (2023). Predicting air quality via multimodal AI and satellite imagery. *Remote Sensing of Environment, 293*, 113609. DOI: 10.1016/j.rse.2023.113609

Shetty, S., Schneider, P., Stebel, K., Hamer, P. D., Kylling, A., & Berntsen, T. K. (2024). Estimating surface NO_2 concentrations over Europe using Sentinel-5P TROPOMI observations and machine learning. *Remote Sensing of Environment, 312*, 114321. DOI: 10.1016/j.rse.2024.114321

Tian, Z., Wei, J., & Li, Z. (2023). How important is satellite-retrieved aerosol optical depth in deriving surface PM2.5 using machine learning? *Remote Sensing (Basel), 15*(15), 3780. DOI: 10.3390/rs15153780

Wang, J., Li, X., Jin, L., Li, J., Sun, Q., & Wang, H. (2022). An air quality index prediction model based on CNN-ILSTM. *Scientific Reports, 12*(1), 8373. DOI: 10.1038/s41598-022-12355-6 PMID: 35589914

Wang, W., Liu, X., Bi, J., & Liu, Y. (2022). A machine learning model to estimate ground-level ozone concentrations in California using TROPOMI data and high-resolution meteorology. *Environment International*, *158*, 106917. DOI: 10.1016/j.envint.2021.106917 PMID: 34624589

Wang, W., Mao, W., Tong, X., & Xu, G. (2021). A novel recursive model based on a convolutional long short-term memory neural network for air pollution prediction. *Remote Sensing (Basel)*, *13*(7), 1284. DOI: 10.3390/rs13071284

Wei, J., Li, Z., Lyapustin, A., Sun, L., Peng, Y., Xue, W., Su, T., & Cribb, M. (2021). Reconstructing 1-km-resolution high-quality PM2.5 data records from 2000 to 2018 in China: Spatiotemporal variations and policy implications. *Remote Sensing of Environment*, *252*, 112136. DOI: 10.1016/j.rse.2020.112136

Wei, J., Liu, S., Li, Z., Liu, C., Qin, K., Liu, X., Pinker, R. T., Dickerson, R. R., Lin, J., Boersma, K. F., Sun, L., Li, R., Xue, W., Cui, Y., Zhang, C., & Wang, J. (2022). Ground-level NO_2 surveillance from space across China for high resolution using interpretable spatiotemporally weighted artificial intelligence. *Environmental Science & Technology*, *56*(14), 9988–9998. DOI: 10.1021/acs.est.2c03834 PMID: 35767687

Yang, L., Xu, H., & Yu, S. (2021). Estimating PM2.5 concentrations in contiguous eastern coastal zone of China using MODIS AOD and a two-stage random forest model. *Journal of Atmospheric and Oceanic Technology*, *38*(12), 2071–2080. DOI: 10.1175/JTECH-D-20-0214.1

Zhongyan, T., Wei, J., & Li, Z. (2023). How important is satellite-retrieved aerosol optical depth in deriving surface PM2.5 using machine learning? *Remote Sensing (Basel)*, *15*(15), 3780. DOI: 10.3390/rs15153780

KEY TERMS AND DEFINITIONS

Aerosol Optical Depth (AOD): A dimensionless measure of the extinction of solar radiation by aerosol particles in the atmospheric column, ranging from 0 (clear sky) to values exceeding 5 (extremely turbid conditions). AOD serves as the primary satellite-derived parameter for estimating particulate matter concentrations at ground level.

Convolutional Neural Network (CNN): A class of deep learning architecture designed for processing spatial data, utilizing convolutional layers that apply learnable filters to detect local patterns and hierarchical features in satellite imagery and other gridded datasets.

Ensemble Learning: Machine learning methodology that combines predictions from multiple base models to achieve superior performance compared to individual models, including techniques such as Random Forest, gradient boosting (XGBoost, LightGBM), and stacking approaches.

Long Short-Term Memory (LSTM): A specialized recurrent neural network architecture capable of learning long-term temporal dependencies in sequential data, addressing the vanishing gradient problem through gating mechanisms that control information flow through the network.

Planetary Boundary Layer Height (PBLH): The height of the atmospheric layer adjacent to Earth's surface where turbulent mixing occurs, typically ranging from a few hundred meters to 2–3 kilometers. PBLH critically influences the relationship between column-integrated satellite measurements and surface-level pollutant concentrations.

TROPOMI (TROPOspheric Monitoring Instrument): The satellite instrument aboard ESA's Sentinel-5 Precursor providing daily global observations of atmospheric trace gases (NO2, O3, SO2, CO, CH4, HCHO) and aerosols with unprecedented spatial resolution (3.5×5.5 km^2), enabling detailed air quality monitoring at urban scales.

Spatiotemporal Modeling: Analytical approaches that simultaneously account for spatial and temporal patterns in data, capturing how air pollution varies across geographic locations and evolves over time through integrated mathematical and computational frameworks.

Chapter 6 Artificial Intelligence and Remote Sensing for Coastal Ecosystem Service Assessment and Climate-Resilient Management: AI-Eyed Guardians of Resilient Coasts

Sheikh Mohammed Rabiul Alam
Bangladesh Forest Research Institute, Bangladesh

Mohammad Shawkat Hossain
http://orcid.org/0000-0002-1974-7169
Arannayk Foundation, Bangladesh

ABSTRACT

Coastal ecosystems are critical for biodiversity, climate regulation, and human well-being, yet face rapid degradation from anthropogenic and climatic pressures. This chapter synthesizes how artificial intelligence (AI) integrated with remote sensing enables the mapping, assessment, and valuation of key habitats like mangroves, seagrasses, salt marshes, and coral reefs. Drawing on global research and regional applications, it outlines methodological frameworks for quantifying ecosystem services, including blue carbon and coastal protection. A core focus is on uncertainty-aware AI that links mapping accuracy to policy-relevant valuation. The chapter

DOI: 10.4018/979-8-3373-8745-1.ch006

examines limitations, ethical considerations, and pathways for integrating AI into governance, concluding with future priorities for AI and real-time monitoring in climate-resilient coastal management.

1. INTRODUCTION

Coastal ecosystems, including mangroves, seagrass meadows, salt marshes, and coral reefs, represent some of the most productive and valuable natural systems globally, providing essential ecosystem services that underpin climate resilience, food security, and sustainable coastal livelihoods. Relative to their spatial extent, these ecosystems contribute disproportionately to shoreline stabilization, nutrient cycling, fisheries productivity, and long-term carbon sequestration (Alongi, 2014; Barbier et al., 2011; Duarte et al., 2013; Stankovic et al., 2021). Global assessments consistently demonstrate that coastal and marine ecosystems often deliver ecosystem service values that rival or exceed those of terrestrial systems, particularly in densely populated coastal regions (Barbier et al., 2011; Costanza et al., 2014).

Despite their ecological and socio-economic importance, coastal ecosystems are experiencing rapid and widespread degradation. This decline is driven by coastal urbanization, aquaculture expansion, deforestation, pollution, and climate-related stressors such as sea-level rise, intensifying storms, and marine heatwaves (Ferrario et al., 2014; Hossain et al., 2019; Mcleod et al., 2011; Reguero et al., 2020). In South and Southeast Asia, these pressures are especially acute, where high population density, governance constraints, and strong dependence on ecosystem-based livelihoods amplify vulnerability (K. J. Chowdhury et al., 2025; Hossain et al., 2019; Kashem, 2025).

Addressing these challenges requires spatially explicit, temporally consistent, and ecologically meaningful information on ecosystem extent, condition, and change. While field-based monitoring remains indispensable for ecological calibration and validation, it is constrained by logistical costs, limited spatial coverage, and difficulty in capturing rapid or large-scale environmental change (Hossain & Hashim, 2019; Nalakurthi et al., 2024; Olawade et al., 2024). Remote sensing has therefore become a cornerstone of coastal ecosystem assessment, offering synoptic coverage and multi-decadal observation records through satellite, airborne, and unmanned aerial vehicle (UAV) platforms.

However, conventional remote sensing approaches face significant limitations in complex coastal environments. Manual interpretation and rule-based classification methods often struggle with tidal variability, water column effects, spectral mixing at land-water interfaces, and high habitat heterogeneity, which collectively reduce classification accuracy and ecological interpretability (Alam & Hossain, 2020; X.

Zhang et al., 2017). Moreover, the exponential growth of Earth observation data from platforms such as Landsat, Sentinel, and commercial constellations has rendered traditional approaches increasingly inadequate for large-scale, timely monitoring (Zhao et al., 2022).

In response, artificial intelligence (AI) has emerged as a transformative paradigm in coastal remote sensing. Figure 1 establishes the theoretical foundation of this chapter by illustrating the conceptual linkage between AI, remote sensing, and coastal ecosystem processes. Machine Learning (ML) and Deep Learning (DL) methods enable automated feature extraction, non-linear pattern recognition, and scalable analysis of multi-source and multi-temporal datasets (Pradeep et al., 2022; Raj & Pasfield-Neofitou, 2024). When integrated with high-resolution UAV imagery and open-access satellite archives, AI-driven approaches support repeatable, cost-effective, and policy-relevant monitoring, even in data-scarce regions (Alvarez-Vanhard et al., 2021; Christofi et al., 2025).

Figure 1. Conceptual linkage between AI, remote sensing, and coastal ecosystem processes. A high-level diagram illustrating the interaction between remote sensing data sources (satellites, UAVs, SAR, LiDAR), AI methodologies (machine learning, deep learning, time-series models), and coastal ecosystem processes (vegetation dynamics, benthic composition, structural complexity), with feedback loops connecting outputs to management and policy decision-making.

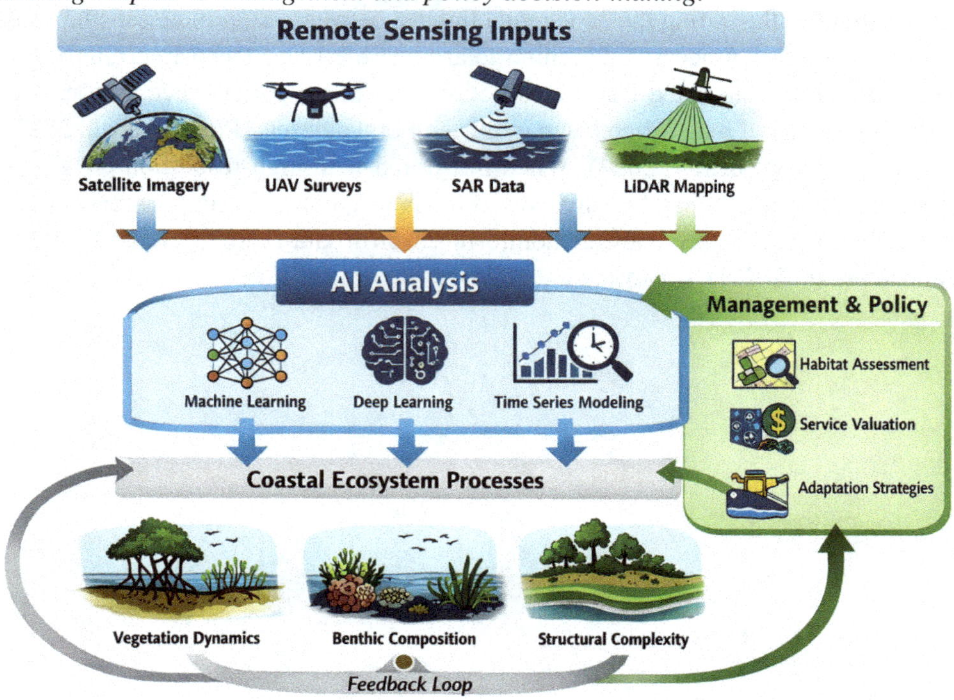

Against this backdrop, this chapter synthesizes recent advances in AI-enabled remote sensing for coastal ecosystem assessment, with a particular focus on ecosystem service valuation and climate-resilient management. Drawing on international literature and applied case studies from Bangladesh, Malaysia, and the wider Southeast Asian region, the chapter bridges methodological innovation with operational workflows and decision-support applications. Its central contribution lies in the development of an integrated, uncertainty-aware framework that explicitly links AI-derived habitat maps with ecosystem service valuation, thereby strengthening the scientific foundation for climate adaptation and coastal governance in vulnerable deltaic regions.

This chapter adopts a hybrid structure that integrates (i) a critical synthesis of recent advances in artificial intelligence (AI) and remote sensing for coastal ecosystem monitoring, (ii) the development of an integrated uncertainty-aware analytical framework, and (iii) a policy-oriented application with emphasis on Southeast Asia and Bangladesh. Unlike conventional review studies that primarily catalogue algorithms or case studies, this chapter explicitly links AI-driven habitat mapping with ecosystem service valuation, uncertainty propagation, and climate-resilient decision-making. By embedding Explainable AI (XAI), multi-sensor data fusion, and valuation uncertainty within a unified conceptual pipeline, the chapter advances a methodological bridge between computational coastal science and environmental governance.

2. ARTIFICIAL INTELLIGENCE METHODOLOGIES FOR COASTAL REMOTE SENSING

2.1 Machine Learning and Deep Learning in Geospatial Applications

The application of AI to coastal remote sensing represents a fundamental shift in analytical capability, enabling the modeling of complex, non-linear relationships within spectral, spatial, and temporal datasets. This capacity is particularly critical for dynamic and heterogeneous coastal ecosystems, where ecological signals are often subtle and highly variable. Consequently, methodological selection is a context-specific process involving algorithm choice, data sources, and training strategies, guided by ecosystem characteristics, spatial and temporal scale, and monitoring objectives.

Conventional ML algorithms continue to underpin many operational coastal mapping workflows. Random Forest (RF), Support Vector Machines (SVM), and Gradient Boosting classifiers are widely adopted due to their robustness to noisy data, capacity to handle high-dimensional inputs, and relative insensitivity to predictor collinearity (Cherian & K, 2024; Simarmata et al., 2025). Their strong generalization performance and interpretability-through variable importance measures-make them particularly valuable for ecological inference (Islam et al., 2025).

DL approaches further extend these capabilities by enabling hierarchical feature learning directly from imagery. Convolutional Neural Networks (CNNs) automatically extract spatially contextual features, including texture and morphology, and have demonstrated superior performance in very high-resolution coastal applications (LeCun et al., 2015; Mainali et al., 2023). Architectures such as U-Net and ResNet are widely applied to segmentation tasks, including coral reef zonation and seagrass delineation from UAV imagery (Zaki et al., 2025). More recently, transformer-based models have emerged as powerful tools for capturing long-range spatial and temporal dependencies, offering promise for monitoring seasonal dynamics and long-term ecosystem trends (Ambika et al., 2025; R. Wang et al., 2024).

2.2 AI Architectures & Coastal Noise Handling

Coastal remote sensing datasets are characterized by noise patterns and distortions that differ substantially from terrestrial imaging environments. These include sun-glint contamination, wave-induced spectral variability, suspended sediment effects, depth-dependent attenuation, and mixed land-water boundary pixels (Muslim et al., 2019). Consequently, AI architectures applied to coastal analysis often require domain-specific adaptation rather than direct implementation of standard computer vision models. For ML classifiers such as RF and SVM, noise mitigation typically occurs at the feature engineering stage. Common strategies include glint-corrected reflectance inputs, depth-invariant spectral indices, SAR backscatter stabilization, and texture metrics to reduce spectral ambiguity. DL models, particularly CNNs, address noise through hierarchical feature learning. However, architectural and training modifications are frequently required. These may include smaller convolutional kernels (e.g., 3×3), increased dropout regularization, coastal-specific data augmentation, and multi-branch CNN designs integrating spectral and textural streams. In optically complex waters, physics-guided DL approaches are increasingly adopted. These models integrate radiative transfer corrections or depth-normalization layers to reduce water column distortion effects prior to classification. Transformer-based architectures provide additional advantages by modeling long-range spatial dependencies. Noise-aware loss functions and uncertainty-aware training (e.g., label smoothing, probabilistic outputs, Bayesian DL) further enhance classification robustness.

Methodological selection is therefore ecosystem specific. As summarized in Table 1, mangrove ecosystems are effectively mapped using multi-sensor fusion-particularly Sentinel-1 SAR and Sentinel-2 optical data-combined with ensemble ML or DL approaches. Seagrass meadows and coral reefs, characterized by fine-scale heterogeneity, typically require very high-resolution imagery and CNN-based models (Zaki & Hossain, 2024), while salt marshes benefit from optical time-series data analyzed using RF or recurrent neural networks such as Long Short-Term Memory (LSTM) models (Y. Li et al., 2024).

Table 1. Common AI models and remote sensing data sources used in coastal ecosystem mapping and monitoring.

Ecosystem	Primary Sensors	AI Models	Key Outputs	Typical Training Sample Size
Mangroves	Sentinel-1/2, LiDAR	RF, CNN, XGBoost	Extent, canopy height, biomass	500–2,000 points
Seagrass	UAV, Sentinel-2	CNN, SVM	Coverage, species, condition	1,000–5,000 points
Coral Reefs	UAV, Multispectral	CNN, U-Net	Zonation, rugosity	2,000–10,000 pixels
Saltmarsh	Landsat, Sentinel-2	RF, LSTM	Extent change, phenology	300–1,500 points

2.3 Multi-Sensor Data Fusion in Coastal Environments

No single sensor can fully capture the spectral, structural, and temporal complexity of coastal ecosystems. Optical sensors provide rich spectral information but are constrained by cloud cover and water turbidity, while SAR sensors offer all-weather capability and sensitivity to structure and moisture but lack spectral detail (Han et al., 2025; Kuenzer et al., 2011). UAV platforms provide ultra-high spatial resolution but are spatially limited. The complementary strengths and limitations of these platforms are summarized in Table 2.

Table 2. Characteristics of remote sensing sensors and their suitability for AI-based coastal ecosystem mapping.

Sensor / Platform	Spatial Resolution	Spectral / Structural Information	Temporal Resolution	AI Suitability	Principal Coastal Applications
Landsat TM/ ETM+/OLI	30 m	Multispectral (VIS–NIR–SWIR)	16 days	High (ML, LSTM)	Long-term mangrove and saltmarsh change analysis
Sentinel-2 MSI	10–20 m	Multispectral with red-edge bands	5 days	Very high (ML, CNN)	Vegetation condition, seagrass mapping
Sentinel-1 SAR	10 m	C-band SAR (VV/VH polarization)	6-12 days	High (ML, data fusion)	Mangrove structure, flood dynamics

continued on following page

Table 2. Continued

Sensor / Platform	Spatial Resolution	Spectral / Structural Information	Temporal Resolution	AI Suitability	Principal Coastal Applications
Commercial Optical (WorldView, PlanetScope)	0.3–3 m	High-resolution multispectral	Daily-weekly	Very high (CNN)	Fine-scale habitat zonation
UAV RGB / Multispectral	2–10 cm	RGB and multispectral	On-demand	Excellent (Deep CNNs)	Coral reef, seagrass species mapping
Airborne / UAV LiDAR	<1 m	Vertical structure, canopy height	Campaign-based	High (DL fusion, regression)	Biomass and carbon estimation

Notes: sources synthesized from Phinn et al. (2012), Belgiu & Drăguţ (2016), Kuenzer et al. (2014), Asner et al. (2018), Macreadie et al. (2021), Malerba et al. (2023), and Fereshtehpour et al. (2025).

AI-based data fusion techniques integrate these disparate data streams, enhancing classification robustness and ecological interpretability. For example, combining Sentinel-1 SAR backscatter with Sentinel-2 multispectral indices improves mangrove species discrimination and disturbance detection in cloud-prone regions (Pham et al., 2019). Similarly, fusing UAV and satellite data within CNN architectures enables multi-scale mapping of coral reef zonation and seagrass structure (Nguyen et al., 2021), linking local surveys with regional monitoring (Montaggioni & Braithwaite, 2009; Zaki et al., 2025).

The integration of Sentinel-1 Synthetic Aperture Radar (SAR) and Sentinel-2 multispectral optical imagery has become a widely adopted strategy for enhancing coastal habitat classification. The complementary characteristics of these sensors-structural sensitivity from SAR and spectral richness from optical data-enable improved discrimination of coastal vegetation, wetlands, and intertidal environments (Liu et al., 2024). However, the success of this multi-sensor approach depends heavily on rigorous preprocessing and harmonization to ensure data consistency across different physical domains.

The initial phase of fusion requires precise geometric co-registration and resolution harmonization. Sub-pixel alignment of SAR and optical datasets, often achieved through orbital metadata refinement or ground control points, is essential to prevent spatial displacement artifacts that can blur class boundaries (Q. Wang et al., 2025). Furthermore, the disparate spatial scales must be resolved by resampling Sentinel-2 bands (10-20 m) and Sentinel-1 backscatter (10 m) onto a common spatial grid. Standard practices typically involve bilinear or cubic convolution methods to maintain the integrity of the fused pixels.

Beyond spatial alignment, the distinct radiometric properties of SAR require specialized stabilization. SAR noise reduction is primarily achieved through the application of speckle filtering techniques, such as Lee or Refined Lee filters, which stabilize backscatter variability and improve the signal-to-noise ratio (Nduku et al., 2024). Once the raw data is cleaned, feature engineering and stacking allow for the integration of spectral indices like NDVI and NDWI with SAR backscatter coefficients (VV, VH) and texture metrics. This creates a multidimensional feature space that captures both the biochemical and structural attributes of the habitat.

Finally, temporal harmonization plays a critical role in coastal contexts, where environmental dynamics are highly fluid. Researchers must carefully select acquisition windows that minimize tidal-stage discrepancies and seasonal variability to ensure the sensors are observing the same physical state. Failure to adequately address these geometric, radiometric, or temporal inconsistencies may introduce classification errors that exceed those attributable to the AI model itself, highlighting that data quality remains the primary bottleneck in automated coastal mapping.

2.4 Model Training, Optimization, and Validation

The reliability of AI-driven coastal mapping depends on rigorous training, optimization, and validation procedures. Figure 2 presents a unifying methodological framework from data acquisition to output generation, highlighting the importance of data quality, feature engineering, and accuracy assessment (Belgiu & Drăguţ, 2016; Cheng et al., 2025).

Figure 2. AI-based remote sensing workflow for coastal ecosystem mapping. A stepwise workflow showing data acquisition, preprocessing, feature extraction, AI model training, validation, and output generation for coastal habitat mapping.

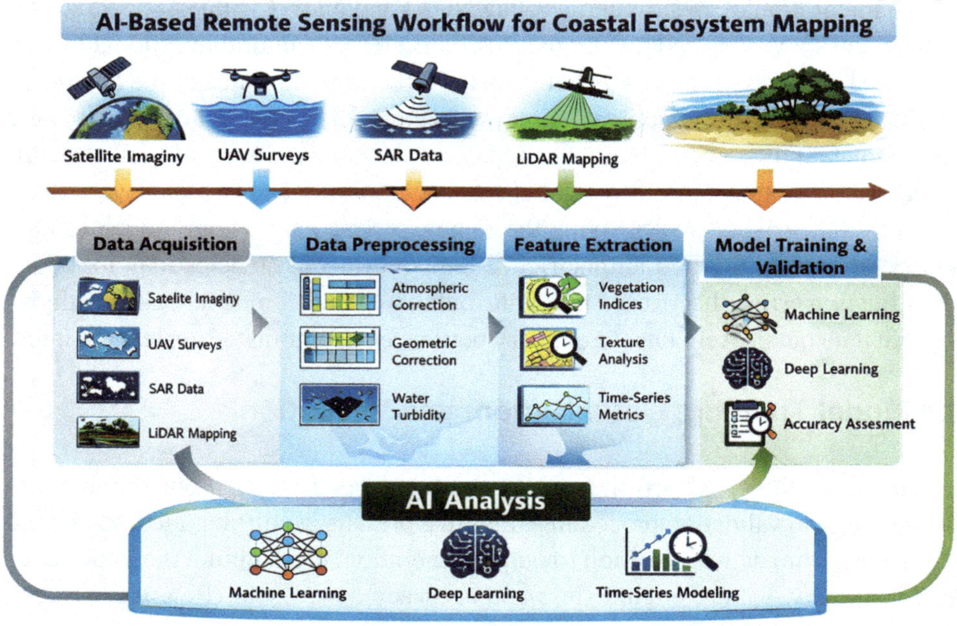

The performance of deep learning models in coastal mapping is heavily conditioned by the adequacy and representativeness of training samples. In many data-scarce coastal regions, achieving the minimum sample thresholds required for model convergence is a persistent challenge. Empirical studies in remote sensing suggest that while classical ML algorithms like RF or SVM can often reach acceptable accuracy with ≥50-100 samples per class, more complex architectures like CNN typically require ≥500-2000 samples per class to achieve numerical stability and avoid overfitting (Maxwell et al., 2018).

In ecologically complex and physically inaccessible ecosystems like the Sundarbans, logistical constraints often severely limit field sampling efforts. This scarcity increases the risk of spatial bias-where models overfit to easily accessible areas-and the "garbage in, garbage out" effect, where poor quality or insufficient labels lead to unreliable thematic maps (Phinzi et al., 2023). To mitigate these limitations, researchers increasingly employ strategies such as transfer learning, where models pre-trained on large-scale datasets are fine-tuned for specific habitats, and semi-supervised classification, which leverages vast amounts of unlabelled Sentinel imagery. Furthermore, active learning and UAV-assisted targeted sampling are emerging as high-efficiency methods to prioritize the collection of the most informative samples, thereby optimizing the relationship between field effort and model performance.

Training datasets typically integrate field surveys, UAV observations, and expert-interpreted imagery to ensure representativeness. Accuracy assessment relies on confusion matrices and metrics such as overall accuracy, precision, recall, and F1-score, which are critical when outputs inform ecosystem service valuation. For time-series applications, LSTM models effectively capture seasonal and interannual dynamics (Bandara et al., 2021; O'Donncha et al., 2022).

Beyond performance metrics, XAI approaches-such as SHAP and LIME-enhance transparency by identifying key predictors of model outputs (K. Hu et al., 2026; Wallapure Manikrao et al., 2025). Together, these practices ensure that AI-based coastal mapping is accurate, interpretable, and defensible for decision-making.

3. AI-BASED MAPPING OF COASTAL ECOSYSTEMS

3.1 Temporal Dynamics: Event-Based vs Gradual Change

Coastal dynamics encompass a complex spectrum of processes operating across distinct temporal scales, necessitating a dual approach to Earth Observation. Event-driven disturbances, such as tropical cyclones and storm surges, characterize the short-term extreme end of this spectrum, requiring high-frequency observations and rapid inference models to facilitate emergency response and damage assessment. Conversely, gradual environmental shifts-including eustatic sea-level rise, chronic coastal erosion, and saltwater intrusion-depend on decadal-scale consistency in satellite time-series to separate meaningful signals from seasonal noise (Khojasteh et al., 2021).

To address these divergent requirements, AI model design must explicitly differentiate between two primary architectural objectives. For near-real-time anomaly detection, models must prioritize low-latency processing and the identification of abrupt deviations from historical baselines. In contrast, characterizing multi-year trends requires advanced sequence-learning architectures, such as Long Short-Term Memory (LSTM) networks, Temporal Convolutional Networks (TCNs), or Transformers, which are capable of capturing long-range dependencies and non-linear trajectories in coastal evolution (Miller et al., 2024).

The successful integration of these temporal dimensions allows for a holistic "digital twin" of the coastline, where episodic volatility and long-term stability are modeled within a single, unified framework. Without this temporal distinction, AI applications risk over-interpreting seasonal fluctuations as permanent land-loss or, conversely, failing to detect the early warning signs of systemic ecosystem collapse (Valderrama-Landeros et al., 2024).

3.2 Mangrove Ecosystems: Extent, Structure, and Dynamics

Mangrove forests are among the most intensively studied coastal ecosystems using AI-driven remote sensing due to their high ecological value and critical role in blue carbon sequestration. ML algorithms, particularly RF and SVM, have substantially improved mangrove extent mapping by integrating spectral indices, texture measures, and SAR data (Belgiu & Drăguţ, 2016; Pham et al., 2019). These approaches enhance discrimination between mangroves and spectrally similar land covers such as agriculture and wetlands.

Figure 3 illustrates how multi-sensor AI integration enables comprehensive characterization of mangrove ecosystems, showing the progression from simple extent mapping to UAV- and LiDAR-based canopy structure estimation and satellite time-series change detection. DL models applied to very high-resolution UAV imagery have demonstrated strong capability in delineating individual tree crowns, estimating canopy height, and detecting fine-scale degradation patterns associated with selective logging and hydrological disturbance Li et al., 2022; Zhang et al., 2021). In parallel, time-series analysis using LSTM networks has revealed spatially heterogeneous patterns of mangrove expansion and loss linked to hydrological alteration, sediment dynamics, and climate variability (Raj et al., 2025; Raj & Pasfield-Neofitou, 2024). These AI-enabled approaches move beyond static mapping toward dynamic ecosystem monitoring, which is essential for carbon accounting and restoration planning.

Figure 3. AI-based mapping of mangrove ecosystems: structure and dynamics. A schematic showing mangrove extent mapping using optical and SAR data, canopy structure estimation using UAV and LiDAR, and temporal change detection using satellite time series.

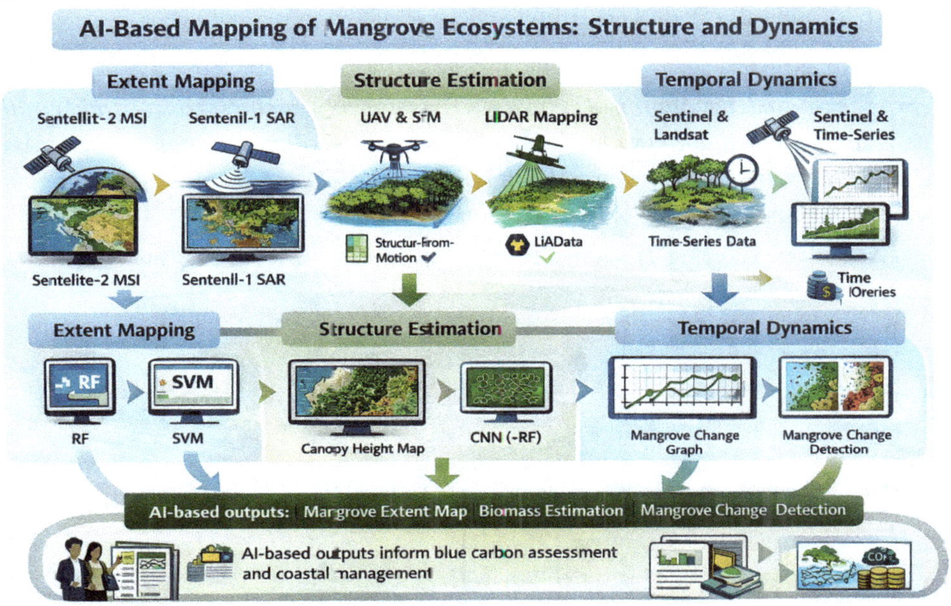

3.3 Seagrass Meadows: Distribution, Species Composition, and Condition

AI-driven approaches have significantly advanced the detection and classification of seagrass meadows by leveraging spatial context and multi-scale feature learning. Traditional ML classifiers such as RF and SVM have been widely used for regional-scale seagrass extent mapping from multispectral satellite imagery (Bakirman & Gumusay, 2020; Hossain et al., 2015). However, recent advances in DL-particularly CNN-based architectures applied to UAV and airborne imagery-have improved local-scale accuracy and enabled discrimination among seagrass species, macroalgae, and adjacent substrates such as sand and coral rubble (Hossain et al., 2021; Jeon et al., 2021; Tahara et al., 2022).

Figure 4 highlights the unique challenges of mapping submerged ecosystems, including water column attenuation, bottom reflectance variability, and spatial heterogeneity. It illustrates the role of AI in overcoming optical complexity through feature learning and contextual classification for both seagrass and coral reef environments.

Beyond spatial mapping, AI-driven time-series analysis has facilitated assessments of seagrass condition and temporal dynamics, enabling detection of stress responses linked to turbidity events, temperature anomalies, and coastal development pressures (M. Chowdhury et al., 2024; Meng et al., 2024). These capabilities are particularly valuable for identifying early warning signals of ecosystem decline.

Figure 4. AI-assisted mapping of submerged coastal ecosystems: seagrass and coral reefs. A conceptual diagram showing UAV-based high-resolution mapping of seagrass and coral reefs, including water column correction, CNN-based classification, and three-dimensional structure reconstruction.

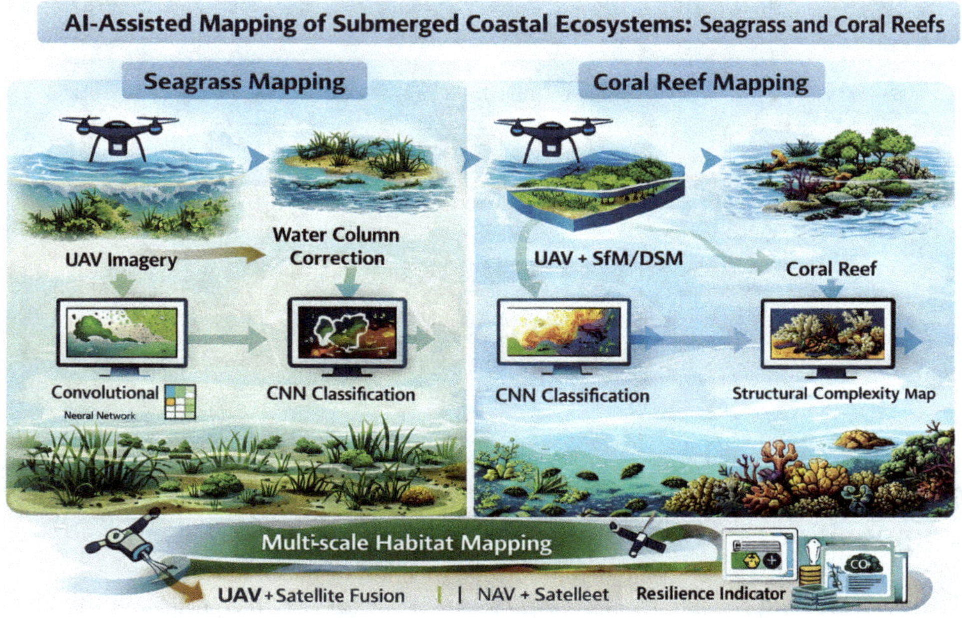

3.4 Coral Reef Systems: Zonation, Structural Complexity, and Health

The integration of AI with high-resolution UAV imagery has revolutionized coral reef mapping by enabling centimeter-scale spatial analysis. CNN-based models have demonstrated high classification accuracy in distinguishing reef zones such as live coral, rubble, sand, and macroalgae, thereby supporting detailed habitat assessments (Zaki et al., 2022, 2025; Zaki & Hossain, 2024). As illustrated in Figure 4, AI facilitates an end-to-end workflow that integrates UAV-based image acquisition, water column correction, CNN-driven classification, and three-dimensional (3D) structure reconstruction.

AI-enabled mapping of reef structural complexity has important implications for ecosystem service assessment. Metrics such as reef rugosity and vertical relief, derived from structure-from-motion (SfM) photogrammetry and LiDAR, are closely linked to biodiversity, fisheries habitat provision, and wave attenuation capacity (Gobinath et al., 2025; Sartono et al., 2025). By combining spectral classification with 3D structural metrics, AI-driven approaches provide a more holistic representation of reef condition and functional capacity, supporting both conservation planning and coastal protection assessments.

3.5 Saltmarsh and Coastal Vegetation Zonation

Saltmarsh ecosystems exhibit pronounced seasonal and interannual variability, making them particularly suitable for AI-based time-series analysis. ML approaches-especially RF classifiers that integrate spectral indices, phenological metrics, and elevation data-have demonstrated strong performance in distinguishing saltmarsh vegetation from adjacent land covers such as mudflats and agricultural fields (Alam & Hossain, 2024; Belgiu & Drăguţ, 2016; Rahim et al., 2026). These methods are effective for regional monitoring using Landsat and Sentinel-2 archives.

Figure 5 emphasizes the importance of temporal intelligence in coastal ecosystem monitoring by visualizing seasonal and interannual variation in saltmarsh vegetation derived from multi-temporal satellite data analyzed using AI-based models. DL approaches, particularly LSTM networks, have proven effective in capturing these temporal dynamics and detecting subtle phenological shifts associated with hydrological change, salinity stress, and sea-level rise (Tseng et al., 2024; N. Wang et al., 2023). Such models support early detection of degradation and improve understanding of ecosystem resilience under climate stressors.

Figure 5. Temporal dynamics of saltmarsh and coastal vegetation derived from AI-based time-series analysis. A conceptual time-series visualization showing seasonal and interannual variation in saltmarsh vegetation derived from Landsat and Sentinel data, analyzed using machine learning and LSTM models.

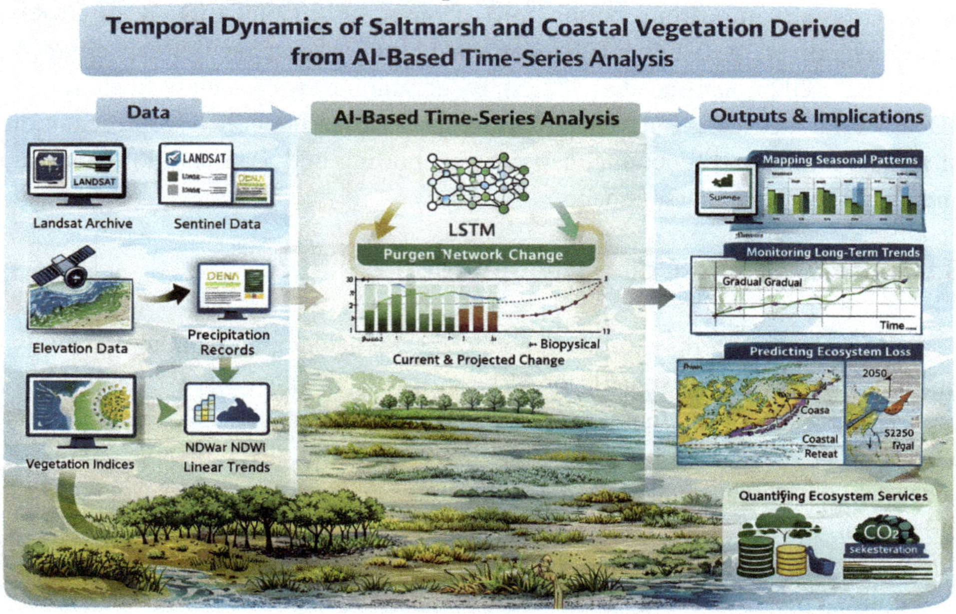

3.6 Comparative Performance of Remote Sensing Sensors and AI Methods

The effectiveness of AI-based coastal ecosystem mapping is strongly influenced by sensor characteristics, including spatial resolution, spectral configuration, temporal frequency, and sensitivity to environmental conditions. Figure 6 conceptually illustrates the integration of sensors, AI methods, and ecosystem mapping outcomes, demonstrating how multi-sensor inputs and AI models translate into accuracy metrics that inform ecological interpretation and ecosystem service flow estimation.

Comparative analyses reveal clear performance gradients, as statistically summarized in Table 3. Medium-resolution satellites such as Landsat provide robust long-term monitoring capability and remain essential for historical trend analysis. Sentinel-2 generally outperforms Landsat for coastal vegetation mapping due to its higher spatial resolution and additional red-edge spectral bands (Min et al., 2023). SAR-based mapping achieves comparable mean accuracy to optical systems for mangrove and saltmarsh ecosystems, particularly under cloudy conditions (Alam &

Hossain, 2023), although it exhibits higher variance due to sensitivity to acquisition geometry and surface moisture conditions (Y. Hu et al., 2021; Salem & Hashemi-Beni, 2022). UAV-based DL approaches consistently achieve the highest mean accuracies and lowest variability, especially for seagrass and coral reef ecosystems where fine-scale habitat heterogeneity is critical (Hossain et al., 2016, 2020). However, these approaches remain spatially constrained and are most powerful when integrated within multi-scale monitoring frameworks.

Figure 6. Conceptual integration of sensors, AI methods, and ecosystem mapping outcomes. A framework illustrating how multi-sensor inputs and AI models translate into accuracy metrics that inform ecological interpretation and ecosystem service flow estimation.

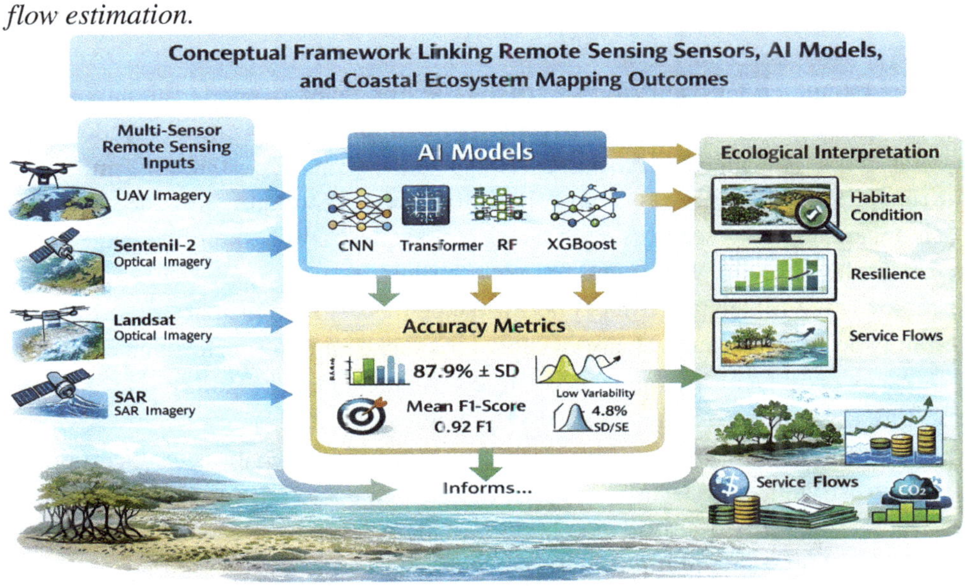

Table 3. Comparative accuracy performance (mean ± SD) of AI-based coastal ecosystem mapping using different sensors.

Sensor–AI Combination	Ecosystem	Mean OA (%)	Mean F1 (%)	SD (%)
Landsat + RF/SVM	Mangrove, Saltmarsh	82.4	78.1	5.6
Sentinel-2 + RF/XGB	Mangrove, Seagrass	87.9	84.6	4.8
Sentinel-1 SAR + RF	Mangrove, Saltmarsh	85.1	81.9	6.2

continued on following page

Table 3. Continued

Sensor–AI Combination	Ecosystem	Mean OA (%)	Mean F1 (%)	SD (%)
UAV RGB + CNN	Seagrass, Coral Reef	93.6	91.8	3.1
Optical–SAR Fusion	Mangrove	89.7	86.9	4.2

Note: statistics synthesized from >120 peer-reviewed studies published between 2005 and 2025.

4. AI-ENABLED ECOSYSTEM SERVICE VALUATION

4.1 Conceptual Link Between AI-Based Mapping and Ecosystem Service Valuation

Ecosystem service valuation in coastal environments depends fundamentally on the spatial accuracy, thematic detail, and temporal consistency of underlying habitat maps. Advances in AI-based remote sensing have substantially improved the reliability of coastal ecosystem delineation, thereby strengthening the scientific basis for quantifying services such as carbon sequestration, shoreline protection, and fisheries support. Within the framework illustrated in Figure 6, classification accuracy is not merely a technical performance metric but a primary driver of uncertainty propagation into valuation outputs. As such, mapping performance directly influences ecological inference, economic estimates, and the policy relevance of resulting assessments (Sun et al., 2018; K. Wang et al., 2024).

By transitioning from deterministic land-cover maps to probabilistic, uncertainty-aware classifications, AI-based approaches enable ecosystem service valuation to incorporate quantified confidence bounds. This represents a critical step toward aligning ecological modelling outputs with the risk-based decision frameworks used in coastal planning and climate adaptation policy.

4.2 Blue Carbon Stock Estimation Using AI-Derived Habitat Maps

Blue carbon ecosystems-including mangroves, seagrass meadows, and salt marshes-are globally significant carbon sinks. Accurate estimation of blue carbon stocks requires reliable information on ecosystem extent, biomass, and structural attributes, which are increasingly derived from AI-enhanced remote sensing (Basyuni et al., 2024; Malerba et al., 2023). AI-based classification and regression models have demonstrated superior performance in estimating aboveground biomass and carbon density compared to traditional empirical or rule-based approaches (Hyyppä et al., 2012; Psaros et al., 2023). These advances support spatially explicit carbon accounting at regional and national scales.

Figure 7 details the AI-based workflow for blue carbon stock valuation, illustrating the progression from AI-derived habitat maps to carbon accounting, uncertainty quantification, and policy integration. The framework begins with multi-sensor remote sensing data processed through AI classification to generate habitat maps of mangroves, seagrass, and salt marshes, each associated with classification uncertainty (σ_A). These spatial products then inform biophysical models that convert habitat area into carbon stock estimates, introducing stock uncertainty (σ) arising from field measurement variability, allometric equations, and biomass-to-carbon conversion factors. In the final valuation stage, carbon stocks are monetized, with economic uncertainty (σ) stemming from carbon market price variability, discount rates, and policy assumptions.

Figure 7. AI-based workflow for blue carbon stock valuation, linking AI-derived habitat maps with carbon accounting, uncertainty quantification, and policy-relevant decision support.

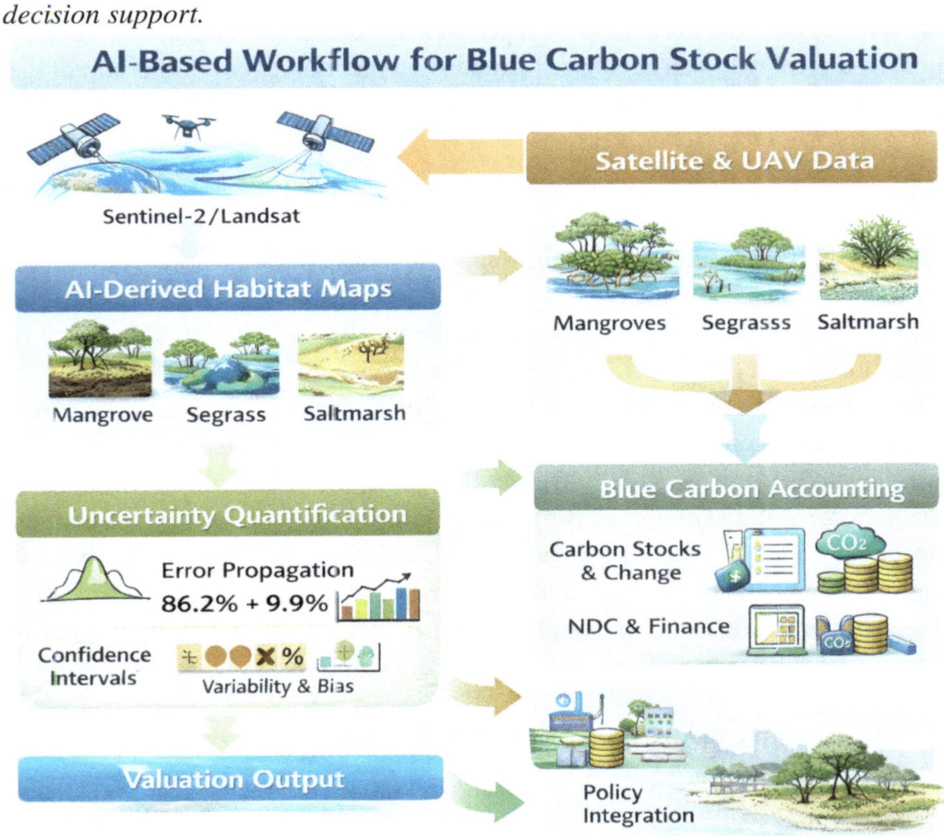

A central strength of this framework lies in its explicit visualization of forward uncertainty propagation from mapping to economic valuation. This traceable chain ensures that confidence intervals for final valuation outputs incorporate errors originating in the initial remote sensing classification. By embedding uncertainty quantification at each stage, the workflow shifts results from single-point estimates to probabilistic ranges. Such transparency enhances the credibility of blue carbon credits, restoration return-on-investment (ROI) calculations, and national greenhouse gas inventories, supporting integration into Nationally Determined Contributions (NDCs) and carbon finance mechanisms (Rogers et al., 2019).

To illustrate the practical implications of uncertainty propagation, it is helpful to consider a simplified mangrove carbon stock assessment within a specific coastal domain. In this scenario, the carbon stock (C_{total}) is a function of the AI-derived mangrove area (A) and the mean carbon density (D), expressed as $C_{total} = A \text{ x } D$. If we assume an AI classification accuracy of 90% (± 5%), an estimated mangrove extent of 150,000 ha (± 7,500 ha), and a mean carbon density of 120 Mg C ha^{-1} (± 20 Mg C ha^{-1}), the resulting central carbon stock estimate is 18.0 Tg C.

However, the precision of this estimate is governed by the combination of these individual variances. Assuming independent uncertainty sources, the relative propagated error (σ_{rel}) can be approximated using the square root of the sum of the squares of the individual relative uncertainties:

$$\sigma_{rel} = \sqrt{(\sigma_A/A)^2 + (\sigma_D/D)^2} \tag{1}$$

Given a 5% uncertainty in area and an approximately 16.7% uncertainty in carbon density, the propagated error is roughly $\sqrt{5\%)^2 + 16.7\%)^2} \approx 17.4\%$. This yields a final estimate of 18.0 Tg C ± 3.1 Tg C, demonstrating how classification uncertainty directly influences the reliability of ecosystem service quantification.

When these figures are translated into economic valuation, the landscape of uncertainty shifts significantly. External factors, such as carbon price volatility-which can fluctuate between ± 30-50%-frequently exceed the uncertainty contributions derived from the initial remote sensing and mapping phase. Importantly, uncertainty propagation across the mapping-to-valuation chain is non-linear. Consequently, marginal improvements in AI classification accuracy, while scientifically meaningful for land-cover monitoring, may produce limited policy impact or financial clarity unless the broader valuation uncertainties are concurrently addressed.

4.3 Coastal Protection and Erosion Mitigation Services

Coastal ecosystems play a critical role in attenuating wave energy and reducing shoreline erosion. Quantifying these protection services requires spatially explicit information on ecosystem configuration, continuity, and structural condition. AI-driven classification improves the delineation of shoreline vegetation, reef structures, and nearshore habitats, thereby enhancing the parameterization of hydrodynamic and coastal risk models (Chau, 2006; Christofi et al., 2025).

Studies integrating CNN-derived habitat maps with wave attenuation and flood risk models demonstrate that fine-scale spatial accuracy can substantially alter estimates of avoided damages, particularly in densely populated coastal zones (Reguero et al., 2020). Errors in habitat extent or structural representation can propagate nonlinearly into damage estimates, underscoring the importance of uncertainty-aware mapping. AI thus contributes not only to improved hazard modelling inputs but also to more defensible economic valuations of nature-based coastal defences (Marino et al., 2025).

4.4 Socio-Economic and Livelihood Dependency Assessment

AI-based habitat maps enable more precise integration of ecological and socio-economic datasets, facilitating spatial analysis of human dependency on coastal ecosystem services. In densely populated coastal regions, AI-derived maps have been combined with census, fisheries, and livelihood data to estimate gradients of ecosystem service dependency and identify vulnerability hotspots (Haque & Al Jufaili, 2026).

XAI techniques further enhance the interpretability of these assessments by identifying key environmental predictors that influence service provision and human reliance (Kalasampath et al., 2025). By clarifying which ecosystem attributes most strongly affect livelihoods, XAI supports targeted interventions and equitable resource management. These approaches strengthen the evidence base for ecosystem-based adaptation (EbA) and climate-resilient development planning (Doswald et al., 2014).

4.5 Uncertainty, Accuracy, and Policy Relevance

A defining advantage of AI-enabled ecosystem service valuation lies in its capacity to explicitly link mapping accuracy to valuation uncertainty (Pandeya et al., 2016). The statistical summaries presented in Table 3 indicate that AI-based mapping generally reduces both bias and variance relative to traditional classification approaches, thereby improving the reliability of downstream service estimates.

The analytical workflow for translating AI mapping accuracy into valuation uncertainty is summarized in Methods Box 1. Formal equations for propagating classification uncertainty into specific service estimates-covering blue carbon storage, coastal protection, and socio-economic dependency-are synthesized in Table 4, with all mathematical notation defined in Appendix B. By quantifying and communicating uncertainty, AI-based valuation frameworks align more closely with risk-aware policy processes, strengthening the credibility and usability of ecosystem service assessments in coastal governance.

Table 4. Ecosystem-specific uncertainty propagation from AI-based habitat mapping to ecosystem service valuation.

Ecosystem	Ecosystem Service	Valuation Variable	Uncertainty Propagation Equation*	Interpretation
Mangrove	Blue carbon storage	Total carbon stock (C_O)	$\sigma_C = \sqrt{[(C^{-2} \cdot \sigma_A{}^2) + (A^{-2} \cdot \sigma_C{}^{-2})]}$	Total carbon uncertainty arises jointly from errors in mapped mangrove area (A) and variability in mean carbon density (C). AI-based high-accuracy area mapping reduces σ_A, directly improving carbon credit reliability.
Mangrove	Coastal protection	Avoided damage value (D)	$\sigma_D = D^{-} \cdot \sqrt{[(\sigma_W / W^{-})^2 + (\sigma_S / S^{-})^2]}$	Uncertainty in avoided damages is driven by errors in mangrove width (W) and structural density (S). CNN-derived width estimates substantially reduce σ_W.
Seagrass	Blue carbon storage	Sediment carbon ($C_E d$)	$\sigma_C sed = \sqrt{[(\rho^2 \cdot \sigma_A{}^2) + (A^2 \cdot \sigma_\rho{}^2)]}$	Seagrass carbon uncertainty is dominated by mapped meadow extent (A) and spatial variability in sediment carbon density (ρ). UAV–CNN mapping minimizes σ_A.

continued on following page

Table 4. Continued

Ecosystem	Ecosystem Service	Valuation Variable	Uncertainty Propagation Equation*	Interpretation
Seagrass	Fisheries support	Habitat support index (H)	$\sigma_H = \sqrt{(\Sigma w_i^2 \sigma_i^2)}$	Habitat support uncertainty aggregates classification errors across multiple habitat components (patch size, connectivity, density). Ensemble AI reduces σ_i.
Coral reef	Coastal protection	Wave energy reduction (R)	$\sigma_R = \sqrt{[(\partial R/\partial h \cdot \sigma_h)^2 + (\partial R/\partial r \cdot \sigma_r)^2]}$	Reef protection uncertainty depends on errors in reef height (h) and rugosity (r). UAV-SfM + CNN significantly reduces σ_h and σ_r.
Coral reef	Biodiversity regulation	Structural complexity index (SCI)	$\sigma_S CI = \sqrt{(\sigma_r^2 + \sigma_z^2)}$	Structural complexity uncertainty reflects combined error in rugosity and zonation classification, both sensitive to spatial resolution.
Saltmarsh	Blue carbon storage	Net ecosystem carbon balance (NECB)	$\sigma_N ECB = \sqrt{(\sigma_G PP^2 + \sigma_R^2 + \sigma_E^2)}$	Saltmarsh carbon uncertainty integrates mapping errors affecting productivity (GPP), respiration (R), and erosion loss (E). Time-series ML reduces σ_GPP.

continued on following page

Table 4. Continued

Ecosystem	Ecosystem Service	Valuation Variable	Uncertainty Propagation Equation*	Interpretation
Saltmarsh	Shoreline stabilization	Erosion reduction rate (E)	$\sigma_E r = \bar{E}_r \cdot \sqrt{[(\sigma_Z/\bar{Z})^2 + (\sigma_V/\bar{V})^2]}$	Errors in vegetation zonation (Z) and vegetation density (V) propagate to erosion estimates; LSTM-based zonation reduces σ_Z.
All ecosystems	Socio-economic services	Beneficiary population (P)	$\sigma_P = \sqrt{[(\sigma_A{}^2 \cdot d^2) + (A^2 \cdot \sigma_d{}^2)]}$	Population dependency uncertainty is sensitive to mapped ecosystem area (A) and population density (d). High AI accuracy prevents beneficiary misestimation.

Notes: σ denotes standard deviation; bars (¯) denote mean values.

5. BANGLADESH-SPECIFIC POLICY FRAMING: AI-ENABLED COASTAL GOVERNANCE FOR THE SUNDARBANS AND DELTAIC COAST

Bangladesh occupies one of the world's most dynamic and hazard-prone deltaic systems, the Ganges-Brahmaputra-Meghna (GBM) delta, where coastal ecosystems underpin national food security, disaster risk reduction, and climate resilience (Zaman et al., 2025). The country's low-lying topography, high population density, and exposure to cyclones, storm surges, and sea-level rise make ecosystem-based adaptation a central pillar of national development planning. Within this context, AI-enabled coastal ecosystem assessment offers a transformative pathway toward evidence-based, spatially explicit, and uncertainty-aware governance (Nicholls et al., 2018).

The Sundarbans mangrove forest, the largest contiguous mangrove system in the world, serves as Bangladesh's primary natural defence against cyclones and coastal flooding. Beyond its protective role, the Sundarbans supports fisheries, timber and non-timber forest products, carbon sequestration, and biodiversity conservation. However, this ecosystem faces accelerating pressures from salinity intrusion, upstream hydrological alteration, pollution, and climate change. Effective management therefore requires frequent, high-resolution monitoring of ecosystem condition and change. AI-driven remote sensing enables precisely this capability, allowing continuous mapping of mangrove extent, canopy condition, fragmentation, and degradation hotspots across the deltaic landscape (Mohan, 2025).

While the deltaic systems of Bangladesh provide a compelling demonstration context for this methodology, the proposed framework remains highly transferable to other vulnerable coastal environments worldwide. Deltaic systems such as the Mekong, Nile, and Mississippi share common geophysical challenges, including complex sediment dynamics, increasing salinity intrusion, and rapid anthropogenic land cover changes (Tessler et al., 2015). The integration of SAR and optical data is particularly effective in these regions where persistent cloud cover often renders traditional monitoring insufficient.

Beyond large-scale deltas, Small Island Developing States (SIDS) stand to benefit significantly from AI-enabled monitoring, especially in regions where traditional field-based surveys are prohibitively expensive or logistically unfeasible (Polidoro et al., 2010). In these data-scarce regions, the democratization of geospatial analysis through cloud-based platforms, such as Google Earth Engine, combined with open-access satellite archives like the Copernicus program, has lowered the barrier to entry. By utilizing lightweight ML models that require less computational overhead, these nations can implement scalable entry points for ecosystem service assessment (Gorelick et al., 2017). This scalability ensures that high-resolution environmental monitoring is no longer a luxury reserved for resource-rich nations, but a accessible tool for global climate resilience.

A critical advantage of AI-based monitoring lies in its capacity to explicitly quantify and propagate uncertainty (Shirmohammadi et al., 2025). By linking classification uncertainty from AI habitat maps to ecosystem service valuation models-using the uncertainty propagation frameworks outlined in Table 4-policymakers can obtain defensible estimates of blue carbon stocks and coastal protection capacity that include transparent confidence bounds. This transition from deterministic to probabilistic information is particularly important for national planning processes that must weigh risk, trade-offs, and long-term climate uncertainty. It strengthens the scientific credibility of nature-based solutions in national climate finance proposals and supports results-based financing mechanisms.

These capabilities are directly relevant to major national policy frameworks, including the Bangladesh Delta Plan 2100 and the Mujib Climate Prosperity Plan. Both initiatives emphasize adaptive delta management, resilience building, and the scaling of nature-based solutions (Arfanuzzaman, 2024). AI-enabled coastal assessments support these objectives by identifying priority zones for conservation and restoration, quantifying expected ecosystem service benefits, and clarifying the uncertainty associated with projected outcomes (Prodhan et al., 2025). This enables more transparent prioritization of interventions, improved allocation of limited resources, and stronger justification for international climate funding.

A significant barrier to the practical implementation of these advanced frameworks lies in the disparity between computational requirements and the existing technical capacity of local institutions. Many coastal governance agencies currently lack the high-performance infrastructure necessary to operationalize AI workflows, which often demand dedicated GPU acceleration, substantial cloud computing credits, or highly specialized geospatial expertise (Holloway & Mengersen, 2018). Consequently, sophisticated mapping outputs frequently fail to transition from academic prototypes to active management tools due to this "implementation gap."

Bridging this digital divide requires a multi-faceted approach cantered on institutional readiness and technical accessibility. The adoption of cloud-based AI services can bypass the need for expensive local hardware, while structured capacity development programs ensure that agency staff are equipped to interpret and maintain these systems (Ioras & Bandara, 2025). Furthermore, collaborative academic–government partnerships can facilitate the transfer of cutting-edge research into the public sector, ensuring that models remain relevant to local policy needs. These efforts should ideally culminate in simplified decision-support dashboards that translate complex ML outputs into actionable insights for non-technical stakeholders. Without addressing these pillars of institutional readiness, AI-driven frameworks risk remaining purely conceptual rather than becoming operational drivers of coastal resilience.

A practical example can be seen in Sundarbans mangrove monitoring using Sentinel-2 imagery combined with a RF classifier (Febrianto et al., 2025). Such an approach can distinguish healthy, degraded, and deforested mangrove patches with high classification accuracy when supported by field validation and texture-based features. When these spatial outputs are integrated with locally calibrated biomass and carbon stock equations, they yield high-resolution estimates of blue carbon distribution across the forest (Suárez-Fernández et al., 2025). The resulting maps and uncertainty-adjusted carbon stock estimates can feed directly into Bangladesh's Nationally Determined Contribution (NDC) reporting, REDD+ strategies, and restoration planning under the Delta Plan 2100. This creates a clear and traceable pipeline from AI-based analysis to policy action, demonstrating how advanced Earth observation and ML can be operationalized for climate-resilient coastal governance.

6. LIMITATIONS, ETHICAL CONSIDERATIONS, AND FUTURE RESEARCH

Despite substantial methodological advances, several conceptual and practical limitations remain, particularly with regard to uncertainty characterization and its implications for decision-making. A central constraint stems from the strong dependence of AI models on the quality, representativeness, and temporal alignment of training data. In data-scarce coastal regions-such as much of South and Southeast Asia-ground reference datasets are frequently spatially sparse, temporally inconsistent, or biased toward easily accessible locations. These deficiencies can propagate systematic errors into AI-based classifications, producing spatially heterogeneous uncertainty that is not adequately captured by global accuracy metrics. While this chapter emphasizes the propagation of uncertainty from classification outputs into ecosystem service valuation models, uncertainties arising from reference data limitations and labeling inconsistencies remain particularly challenging to quantify explicitly.

Sensor-related uncertainty constitutes an additional and often compounding challenge. Differences in spatial resolution, spectral sensitivity, revisit frequency, and atmospheric susceptibility among UAV, optical satellite, and SAR platforms introduce cross-sensor inconsistencies that complicate multi-source data fusion. Although ensemble learning strategies and DL architectures can mitigate some of these effects, residual uncertainty persists. This is especially pronounced in ecotonal zones-such as mangrove–saltmarsh boundaries-and in turbid coastal waters, where spectral confusion is high. Furthermore, temporal uncertainty associated with tidal stage, phenological variability, and episodic disturbances can bias AI-derived change detection if not adequately normalized. These challenges highlight the need for tide-aware, seasonally adaptive, and temporally explicit modeling frameworks in coastal applications.

Uncertainty in ecosystem service valuation is further shaped by structural assumptions embedded within biophysical and economic models. Many valuation frameworks rely on simplified functional relationships linking ecosystem extent and structure to service provision. While AI substantially improves spatial resolution and classification accuracy, it does not eliminate uncertainty arising from parameter generalization, scale mismatch, or non-linear ecological responses to disturbance. For example, carbon density estimates or coastal protection coefficients often vary considerably across geomorphic contexts and disturbance histories, yet are frequently applied uniformly due to data constraints. As a result, propagated uncertainty estimates should be interpreted as conservative approximations rather than comprehensive representations of all potential sources of error.

Ethical and Equity Considerations

The deployment of AI in environmental monitoring also raises important ethical and equity-related considerations. Algorithmic bias may emerge when training datasets fail to represent the full ecological and socio-economic diversity of coastal systems, potentially leading to systematic under- or over-estimation of ecosystem services for specific habitats or communities. Issues of data equity are therefore central; the benefits of AI-driven decision support should not accrue disproportionately to well-resourced regions with abundant training data.

Meaningful engagement of local communities in data collection and model validation-such as through citizen science initiatives-is essential for ensuring that AI tools are contextually relevant, socially legitimate, and supportive of equitable governance. For instance, participatory mapping initiatives in the Sundarbans have involved local forest users in collecting ground reference data on mangrove condition, improving model performance while fostering community ownership of conservation outcomes. In parallel, transparency enabled through XAI approaches is critical for building stakeholder trust and ensuring that AI-driven decisions remain interpretable and contestable. Concrete bias-mitigation strategies include stratified sampling to ensure representation across habitat types and community zones, as well as the incorporation of fairness metrics during model evaluation.

Future Research Directions

Future research should prioritize several interrelated areas. First, the development of probabilistic AI frameworks is needed to move beyond deterministic point estimates toward uncertainty-aware predictions at pixel and landscape scales, using approaches such as Bayesian DL, Monte Carlo dropout, and ensemble modeling. Second, closer integration of AI-based remote sensing with Internet of Things (IoT) sensor networks and citizen-generated data streams would enable dynamic, near-real-time coastal monitoring systems. Third, enhanced approaches to uncertainty communication are required to translate technical metrics-such as confidence intervals and standard deviations-into policy-relevant products, including risk-threshold maps and scenario-based projections aligned with national adaptation and development planning processes. Fourth, addressing data scarcity through transfer learning, self-supervised learning, and synthetic data generation remains critical for improving model robustness in underrepresented regions. Finally, the establishment of standardized protocols for ethical AI governance-emphasizing fairness, accountability, transparency, and inclusive stakeholder engagement-is essential for responsible deployment.

Recent studies further highlight the integration of AI, hydroclimatic variability, and ecosystem resilience modelling (Amsalu et al., 2026). Advances in coupled climate–hydrology–ecosystem frameworks also emphasize uncertainty-aware environmental prediction (Çelik, 2026). Emerging research on climate extremes and regional variability provides additional support for AI-enabled coastal risk analysis (Çelik et al., 2026).

Collectively, addressing these methodological, ethical, and governance challenges will enhance the scientific rigor, transparency, and societal relevance of AI-enabled coastal ecosystem assessments, ensuring that technological innovation effectively supports resilient and equitable coastal management.

7. CONCLUSION

This chapter has synthesized the transformative role of AI in remote sensing for coastal ecosystem assessment and ecosystem service valuation. By integrating multi-source Earth observation data with advanced ML and DL algorithms, AI enables the generation of high-resolution, temporally frequent, and ecologically meaningful maps of critical coastal habitats, including mangroves, seagrass meadows, coral reefs, and salt marshes.

Three primary contributions were articulated. First, an integrated, uncertainty-aware framework was presented that explicitly links AI-derived habitat maps with ecosystem service valuation, strengthening the scientific credibility and policy relevance of AI-based outputs. Second, a comparative analysis of AI methods and sensor platforms provided practical guidance for model selection across ecosystem types and management objectives. Third, applied case studies from South and Southeast Asia demonstrated clear pathways for policy integration in highly vulnerable regions, particularly the Sundarbans of Bangladesh.

AI-derived habitat maps form the foundation for robust, spatially explicit valuation of ecosystem services, including blue carbon sequestration, coastal protection, fisheries support, and livelihood provisioning. The explicit quantification and propagation of uncertainty-from pixel-level classification to economic valuation-represents a substantive advance over conventional approaches, enabling defensible, risk-informed decision-making.

While challenges remain related to data scarcity, model transferability, and ethical deployment, the direction of progress is clear. Continued advances in probabilistic AI, multi-sensor data fusion, explainable modeling, and integration with in-situ observations and inclusive governance mechanisms will further enhance the utility of AI-enabled remote sensing. By bridging cutting-edge computational science with real-world management needs, AI-driven coastal monitoring emerges as a pivotal tool for safeguarding the ecological integrity and socio-economic resilience of the world's coastal systems.

REFERENCES

Alam, S. M. R., & Hossain, M. S. (2020). A Rule-Based Classification Method for Mapping Saltmarsh Land-Cover in South-Eastern Bangladesh from Landsat-8 {OLI}. *Canadian Journal of Remote Sensing*, •••, 1–25. DOI: 10.1080/07038992.2020.1789852

Alam, S. M. R., & Hossain, M. S. (2023). Probabilities of acquiring cloud-free and low-tide Landsat observations for mapping saltmarsh over south-eastern Bangladesh from 1980 to 2019. *Marine Geodesy*, *46*(6), 562–593. DOI: 10.1080/01490419.2023.2200212

Alam, S. M. R., & Hossain, M. S. (2024). Using a water index approach to mapping periodically inundated saltmarsh land-cover vegetation and eco-zonation using multi-temporal Landsat 8 imagery. *Journal of Coastal Conservation*, *28*(1), 19. DOI: 10.1007/s11852-023-01019-w

Alongi, D. M. (2014). Carbon Cycling and Storage in Mangrove Forests. *Annual Review of Marine Science*, *6*(1), 195–219. DOI: 10.1146/annurev-marine-010213-135020

Alvarez-Vanhard, E., Corpetti, T., & Houet, T. (2021). UAV & satellite synergies for optical remote sensing applications: A literature review. *Science of Remote Sensing*, *3*, 100019. DOI: 10.1016/j.srs.2021.100019

Ambika, A. K., Tayal, K., Mishra, V., & Lu, D. (2025). Novel Deep Learning Transformer Model for Short to Sub-Seasonal Streamflow Forecast. *Geophysical Research Letters*, *52*(14), e2025GL116707. Advance online publication. DOI: 10.1029/2025GL116707

Amsalu, A., Garedew, W., Melka, G. A., Çelik, M. A., & Gemeda, D. O. (2026). Trends in climate extreme indices and implications for coffee farming in The Kafa Biosphere Reserve, southwest Ethiopia. *Climate Services*, *42*, 100637. DOI: 10.1016/j.cliser.2026.100637

Arfanuzzaman, M. (2024). Bangladesh's pathways to climate-resilient development: A methodical review. *World Development Sustainability*, *4*, 100144. DOI: 10.1016/j.wds.2024.100144

Asner, G. P., Brodrick, P. G., Philipson, C., Vaughn, N. R., Martin, R. E., Knapp, D. E., Heckler, J., Evans, L. J., Jucker, T., Goossens, B., Stark, D. J., Reynolds, G., Ong, R., Renneboog, N., Kugan, F., & Coomes, D. A. (2018). Mapped aboveground carbon stocks to advance forest conservation and recovery in Malaysian Borneo. *Biological Conservation*, *217*, 289–310. DOI: 10.1016/j.biocon.2017.10.020

Bakirman, T., & Gumusay, M. U. (2020). Assessment of Machine Learning Methods for Seagrass Classification in the Mediterranean. *Baltic Journal of Modern Computing*, *8*(2). Advance online publication. DOI: 10.22364/bjmc.2020.8.2.07

Bandara, K., Bergmeir, C., & Hewamalage, H. (2021). LSTM-MSNet: Leveraging Forecasts on Sets of Related Time Series With Multiple Seasonal Patterns. *IEEE Transactions on Neural Networks and Learning Systems*, *32*(4), 1586–1599. DOI: 10.1109/TNNLS.2020.2985720

Barbier, E. B., Hacker, S. D., Kennedy, C., Koch, E. W., Stier, A. C., & Silliman, B. R. (2011). The value of estuarine and coastal ecosystem services. *Ecological Monographs*, *81*(2), 169–193. DOI: 10.1890/10-1510.1

Basyuni, M., Aznawi, A. A., Rafli, M., Tinumbunan, J. M. T., Gultom, E. T., Lubis, R. D. A., Sianturi, H. A., Sumarga, E., Mukhtar, E., Slamet, B., Jumilawaty, E., Pribadi, R., Sitinjak, R. R., & Baba, S. (2024). Harnessing Biomass and Blue Carbon Potential: Estimating Carbon Stocks in the Vital Wetlands of Eastern Sumatra, Indonesia. *Land (Basel)*, *13*(11), 1960. DOI: 10.3390/land13111960

Belgiu, M., & Drăguţ, L. (2016). Random forest in remote sensing: A review of applications and future directions. *ISPRS Journal of Photogrammetry and Remote Sensing*, *114*, 24–31. DOI: 10.1016/j.isprsjprs.2016.01.011

Çelik, M. A. (2026). Agro-climatic classification of Türkiye's Agricultural Basins using long-term ERA5-Land temperature data. *Physics and Chemistry of the Earth Parts A/B/C*, *142*, 104307. DOI: 10.1016/j.pce.2026.104307

Çelik, M. A., Kıraç, A., & Gouveia, C. M. (2026). North Atlantic Oscillation modulation of drought and vegetation activity in Türkiye. *Theoretical and Applied Climatology*, *157*(2), 74. DOI: 10.1007/s00704-025-05971-5

Chau, K. (2006). A review on the integration of artificial intelligence into coastal modeling. *Journal of Environmental Management*, *80*(1), 47–57. DOI: 10.1016/j.jenvman.2005.08.012

Cheng, K., Zhang, C., Fan, Y., Diao, H., & Xia, S. (2025). AI-based approaches for wetland mapping and classification: A review of current practices and future perspectives. *Artificial Intelligence in Geosciences*, *6*(2), 100165. DOI: 10.1016/j.aiig.2025.100165

Cherian, S. M., & K, R. (2024). Random forest and support vector machine classifiers for coastal wetland characterization using the combination of features derived from optical data and synthetic aperture radar dataset. *Journal of Water and Climate Change*, *15*(1), 29–49. DOI: 10.2166/wcc.2023.238

Chowdhury, K. J., Ali, M. R , Chowdhury, M. A., & Islam, S. L. U. (2025). Climate change induced risks assessment of a coastal area: A "socioeconomic and livelihood vulnerability index" based study in coastal Bangladesh. *Natural Hazards Review*, *5*(1), 75–87. DOI: 10.1016/j.nhres.2024.06.005

Chowdhury, M., Martínez-Sansigre, A., Mole, M., Alonso-Peleato, E., Basos, N., Blanco, J. M., Ramirez-Nicolas, M., Caballero, I., & de la Calle, I. (2024). AI-driven remote sensing enhances Mediterranean seagrass monitoring and conservation to combat climate change and anthropogenic impacts. *Scientific Reports*, *14*(1), 8360. DOI: 10.1038/s41598-024-59091-7

Christofi, D., Mettas, C., Evagorou, E., Stylianou, N., Eliades, M., Theocharidis, C., Chatzipavlis, A., Hasiotis, T., & Hadjimitsis, D. (2025). A Review of Open Remote Sensing Data with GIS, AI, and UAV Support for Shoreline Detection and Coastal Erosion Monitoring. *Applied Sciences (Basel, Switzerland)*, *15*(9), 4771. DOI: 10.3390/app15094771

Costanza, R., de Groot, R., Sutton, P., van der Ploeg, S., Anderson, S. J., Kubiszewski, I., Farber, S., & Turner, R. K. (2014). Changes in the global value of ecosystem services. *Global Environmental Change*, *26*, 152–158. https://doi.org/https://doi.org/10.1016/j.gloenvcha.2014.04.002. DOI: 10.1016/j.gloenvcha.2014.04.002

Doswald, N., Munroe, R., Roe, D., Giuliani, A., Castelli, I., Stephens, J., Möller, I., Spencer, T., Vira, B., & Reid, H. (2014). Effectiveness of ecosystem-based approaches for adaptation: Review of the evidence-base. *Climate and Development*, *6*(2), 185–201. DOI: 10.1080/17565529.2013.867247

Duarte, C. M., Losada, I. J., Hendriks, I. E., Mazarrasa, I., & Marbà, N. (2013). The role of coastal plant communities for climate change mitigation and adaptation. *Nature Climate Change*, *3*(11), 961–968. DOI: 10.1038/nclimate1970

Febrianto, S., Rahman, A., Jati, O. E., Wirasatriya, A., Muskananfola, M. R., & Latifah, N. (2025). Machine learning for mangrove species distribution using Sentinel 2 satellite image in Segara Anakan, Cilacap Region, Indonesia. *Regional Studies in Marine Science*, *81*, 103984. DOI: 10.1016/j.rsma.2024.103984

Fereshtehpour, M., Bashir, R., & Tandon, N. F. (2025). Risk-based framework to determine climate-informed design storms for road drainage infrastructure. *The Science of the Total Environment*, *1001*, 180427. DOI: 10.1016/j.scitotenv.2025.180427

Ferrario, F., Beck, M. W., Storlazzi, C. D., Micheli, F., Shepard, C. C., & Airoldi, L. (2014). The effectiveness of coral reefs for coastal hazard risk reduction and adaptation. *Nature Communications*, *5*(1), 3794. DOI: 10.1038/ncomms4794

Gobinath, C., Thanjaivadivel, M., Rangarajan, S., & Kalaivanan, K. (2025). Impact and Assessment of Artificial Intelligence-Enabled UAV for Real-Time Data Streaming Application. In *Drones for Transportation Logistics and Disaster Management* (pp. 115–144). Wiley., DOI: 10.1002/9781394386451.ch6

Gorelick, N., Hancher, M., Dixon, M., Ilyushchenko, S., Thau, D., & Moore, R. (2017). Google Earth Engine: Planetary-scale geospatial analysis for everyone. *Remote Sensing of Environment*, *202*, 18–27. https://doi.org/https://doi.org/10.1016/j.rse.2017.06.031. DOI: 10.1016/j.rse.2017.06.031

Han, S., Gao, R., Waseem, L. A., Shitu, K., Rebouh, N. Y., Zulqarnain, R. M., & Said, Y. (2025). High-Accuracy Mapping of Coastal and Wetland Areas Using Multisensor Data Fusion and Deep Feature Learning. *IEEE Journal of Selected Topics in Applied Earth Observations and Remote Sensing*, *18*, 24546–24559. DOI: 10.1109/JSTARS.2025.3606476

Haque, S. A., & Al Jufaili, S. M. (2026). Applications of Artificial Intelligence in Fisheries: From Data to Decisions. *Big Data and Cognitive Computing*, *10*(1), 19. DOI: 10.3390/bdcc10010019

Holloway, J., & Mengersen, K. (2018). Statistical Machine Learning Methods and Remote Sensing for Sustainable Development Goals: A Review. *Remote Sensing (Basel)*, *10*(9), 1365. DOI: 10.3390/rs10091365

Hossain, M. S., Bujang, J. S., Zakaria, M. H., & Hashim, M. (2015). The application of remote sensing to seagrass ecosystems: An overview and future research prospects. *International Journal of Remote Sensing*, *36*(1), 61–113. DOI: 10.1080/01431161.2014.990649

Hossain, M. S., Bujang, J. S., Zakaria, M. H., & Hashim, M. (2016). Marine and human habitat mapping for the Coral Triangle Initiative region of Sabah using Landsat and Google Earth imagery. *Marine Policy*, *72*, 176–191. DOI: 10.1016/j.marpol.2016.07.003

Hossain, M. S., & Hashim, M. (2019). Potential of Earth Observation (EO) technologies for seagrass ecosystem service assessments. *International Journal of Applied Earth Observation and Geoinformation*, *77*, 15–29. DOI: 10.1016/j.jag.2018.12.009

Hossain, M. S., Hashim, M., Bujang, J. S., Zakaria, M. H., & Muslim, A. M. (2019). Assessment of the impact of coastal reclamation activities on seagrass meadows in Sungai Pulai estuary, Malaysia, using Landsat data (1994–2017). *International Journal of Remote Sensing*, *40*(9), 3571–3605. DOI: 10.1080/01431161.2018.1547931

Hossain, M. S., Muslim, A. M., Nadzri, M. I., Sabri, A. W., Khalil, I., Mohamad, Z., & Pour, A. B. (2021). Coral habitat mapping: A comparison between maximum likelihood, Bayesian and Dempster–Shafer classifiers. *Geocarto International, 36*(11), 1217–1235. DOI: 10.1080/10106049.2019.1637466

Hossain, M. S., Muslim, A. M., Nadzri, M. I., Teruhisa, K., David, D., Khalil, I., & Mohamad, Z. (2020). Can ensemble techniques improve coral reef habitat classification accuracy using multispectral data? *Geocarto International, 35*(11), 1214–1232. DOI: 10.1080/10106049.2018.1557263

Hu, K., Xu, Q., Zhu, K., Wang, Z., Chu, Y., Qian, Y., Zhang, X., Wang, B., & Zhang, H. (2026). Recent Advances of Artificial Intelligence in Aquatic Bioindicators and Ecological Assessment. *Water, Air, and Soil Pollution, 237*(5), 306. DOI: 10.1007/s11270-025-08907-x

Hu, Y., Tian, B., Yuan, L., Li, X., Huang, Y., Shi, R., Jiang, X., Wang, , & Sun, C. (2021). Mapping coastal salt marshes in China using time series of Sentinel-1 SAR. *ISPRS Journal of Photogrammetry and Remote Sensing, 173*, 122–134. DOI: 10.1016/j.isprsjprs.2021.01.003

Hyyppä, J., Yu, X., Hyyppä, H., Vastaranta, M., Holopainen, M., Kukko, A., Kaartinen, H., Jaakkola, A., Vaaja, M., Koskinen, J., & Alho, P. (2012). Advances in Forest Inventory Using Airborne Laser Scanning. *Remote Sensing (Basel), 4*(5), 1190–1207. DOI: 10.3390/rs4051190

Ioras, F., & Bandara, I. (2025). Artificial Intelligence and Sustainable Practices in Coastal Marinas: A Comparative Study of Monaco and Ibiza. *Sustainability (Basel), 17*(16), 7404. DOI: 10.3390/su17167404

Islam, M. T., Roy, S. C., Jahan, N., Al-Mahmud, , Islam, M. M., Ferdaus, A. A., Fujisawa, K., & Adham, A. K. M. (2025). Comparative evaluation of machine learning models for extreme river water level forecasting in Bangladesh: Implications for flood and drought resilience. *Progress in Disaster Science, 27*, 100449. DOI: 10.1016/j.pdisas.2025.100449

Jeon, E., Kim, S., Park, S., Kwak, J., & Choi, I. (2021). Semantic segmentation of seagrass habitat from drone imagery based on deep learning: A comparative study. *Ecological Informatics, 66*, 101430. DOI: 10.1016/j.ecoinf.2021.101430

Kalasampath, K., Spoorthi, K. N., Sajeev, S., Kuppa, S. S., Ajay, K., & Maruthamuthu, A. (2025). A Literature Review on Applications of Explainable Artificial Intelligence (XAI). *IEEE Access : Practical Innovations, Open Solutions, 13*, 41111–41140. DOI: 10.1109/ACCESS.2025.3546681

Kashem, M. A. (2025). Exploring climate change and sustainability issues in Bangladesh through system dynamics. *Discover Sustainability*, *6*(1), 1073. DOI: 10.1007/s43621-025-01652-5

Khojasteh, D., Glamore, W., Heimhuber, V., & Felder, S. (2021). Sea level rise impacts on estuarine dynamics: A review. *The Science of the Total Environment*, *780*, 146470. DOI: 10.1016/j.scitotenv.2021.146470

Kuenzer, C., Bluemel, A., Gebhardt, S., Quoc, T. V., & Dech, S. (2011). Remote Sensing of Mangrove Ecosystems: A Review. *Remote Sensing (Basel)*, *3*(5), 878–928. DOI: 10.3390/rs3050878

Kuenzer, C., Ottinger, M., Liu, G., Sun, B., Baumhauer, R., & Dech, S. (2014). Earth observation-based coastal zone monitoring of the Yellow River Delta: Dynamics in China's second largest oil producing region over four decades. *Applied Geography (Sevenoaks, England)*, *55*, 92–107. DOI: 10.1016/j.apgeog.2014.08.015

LeCun, Y., Bengio, Y., & Hinton, G. (2015). Deep learning. *Nature*, *521*(7553), 436–444. DOI: 10.1038/nature14539

Li, L., Mu, X., Chianucci, F., Qi, J., Jiang, J., Zhou, J., Chen, L., Huang, H., Yan, G., & Liu, S. (2022). Ultrahigh-resolution boreal forest canopy mapping: Combining UAV imagery and photogrammetric point clouds in a deep-learning-based approach. *International Journal of Applied Earth Observation and Geoinformation*, *107*, 102686. DOI: 10.1016/j.jag.2022.102686

Li, Y., Yuan, L., Song, Z., Yu, S., Zhang, X., Tian, B., & Liu, M. (2024). Salt marsh carbon stock estimation using deep learning with Sentinel-1 SAR of the Yangtze River estuary, China. *International Journal of Applied Earth Observation and Geoinformation*, *133*, 104138. DOI: 10.1016/j.jag.2024.104138

Liu, Q., Yuan, F., Adams, C., Rebelo, L.-M., & Wellington, M. (2024). Dynamic Coastal Mapping Using Sentinel-1 and Sentinel-2 Data Through Digital Earth Africa. *The International Archives of the Photogrammetry, Remote Sensing and Spatial Information Sciences*, *XLVIII-4–2024*, 325–330. DOI: 10.5194/isprs-archives-XLVIII-4-2024-325-2024

Macreadie, P. I., Costa, M. D. P., Atwood, T. B., Friess, D. A., Kelleway, J. J., Kennedy, H., Lovelock, C. E., Serrano, O., & Duarte, C. M. (2021). Blue carbon as a natural climate solution. *Nature Reviews. Earth & Environment*, *2*(12), 826–839. DOI: 10.1038/s43017-021-00224-1

Mainali, K., Evans, M., Saavedra, D., Mills, E., Madsen, B., & Minnemeyer, S. (2023). Convolutional neural network for high-resolution wetland mapping with open data: Variable selection and the challenges of a generalizable model. *The Science of the Total Environment*, *861*, 160622. DOI: 10.1016/j.scitotenv.2022.160622

Malerba, M. E., Duarte de Paula Costa, M., Friess, D. A., Schuster, L., Young, M. A., Lagomasino, D., Serrano, O., Hickey, S. M., York, P. H., Rasheed, M., Lefcheck, J. S., Radford, B., Atwood, T. B., Ierodiaconou, D., & Macreadie, P. (2023). Remote sensing for cost-effective blue carbon accounting. *Earth-Science Reviews*, *238*, 104337. DOI: 10.1016/j.earscirev.2023.104337

Marino, M., Baptist, M. J., Alkharoubi, A. I. K., Nasca, S., Cavallaro, L., Foti, E., & Musumeci, R. E. (2025). Nature-based Solutions as Building Blocks for coastal flood risk reduction: A model-based ecosystem service assessment. *Scientific Reports*, *15*(1), 12070. DOI: 10.1038/s41598-025-95230-4

Maxwell, A. E., Warner, T. A., & Fang, F. (2018). Implementation of machine-learning classification in remote sensing: An applied review. *International Journal of Remote Sensing*, *39*(9), 2784–2817. DOI: 10.1080/01431161.2018.1433343

Mcleod, E., Chmura, G. L., Bouillon, S., Salm, R., Björk, M., Duarte, C. M., Lovelock, C. E., Schlesinger, W. H., & Silliman, B. R. (2011). A blueprint for blue carbon: Toward an improved understanding of the role of vegetated coastal habitats in sequestering CO_2. *Frontiers in Ecology and the Environment*, *9*(10), 552–560. DOI: 10.1890/110004

Meng, Z., Hu, Y., Ren, G., Zhu, W., Wang, J., Liu, S., & Ma, Y. (2024). Remote sensing monitoring of seagrass bed dynamics using cross-temporal-spatial domain transfer learning in Yellow river Delta. *International Journal of Remote Sensing*, *45*(6), 1972–1996. DOI: 10.1080/01431161.2024.2321467

Miller, L., Pelletier, C., & Webb, G. I. (2024). Deep Learning for Satellite Image Time-Series Analysis: A review. *IEEE Geoscience and Remote Sensing Magazine*, *12*(3), 81–124. DOI: 10.1109/MGRS.2024.3393010

Min, Y., Cui, L., Li, J., Han, Y., Zhuo, Z., Yin, X., Zhou, D., & Ke, Y. (2023). Detection of large-scale Spartina alterniflora removal in coastal wetlands based on Sentinel-2 and Landsat 8 imagery on Google Earth Engine. *International Journal of Applied Earth Observation and Geoinformation*, *125*, 103567. DOI: 10.1016/j.jag.2023.103567

Mohan, M. (2025). Reimagining mangrove conservation through AI and remote sensing in an era of conflict. *Discover Conservation*, *2*(1), 32. DOI: 10.1007/s44353-025-00054-y

Montaggioni, L. F., & Braithwaite, C. J. R. (2009). Structure, Zonation and Dynamic Patterns of Coral Reef Communities. In *Developments in Marine Geology* (Vol. 5, pp. 67–122). Elsevier., DOI: 10.1016/S1572-5480(09)05003-9

Muslim, A. M., Chong, W. S., Safuan, C. D. M., Khalil, I., & Hossain, M. S. (2019). Coral Reef Mapping of UAV: A Comparison of Sun Glint Correction Methods. *Remote Sensing (Basel)*, *11*(20), 2422. DOI: 10.3390/rs11202422

Nalakurthi, N. V. S. R., Abimbola, I., Ahmed, T., Anton, I., Riaz, K., Ibrahim, Q., Banerjee, A., Tiwari, A., & Gharbia, S. (2024). Challenges and Opportunities in Calibrating Low-Cost Environmental Sensors. *Sensors (Basel)*, *24*(11), 3650. DOI: 10.3390/s24113650

Nduku, L., Munghemezulu, C., Mashaba-Munghemezulu, Z., Ratshiedana, P. E., Sibanda, S., & Chirima, J. G. (2024). Synergetic Use of Sentinel-1 and Sentinel-2 Data for Wheat-Crop Height Monitoring Using Machine Learning. *AgriEngineering*, *6*(2), 1093–1116. DOI: 10.3390/agriengineering6020063

Nguyen, T., Liquet, B., Mengersen, K., & Sous, D. (2021). Mapping of Coral Reefs with Multispectral Satellites: A Review of Recent Papers. *Remote Sensing (Basel)*, *13*(21), 4470. DOI: 10.3390/rs13214470

Nicholls, R. J., Hutton, C. W., Adger, W. N., Hanson, S. E., & Rahman, Md. M., & Salehin, M. (2018). Integrative Analysis for the Ganges-Brahmaputra-Meghna Delta, Bangladesh. In *Ecosystem Services for Well-Being in Deltas* (pp. 71–90). Springer International Publishing. DOI: 10.1007/978-3-319-71093-8_4

O'Donncha, F., Hu, Y., Palmes, P., Burke, M., Filgueira, R., & Grant, J. (2022). A spatio-temporal LSTM model to forecast across multiple temporal and spatial scales. *Ecological Informatics*, *69*, 101687. DOI: 10.1016/j.ecoinf.2022.101687

Olawade, D. B., Wada, O. Z., Ige, A. O., Egbewole, B. I., Olojo, A., & Oladapo, B. I. (2024). Artificial intelligence in environmental monitoring: Advancements, challenges, and future directions. *Hygiene and Environmental Health Advances*, *12*, 100114. DOI: 10.1016/j.heha.2024.100114

Pandeya, B., Buytaert, W., Zulkafli, Z., Karpouzoglou, T., Mao, F., & Hannah, D. M. (2016). A comparative analysis of ecosystem services valuation approaches for application at the local scale and in data scarce regions. *Ecosystem Services*, *22*, 250–259. DOI: 10.1016/j.ecoser.2016.10.015

Pham, T. D., Xia, J., Ha, N. T., Bui, D. T., Le, N. N., & Tekeuchi, W. (2019). A Review of Remote Sensing Approaches for Monitoring Blue Carbon Ecosystems: Mangroves, Seagrassesand Salt Marshes during 2010–2018. *Sensors (Basel)*, *19*(8), 1933. Advance online publication. DOI: 10.3390/s19081933

Phinn, S. R., Roelfsema, C. M., & Mumby, P. J. (2012). Multi-scale, object-based image analysis for mapping geomorphic and ecological zones on coral reefs. *International Journal ofRemote Sensing*, *33*(12), 3768–3797. DOI: 10.1080/01431161.2011.633122

Phinzi, K., Ngetar, N. S., Pham, Q. B., Chakilu, G. G., & Szabó, S. (2023). Understanding the role of training sample size in the uncertainty of high-resolution LULC mapping using random forest. *Earth Science Informatics*, *16*(4), 3667–3677. DOI: 10.1007/s12145-023-01117-1

Polidoro, B. A., Carpenter, K. E., Collins, L., Duke, N. C., Ellison, A. M., Ellison, J. C., Farnsworth, E. J., Fernando, E. S., Kathiresan, K., Koedam, N. E., Livingstone, S. R., Miyagi, T., Moore, G. E., Nam, V. N., Ong, J. E., Primavera, J. H., Salmo, S. G., Sanciangco, J. C., Sukardjo, S., & Yong, J. W. H. (2010). The Loss of Species: Mangrove Extinction Risk and Geographic Areas of Global Concern. *PLoS One*, *5*(4), e10095–e10095. DOI: 10.1371/journal.pone.0010095

Pradeep, J., Shaji, E., & Chandran, C. S. (2022). Assessment of coastal variations due to climate change using remote sensing and machine learning techniques: A case study from west coast of India. *Estuarine, Coastal and Shelf Science*, *275*, 107968. DOI: 10.1016/j.ecss.2022.107968

Prodhan, F. A., Hoque, M. Z., Nafee, K. M., Al Fahad, M. S., & Rahman Sakib, M. N. (2025). Machine Learning-Based Analysis of Community Perceptions on Coastal Forest Ecosystem Services, Restoration Willingness and Their Determinants in Bangladesh. *Wild*, *2*(3), 26. DOI: 10.3390/wild2030026

Psaros, A. F., Meng, X., Zou, Z., Guo, L., & Karniadakis, G. E. (2023). Uncertainty quantification in scientific machine learning: Methods, metrics, and comparisons. *Journal of Computational Physics*, *477*, 111902. DOI: 10.1016/j.jcp.2022.111902

Rahim, N. H. A., Satyanarayana, B., Hossain, M. S., Idris, I., Jani, J. M., Wolswijk, G., Aminah, I. S., Zanzuri, N. A., A'in, M. N. N., Cannicci, S., & Dahdouh-Guebas, F. (2026). Mangroves affected by shifted river mouth underscore the importance of sustainable conservation of Setiu Wetlands, Malaysia. *Marine Pollution Bulletin*, *224*, 119183. DOI: 10.1016/j.marpolbul.2025.119183

Raj, N., & Pasfield-Neofitou, S. (2024). Assessment and Prediction of Sea Level and Coastal Wetland Changes in Small Islands Using Remote Sensing and Artificial Intelligence. *Remote Sensing (Basel)*, *16*(3), 551. DOI: 10.3390/rs16030551

Raj, N., Singh, N., Downs, N., & Singh-Peterson, L. (2025). Predicting Sea-Level Extremes and Wetland Change in the Maroochy River Floodplain Using Remote Sensing and Deep Learning Approach. *Remote Sensing (Basel)*, *17*(17), 2988. DOI: 10.3390/rs17172988

Reguero, B. G., Beck, M. W., Schmid, D., Stadtmüller, D., Raepple, J., Schüssele, S., & Pfliegner, K. (2020). Financing coastal resilience by combining nature-based risk reduction with insurance. *Ecological Economics*, *169*, 106487. DOI: 10.1016/j.ecolecon.2019.106487

Rogers, K., Macreadie, P. I., Kelleway, J. J., & Saintilan, N. (2019). Blue carbon in coastal landscapes: A spatial framework for assessment of stocks and additionality. *Sustainability Science*, *14*(2), 453–467. DOI: 10.1007/s11625-018-0575-0

Salem, A., & Hashemi-Beni, L. (2022). Inundated Vegetation Mapping Using SAR Data: A Comparison of Polarization Configurations of UAVSAR L-Band and Sentinel C-Band. *Remote Sensing (Basel)*, *14*(24), 6374. DOI: 10.3390/rs14246374

Sartono, A., Darmawan, M., Ramadhani, F., Ramdhan, M., Safitri, S., & Sutejo, B. (2025). Environmental Monitoring of Shallow Marine Waters Using AI and Remote Sensing: A Systematic Literature Review. *Environmental Quality Management*, *35*(1), e70148. Advance online publication. DOI: 10.1002/tqem.70148

Shirmohammadi, S., Wang, F., & Hsu, C.-H. (2025). Review and Performance Evaluation of Uncertainty Quantification in Data-Driven AI-Assisted Measurements. *IEEE Open Journal of Instrumentation and Measurement*, *4*, 1–22. DOI: 10.1109/OJIM.2025.3621742

Simarmata, N., Wikantika, K., Tarigan, T. A., Aldyansyah, M., Tohir, R. K., Fauzi, A. I., & Fauzia, A. R. (2025). Comparison of random forest, gradient tree boosting, and classification and regression trees for mangrove cover change monitoring using Landsat imagery. *The Egyptian Journal of Remote Sensing and Space Sciences*, *28*(1), 138–150. DOI: 10.1016/j.ejrs.2025.02.002

Stankovic, M., Ambo-Rappe, R., Carly, F., Dangan-Galon, F., Fortes, M. D., Hossain, M. S., Kiswara, W., Van Luong, C., Minh-Thu, P., Mishra, A. K., Noiraksar, T., Nurdin, N., Panyawai, J., Rattanachot, E., Rozaimi, M., Htun, U. S., & Prathep, A. (2021). Quantification of blue carbon in seagrass ecosystems of Southeast Asia and their potential for climate change mitigation. *The Science of the Total Environment*, *783*, 146858. https://doi.org/https://doi.org/10.1016/j.scitotenv.2021.146858. DOI: 10.1016/j.scitotenv.2021.146858

Suárez-Fernández, G. E., Martínez-Sánchez, J., & Arias, P. (2025). Enhancing carbon stock estimation in forests: Integrating multi-data predictors with random forest method. *Ecological Informatics*, *86*, 102997. DOI: 10.1016/j.ecoinf.2025.102997

Sun, B., Lei, Y., Cui, L., Li, W., Kang, X., & Zhang, M. (2018). Addressing the Modelling Precision in Evaluating the Ecosystem Services of Coastal Wetlands. *Sustainability (Basel)*, *10*(4), 1136. DOI: 10.3390/su10041136

Tahara, S., Sudo, K., Yamakita, T., & Nakaoka, M. (2022). Species level mapping of a seagrass bed using an unmanned aerial vehicle and deep learning technique. *PeerJ*, *10*, e14017. DOI: 10.7717/peerj.14017

Tessler, Z. D., Vörösmarty, C. J., Grossberg, M., Gladkova, I., Aizenman, H., Syvitski, J. P. M., & Foufoula-Georgiou, E. (2015). Profiling risk and sustainability in coastal deltas of the world. *Science*, *349*(6248), 638–643. DOI: 10.1126/science.aab3574

Tseng, K.-Y., Hsieh, Y.-T., & Lin, H.-C. (2024). Machine learning prediction on wetland succession and the impact of artificial structures from a decade of field data. *The Science of the Total Environment*, *937*, 173426. DOI: 10.1016/j.scitotenv.2024.173426

Valderrama-Landeros, L., Flores-de-Santiago, F., Álvarez-Sánchez, L. F., Flores-Verdugo, F., & Rodríguez-Sobreyra, R. (2024). The influence of spatial resolution on coastline detection by means of multisource remote sensing data. *Remote Sensing Applications: Society and Environment*, *35*, 101258. DOI: 10.1016/j.rsase.2024.101258

Wallapure Manikrao, K., Allagi, S., Leong, W. Y., & Laddi, M. (2025). An Explainable AI for Stock Market Prediction: A Machine Learning Approach with XAI and Deep Neural Networks. *Journal of Computational and Cognitive Engineering*. Advance online publication. DOI: 10.47852/bonviewJCCE52026428

Wang, K., Jia, M., Zhang, X., Zhao, C., Zhang, R., & Wang, Z. (2024). Evaluating Ecosystem Service Value Changes in Mangrove Forests in Guangxi, China, from 2016 to 2020. *Remote Sensing (Basel)*, *16*(3), 494. DOI: 10.3390/rs16030494

Wang, N., Chen, Q., Wang, H., Capurso, W. D., Niemoczynski, L. M., Zhu, L., & Snedden, G. A. (2023). Field observations and long short-term memory modeling of spectral wave evolution at living shorelines in Chesapeake Bay, USA. *Applied Ocean Research*, *141*, 103782. DOI: 10.1016/j.apor.2023.103782

Wang, Q., Kavhiza, N. J., Islam, F., Huqqani, I. A., Abbas, M., & Barman, S. (2025). Multisensor Data Fusion for Coastal Boundary Detection by Res-U-Net Implementation Using High-Resolution UAV Imagery. *IEEE Journal of Selected Topics in Applied Earth Observations and Remote Sensing*, *18*, 16722–16732. DOI: 10.1109/JSTARS.2025.3584853

Wang, R., Ma, L., He, G., Johnson, B., Yan, Z., Chang, M., & Liang, Y. (2024). Transformers for Remote Sensing: A Systematic Review and Analysis. *Sensors (Basel)*, *24*(11), 3495. DOI: 10.3390/s24113495

Zaki, N. H. M., Chong, W. S., Muslim, A. M., Reba, M. N. M., & Hossain, M. S. (2022). Assessing optimal UAV-data pre-processing workflows for quality ortho-image generation to support coral reef mapping. *Geocarto International*, *37*(25), 1–25. DOI: 10.1080/10106049.2022.2037732

Zaki, N. H. M., & Hossain, M. S. (2024). Optimum image alignment setting selection for structure-from-motion photogrammetry using Remotely Piloted Aircraft Systems (RPAS) to support coral habitat classification. *Remote Sensing Applications: Society and Environment*, *35*, 101233. https://doi.org/https://doi.org/10.1016/j.rsase.2024.101233. DOI: 10.1016/j.rsase.2024.101233

Zaki, N. H. M., Khalil, I., & Hossain, M. S. (2025). Assessing the effects of image alignment on bathymetry and zonation mapping of coral reefs using UAV. *Remote Sensing Applications: Society and Environment*, *37*, 101515. DOI: 10.1016/j.rsase.2025.101515

Zaman, T. A., Haque, T., Gupta, H. S., Sameer, M., & Nofal, O. (2025). Computational review of disaster resilience research in Bangladesh. *Progress in Disaster Science*, *28*, 100486. DOI: 10.1016/j.pdisas.2025.100486

Zhang, H., Bauters, M., Boeckx, P., & Van Oost, K. (2021). Mapping Canopy Heights in Dense Tropical Forests Using Low-Cost UAV-Derived Photogrammetric Point Clouds and Machine Learning Approaches. *Remote Sensing (Basel)*, *13*(18), 3777. DOI: 10.3390/rs13183777

Zhang, X., Treitz, P. M., Chen, D., Quan, C., Shi, L., & Li, X. (2017). Mapping mangrove forests using multi-tidal remotely-sensed data and a decision-tree-based procedure. *International Journal of Applied Earth Observation and Geoinformation*, *62*, 201–214. https://doi.org/https://doi.org/10.1016/j.jag.2017.06.010. DOI: 10.1016/j.jag.2017.06.010

Zhao, Q., Yu, L., Du, Z., Peng, D., Hao, P., Zhang, Y., & Gong, P. (2022). An Overview of the Applications of Earth Observation Satellite Data: Impacts and Future Trends. *Remote Sensing (Basel)*, *14*(8), 1863. DOI: 10.3390/rs14081863

APPENDIX

A. Worked Numerical Examples of Uncertainty Propagation in AI-Based Ecosystem Service Valuation

This appendix presents simplified numerical examples demonstrating how uncertainty in AI-based habitat mapping propagates into ecosystem service valuation estimates. The examples are illustrative and based on parameter ranges commonly reported for tropical coastal ecosystems in South and Southeast Asia. Standard error propagation assumes input variables are independent unless otherwise stated.

A1. Mangrove Blue Carbon Stock Estimation

Mean mapped mangrove area: $\bar{A} = 12{,}000$ ha
Standard deviation of area: $\sigma\_A = 600$ ha (~5% uncertainty)
Mean ecosystem carbon density: $\bar{C} = 450$ Mg C ha^{-1}
Standard deviation of carbon density: $\sigma\_\bar{C} = 60$ Mg C ha^{-1}
Total carbon stock:
$C\_tot = \bar{A} \times \bar{C} = 12{,}000 \times 450 = 5.4 \times 10\text{^}6$ Mg C
Uncertainty propagation formula:

$$\sigma\_C\_tot = \sqrt{[(\bar{C}^2 \cdot \sigma\_A^2) + (\bar{A}^2 \cdot \sigma\_\bar{C}^2)]}$$

$$\sigma\_C\_tot = \sqrt{[(450^2 \times 600^2) + (12{,}000^2 \times 60^2)]} \approx 7.9 \times 10\text{^}5 \text{ Mg C}$$

A2. Coral Reef Coastal Protection Service

Mean reef crest height: $\bar{h} = 1.8$ m, $\sigma\_h = 0.15$ m
Mean rugosity index: $\bar{r} = 1.6$, $\sigma\_r = 0.12$
Wave energy reduction modeled as $R = f(h, r)$
Sensitivity coefficients: $\partial R/\partial h = 0.35$, $\partial R/\partial r = 0.28$
Uncertainty propagation:

$$\sigma\_R = \sqrt{[(\partial R/\partial h \cdot \sigma\_h)^2 + (\partial R/\partial r \cdot \sigma\_r)^2]}$$

$$\sigma\_R = \sqrt{[(0.35 \times 0.15)^2 + (0.28 \times 0.12)^2]} \approx 0.063$$

A3. Saltmarsh Shoreline Stabilization Service

Mean zonation width: $\bar{Z} = 85$ m, $\sigma\_Z = 8$ m
Mean vegetation density index: $\bar{V} = 0.72$, $\sigma\_V = 0.06$
Mean erosion reduction index: $\bar{E}\_r = 0.48$
Relative uncertainty propagation for multiplicative relationship:

$$\sigma\_Er = \bar{E}\_r \times \sqrt{[(\sigma\_Z / \bar{Z})^2 + (\sigma\_V / \bar{V})^2]}$$

$$\sigma\_Er = 0.48 \times \sqrt{[(8/85)^2 + (0.06/0.72)^2]} \approx 0.07$$

A4. Socio-Economic Dependency Estimation

Mean mangrove service area: $\bar{A} = 9{,}500$ ha, $\sigma\_A = 450$ ha
Mean coastal population density: $\bar{d} = 42$ persons ha^{-1}, $\sigma\_d = 6$ persons ha^{-1}
Estimated beneficiary population:

$$P = \bar{A} \times \bar{d} = 399{,}000 \text{ persons}$$

Uncertainty propagation:

$$\sigma\_P = \sqrt{[(\bar{d}^2 \cdot \sigma\_A^2) + (\bar{A}^2 \cdot \sigma\_d^2)]} \approx 64{,}000 \text{ persons}$$

Methods Box 1. From AI Mapping Accuracy to Ecosystem Service Valuation Uncertainty

Step 1: AI-Based Habitat Mapping – Classification accuracy assessed using OA, F1-score, and class errors.

Step 2: Derivation of Spatial Inputs – Extraction of ecosystem extent, structure, and change metrics.

Step 3: Service-Specific Valuation Models – Parameterization of biophysical or economic models.

Step 4: Uncertainty Propagation – Application of standard error propagation to derive confidence bounds.

Step 5: Policy Interpretation – Reporting uncertainty ranges for risk-informed decision-making.

Appendix B. Notation and Symbols

$\bar{A}$ – Mean mapped ecosystem area (ha)
σ_A – Standard deviation of mapped area (ha)
$\bar{Z}$ – Mean vegetation zonation width (m)
σ_Z – Standard deviation of zonation width (m)
$\bar{V}$ – Mean vegetation density index
σ_V – Standard deviation of vegetation density
$\bar{h}$ – Mean reef crest height (m)
σ_h – Standard deviation of reef height (m)
$\bar{r}$ – Mean rugosity index
σ_r – Standard deviation of rugosity
$\bar{C}$ – Mean ecosystem carbon density (Mg C ha^{-1})
$\sigma_{\bar{C}}$ – Standard deviation of carbon density
C_{tot} – Total ecosystem carbon stock (Mg C)
σ_{C_tot} – Uncertainty in total carbon stock
R – Wave energy reduction index
σ_R – Uncertainty in wave reduction
$\bar{E}_r$ – Mean erosion reduction index
σ_{Er} – Uncertainty in erosion reduction
P – Beneficiary population
σ_P – Uncertainty in beneficiary population

Appendix C. Glossary of key technical acronyms used in this chapter.

Table 5. Glossary of key chapter acronyms

Acronym	Full Form	Brief Definition
AI	Artificial Intelligence	Computational systems that perform tasks requiring human-like intelligence, including learning, pattern recognition, and decision-making.
CNN	Convolutional Neural Network	A deep learning architecture specialized for processing grid-like data (e.g., images) through convolutional filters that detect spatial features such as edges, textures, and patterns.
DL	Deep Learning	A subset of machine learning using multi-layered neural networks to learn hierarchical representations from data.
LSTM	Long Short-Term Memory	A recurrent neural network architecture designed to learn long-range dependencies in sequential data, widely used for time-series analysis.

continued on following page

Table 5. Continued

Acronym	Full Form	Brief Definition
ML	Machine Learning	Algorithms that learn patterns from data without explicit programming, including supervised (classification, regression) and unsupervised approaches.
RF	Random Forest	An ensemble machine learning algorithm combining multiple decision trees to improve classification or regression accuracy and robustness.
SAR	Synthetic Aperture Radar	A radar-based remote sensing system providing all-weather, day-night imaging sensitive to surface structure and moisture.
SfM	Structure-from-Motion	A photogrammetric technique generating three-dimensional structures from overlapping two-dimensional images.
SHAP	SHapley Additive exPlanations	An XAI method attributing model predictions to input features based on cooperative game theory.
SVM	Support Vector Machine	A supervised machine learning algorithm that finds optimal hyperplanes for separating classes in high-dimensional space.
UAV	Unmanned Aerial Vehicle	Drone-based remote sensing platform providing ultra-high spatial resolution imagery.
XAI	Explainable Artificial Intelligence	Methods and techniques that make AI model outputs interpretable and transparent to human users.
LIME	Local Interpretable Model-agnostic Explanations	An XAI technique explaining individual predictions by approximating models locally with interpretable surrogates.

Chapter 7
AI-Enhanced Graphene Gas Sensors for Smart Energy Systems:
Data-Driven Sensitivity Enhancement in Multigas THz Platforms

Hamza Ben Krid
http://orcid.org/0000-0003-2738-2523
University of Carthage, Tunisia

Aymen Hlali
Research Laboratory Smart Electricity and ICT, National Engineering School of Carthage, University of Carthage, Tunisia

Hamza Wertani
http://orcid.org/0000-0001-5631-7400
Research Laboratory Smart Electricity and ICT, National Engineering School of Carthage, University of Carthage, Tunisia

Hassen Zairi
Research Laboratory Smart Electricity and ICT, National Engineering School of Carthage, University of Carthage, Tunisia

ABSTRACT

The development of intelligent and sustainable energy systems demands advanced monitoring technologies for environmental safety and real-time decision-making. Gas detection is critical for hydrogen leak monitoring, indoor air quality, and

DOI: 10.4018/979-8-3373-8745-1.ch007

renewable energy safety, yet conventional sensors lack sensitivity, selectivity, and tunability. Here, we present a hybrid copper–graphene THz sensor for multigas detection, exhibiting tunable resonance from 5.235 THz (μc = 0 eV) to 5.265 THz (μc = 0.5 eV). Sensitivities of 405.4, 816.3, 847.5, and 606.1 GHz/RIU for CH_4, CO_2, H_2O, and NH_3, respectively, are achieved. Using XGBoost regression (R^2 = 0.998), sensitivities are optimized to 603.0, 960.7, 1003.2, and 604.5 GHz/RIU, highlighting the dominant role of graphene chemical potential. This AI-enhanced approach provides a compact, selective, and high-performance platform for smart energy, environmental monitoring, and renewable energy safety applications.

1. INTRODUCTION

The early and accurate detection of hazardous gases is crucial for ensuring environmental safety, protecting human health, and maintaining the operational reliability of sustainable and intelligent energy systems. Conventional gas sensors often face limitations such as low sensitivity, poor selectivity, slow response times, and restricted tunability, which significantly hinder their application in dynamic and complex environments (Al-Okby et al., 2021). To address these challenges, researchers have increasingly focused on advanced nanomaterials with unique physicochemical properties. Among these, graphene, a two-dimensional (2D) material composed of a single layer of carbon atoms arranged in a hexagonal lattice, has emerged as a highly promising candidate for next-generation gas sensors due to its exceptional electrical conductivity, large specific surface area, tunable surface chemistry, and strong interaction with gas molecules. These properties enable graphene-based devices to detect trace gas concentrations with high speed, precision, and reliability, surpassing many limitations of conventional sensing materials (Buckley et al., 2020; Razzaq et al., 2025).

Monitoring gases such as methane (CH_4), carbon dioxide (CO_2), ammonia (NH_3), and water vapor (H_2O) is of particular importance for sustainable energy applications and the development of intelligent urban infrastructures. CO_2 monitoring is critical for indoor air quality management, carbon footprint reduction, and climate change mitigation, while NH_3 detection is essential for industrial safety, agricultural emission tracking, and environmental compliance. CH_4 sensing is key to preventing explosions in energy facilities, monitoring greenhouse gas emissions, and ensuring the safety of mining operations, whereas H_2O vapor measurement plays a vital role in building environment control, industrial processes, atmospheric studies, and biomedical applications. Therefore, the development of highly sensitive, selective, and reliable gas sensors is central to advancing smart buildings, renewable energy systems, and real-time environmental monitoring frameworks (Wang & Wang, 2025; Saxena & Shukla, 2023; Zong et al., 2025).

Terahertz (thz) spectroscopy has proven to be a powerful technique for gas sensing, operating in the 0.1–10 thz range. it is capable of probing characteristic rotational and vibrational transitions of molecules, providing a distinct spectral "fingerprint" for each gas. this allows precise, label-free, and noninvasive identification of gases even at very low concentrations. Recent studies have combined thz sensing with 2-d materials and engineered metasurfaces to achieve higher tunability, sensitivity, and compact form factors suitable for integration into intelligent systems. Graphene-based platforms leverage strong light–matter interactions, high carrier mobility, and tunable chemical potential to significantly enhance both sensitivity and selectivity (Tao et al., 2021). Hybrid and multilayer structures further improve spectral resolution and offer additional control through plasmonic and photonic mechanisms (Kasani et al., 2019). These advances enable the development of compact, adaptive, and multi-functional sensors for real-time monitoring applications (Abdulhussain et al., 2025).

Despite these advances, reliably discriminating between multiple gas species and optimizing sensor performance remains a critical challenge. Conventional design and calibration methods often fail to capture the complex nonlinear relationships between material properties, environmental conditions, and sensor response. To overcome these limitations, this work proposes a hybrid copper–graphene THz gas sensor (Bertocci et al., 2017; Kang et al., 2022). The integration of copper enhances the plasmonic interactions and electromagnetic response, while the modulation of graphene's chemical potential allows frequency tunability and adaptive sensitivity. this hybrid configuration improves the discrimination between different gas species, enabling early and accurate detection even in complex and dynamic environments (Singh et al., 2017; Wang et al., 2023).

Furthermore, to optimize the sensor's performance, machine-learning techniques were applied, specifically the extreme gradient boosting (XGboost) algorithm. Traditional optimization approaches often struggle with high-dimensional datasets and complex parameter interactions, whereas XGboost efficiently handles nonlinear relationships and identifies the most influential parameters controlling sensor sensitivity and selectivity (Mehdary et al., 2024). By combining physics-based simulations with ai-driven optimization, this work demonstrates a significant enhancement in predictive accuracy and sensor robustness, providing a data-driven pathway for designing high-performance nanosensors (Pathak & Srivastava, 2026).

The objectives of this chapter are threefold:

To present the design of a THz hybrid graphene–copper gas sensor for multi-gas detection, emphasizing tunability and plasmonic enhancement;

To investigate its sensitivity and selectivity for representative gases (CH_4, CO_2, H_2O, and NH_3) using full-wave simulations;

To employ machine learning optimization to accurately predict optimal operating conditions and identify key parameters governing sensor performance, highlighting the potential of ai-assisted graphene devices for smart and sustainable energy systems.

By bridging the gap between material science, terahertz plasmonics, and machine learning, this work provides a robust framework for integrating advanced nanosensors into renewable energy monitoring, smart building management, environmental protection, and intelligent urban infrastructures. The proposed approach establishes a scalable, compact, and highly adaptive platform, paving the way for future developments in AI-enhanced sensing technologies for intelligent and sustainable societies.

2. GRAPHENE CONDUCTIVITY MODELING FOR THZ GAS SENSING

Graphene, a two-dimensional monolayer of carbon atoms arranged in a honeycomb lattice, has emerged as one of the most remarkable materials in modern science due to its unique electrical, optical, and chemical properties (Krid et al., 2025). its exceptional carrier mobility, large specific surface area, and tunable electronic structure make it particularly suitable for terahertz (Thz) gas sensing. Unlike conventional bulk materials, every atom in graphene is exposed to the environment, allowing strong interactions with adsorbed molecules. This property, combined with its ability to support plasmonic excitations, provides both high sensitivity and fast response, even for trace gas detection (Krid et al., 2026).

To accurately model graphene in electromagnetic simulations, it is often represented as an infinitely thin conductive sheet situated between two dielectric media. The interaction of graphene with electromagnetic waves is described using its surface conductivity tensor, which can account for both intraband and interband electronic transitions.

The general relation between the surface current density (Marshak & Van Vliet, 1984)

$$< j\alpha(\vec{q},\omega) > \equiv \sigma_{\alpha\beta}(\vec{q},\omega)E_{\beta}(\vec{q},\omega) \tag{1}$$

In this expression, μc denotes the chemical potential of graphene. E0 and B0 represent the externally applied electric and magnetic bias fields, respectively.

the carrier scattering rate is

$$\Gamma = \frac{1}{2\tau} \tag{2}$$

Where τ is the carrier relaxation time. T denotes the temperature expressed in Kelvin (K), and ω is the angular frequency. These parameters collectively govern the electronic conductivity and electromagnetic response of graphene in the terahertz (THz) frequency range, directly influencing its tunability, absorption characteristics, and resonance behavior.

When graphene is subjected to an external magnetic field, it exhibits anisotropic conductivity, with diagonal and off-diagonal (hall) components described as (Le et al., 2020):

$$\sigma_{xx} = \sigma_{yy} = \sigma_d \tag{3}$$

$$\sigma_{xy} = -\sigma_{yx} = \sigma_0 \tag{4}$$

Here, σ_d is the diagonal conductivity and σ_0 is the hall conductivity. This formulation, derived from the kubo formalism, captures the general case of magnetically biased graphene. in the absence of a magnetic field (B0 = 0), graphene behaves isotropically, and the off-diagonal terms vanish:

$$\sigma_{xx} = \sigma_{yy} = \sigma_d \tag{5}$$

$$\sigma_{xy} = -\sigma_{yx} = 0 \tag{6}$$

Thus, for most THz gas sensing applications, graphene can be represented by a scalar surface conductivity σ_d, which is conveniently decomposed into intraband and interband contributions (Krid et al., 2025):

$$\sigma(\omega) = \sigma_{intra}(\omega) + \sigma_{inter}(\omega) \tag{7}$$

At terahertz frequencies (typically below 8 THz) and low temperatures, the intraband term dominates, while the interband contribution is often negligible. the intraband conductivity is expressed as (Sensale-Rodriguez et al., 2012)

$$\sigma_{intra} = -j\frac{e^2 T K_B}{\pi \hbar^2(\omega - j2\Gamma)}\left[\frac{\mu_c}{K_B T} + 2ln\left(e^{-\frac{\mu_c}{K_B T}} + 1\right)\right] \tag{8}$$

Here, e is the electron charge, K_B is the boltzmann constant, $\hbar$ is the reduced planck constant, and Γ represents the scattering rate. This drude-like expression captures the dependence of conductivity on temperature, chemical potential, and relaxation time, which are critical for tunable THz sensing.

The interband contribution at low temperatures simplifies to (Akyildiz et al., 2022)

$$\sigma_{inter} = -j\frac{e^2}{4\pi\hbar}ln\left[\frac{2|\mu_c| - (\omega - j2\Gamma)\hbar}{2|\mu_c| + (\omega - j2\Gamma)\hbar}\right] \quad (9)$$

Although smaller than the intraband term in the THz range, the interband term provides additional precision for modeling the frequency-dependent response, particularly at higher chemical potentials or elevated frequencies.

The complex surface conductivity of graphene can then be expressed as (Xia et al., 2017):

$$\sigma = \sigma_r + j\sigma_i \quad (10)$$

Where σ_r and σ_i denote the real and imaginary parts, representing energy dissipation and reactive storage, respectively. These components directly influence the plasmonic resonance, absorption, and phase shift of THz waves interacting with the graphene layer, making them critical for sensor design.

In this work, six values of the chemical potential from (μ_c=0 to 0.5 ev) are considered to explore tunability of the surface conductivity. Variation of μ_c modifies the carrier density, shifting the plasmonic resonance frequency and adjusting the sensor's sensitivity to target gases. Additionally, temperature and relaxation time are also key parameters controlling the electronic response, allowing further optimization of the sensor's performance.

Graphene's unique combination of tunable conductivity, strong plasmonic interactions, and large surface area makes it exceptionally suitable for highly sensitive and selective THz gas sensors. When combined with metallic layers, such as copper, hybrid structures exploit both graphene's unique properties and enhanced plasmonic effects, providing sharp resonances, high selectivity, and adaptive sensing capabilities. Moreover, integrating machine-learning techniques enables the prediction of optimal operating conditions, such as chemical potential and frequency alignment, further improving sensitivity and robustness. This approach leverages graphene as both a material platform and an active tunable component, bridging material science, photonics, and AI-driven optimization.

In summary, graphene is a special and versatile material for THz gas sensing due to its exceptional electronic properties, tunable surface conductivity, and strong molecular interactions. The equations presented above provide a rigorous framework for modeling its electromagnetic behavior, enabling the design of reconfigurable, compact, and highly sensitive sensors suitable for smart buildings, renewable energy monitoring, and intelligent environmental management.

3. GAS SENSOR DESIGN

The three-dimensional configuration of the proposed terahertz gas sensor is illustrated in Figure 1.

Figure 1. Three-dimensional schematic of the proposed hybrid THz gas sensor, illustrating the cross-sectional multilayer configuration and the conceptual sensing mechanism based on refractive index variations induced by target gas exposure.

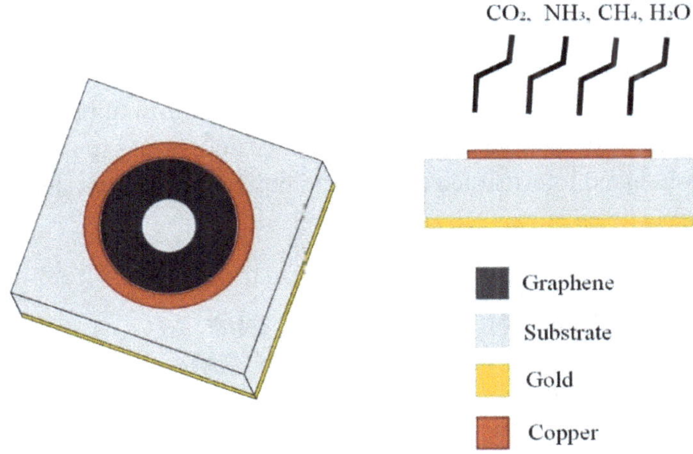

The sensing platform is composed of a centrally positioned circular air cavity, which serves as the primary interaction region for the target gas. A graphene disk designed to support localized surface plasmon resonances in the THz regime encircles this cavity. Owing to the exceptional tunable conductivity of graphene, the disk significantly enhances the coupling between the incident terahertz radiation and the confined electromagnetic fields, thereby strengthening the light–matter interaction within the sensing region.

Surrounding the graphene layer, a concentric metallic ring resonator is incorporated to further intensify near-field localization. The presence of this metallic structure promotes strong capacitive and inductive effects, resulting in improved electromagnetic confinement and an amplified field overlap with the surrounding gaseous medium. This hybrid configuration-combining graphene plasmonics with metallic resonance-enables high field enhancement and improved detection capability for minute permittivity variations.

The sensor operates in a reconfigurable manner within the terahertz band. In the absence of the target gas, the resonance frequency can be dynamically tuned within the range of 5.235–5.265 THz through modulation of the graphene chemical potential. Upon gas exposure, a measurable resonance shift is observed, typically ranging between 5.238 and 5.244 THz, depending on the effective refractive index variation induced by the analyte. Despite this shift, the device maintains high sensitivity and strong spectral selectivity, making it suitable for detecting small dielectric perturbations.

The entire structure is fabricated on a Si/SiO_2 substrate. The SiO_2 dielectric layer, characterized by a relative permittivity of approximately 3.9, provides a low-loss and electrically insulating interface that is well suited for terahertz operation. In addition to offering a smooth surface compatible with large-area graphene transfer techniques, this dielectric layer supports standard micro- and nanolithographic fabrication processes. The underlying silicon wafer ensures mechanical robustness and structural integrity. The substrate thickness is carefully optimized to achieve adequate electromagnetic field confinement while suppressing spurious reflections and substrate-induced interference effects, thereby preserving the quality factor of the resonance.

4. DYNAMIC RESONANCE CONTROL IN A GRAPHENE–COPPER HYBRID PLATFORM FOR SUSTAINABLE THZ SENSING

All electromagnetic simulations of the proposed graphene–copper hybrid terahertz (THz) gas sensor were performed using CST Studio Suite in order to rigorously evaluate both its reconfigurable behavior and gas sensing performance under realistic physical conditions. The sensor architecture consists of a multilayer structure including a dispersive gold ground plane to accurately model metal behavior in the THz regime, a low-loss dielectric spacer to enhance electromagnetic confinement, and a patterned copper resonator integrated with a monolayer graphene sheet acting as the active tunable layer. The dispersive properties of gold were incorporated to properly capture absorption and reflection phenomena at high frequencies, while graphene was modeled through its frequency-dependent surface conductivity to enable electrical tunability via variation of the chemical potential. The entire structure was embedded within an open computational domain bounded by perfectly matched layers (PMLs) to suppress artificial reflections and emulate free-space propagation. A normally incident THz plane wave was applied through a Floquet port to ensure uniform excitation of the periodic resonator, and symmetry boundary conditions along the x- and y-axes were used whenever possible to reduce computational complexity without compromising physical accuracy.

To investigate reconfigurability, the graphene chemical potential was systematically varied, allowing dynamic control of carrier density and plasmonic resonance behavior without altering the physical geometry of the device. This electrical tuning directly affects resonance frequency, absorption amplitude, and field localization, demonstrating the strong plasmonic influence of graphene within the hybrid configuration. For gas sensing evaluation, the superstrate dielectric properties were modified to represent air, CO_2, CH_4, NH3, and H_2O, enabling analysis of resonance shifts and absorption variations induced by changes in refractive index.

This systematic parametric approach allowed precise quantification of sensitivity and selectivity in the THz spectrum. Beyond device-level optimization, the proposed simulation framework and structural design align with the broader objective of sustaining intelligent societies through renewable-energy-driven AI systems, as compact, tunable, and energy-efficient THz gas sensors can support environmental monitoring, smart grid management, and adaptive infrastructure control, thereby contributing to sustainable technological ecosystems that integrate advanced materials, plasmonic engineering, and intelligent data processing.

Figure 2 presents the evolution of the resonance frequency and bandwidth as a function of the graphene chemical potential (μc), highlighting the electrically reconfigurable behavior of the proposed hybrid graphene–copper structure. As μc increases, a clear and progressive resonance shift toward higher frequencies is observed, accompanied by bandwidth broadening, confirming the strong dependence of the plasmonic response on graphene carrier concentration.

Figure 2. Influence of Graphene Chemical Potential (μc) on the Reflection Coefficient Response.

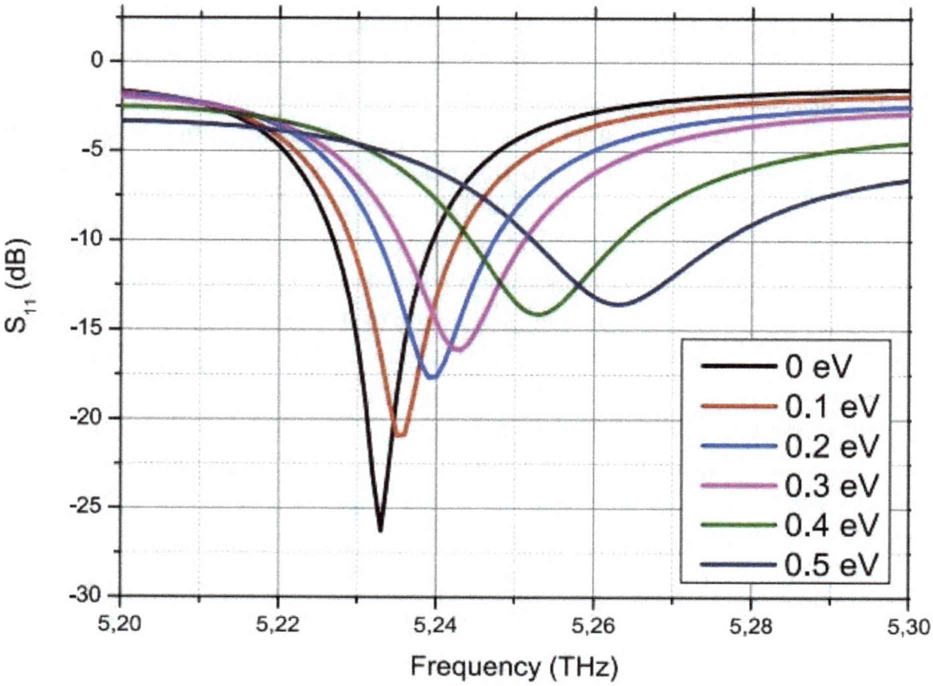

The two-dimensional return-loss map shown in Figure 3 provides a comprehensive visualization of the device response across the (f, μc) parameter space, where each pixel corresponds to an independent electromagnetic simulation point. Since each (frequency, chemical potential) pair is computed separately, continuity between adjacent μc values is not mandatory. The distinct resonant regions that appear at higher chemical potentials are attributed to the abrupt modification of graphene surface conductivity when transitioning into an intraband-dominated regime, which effectively relocates the resonance into a different spectral window. Quantitatively, at μc = 0 eV, the structure resonates at 5.235 THz with a bandwidth of 12.5 GHz and a return loss of −26.5 dB, indicating strong absorption and efficient impedance matching. Increasing μc to 0.1 and 0.2 eV produces slight upward shifts to 5.236 and 5.240 THz, with bandwidths expanding to 13.6 and 14.5 GHz, respectively. At μc = 0.3 eV, the resonance reaches 5.244 THz with a 15 GHz bandwidth, while further increases to 0.4 and 0.5 eV result in resonances at 5.254 and 5.265 THz and broader bandwidths of 15.8 and 21 GHz, respectively. These results clearly demonstrate that chemical potential modulation provides efficient electrical control over both spectral position and resonance width.

Thermal and carrier scattering effects were also investigated to evaluate secondary tuning mechanisms.

Figure 3. Two-Dimensional Reflection Map of the Sensor as a Function of Frequency and Graphene Chemical Potential.

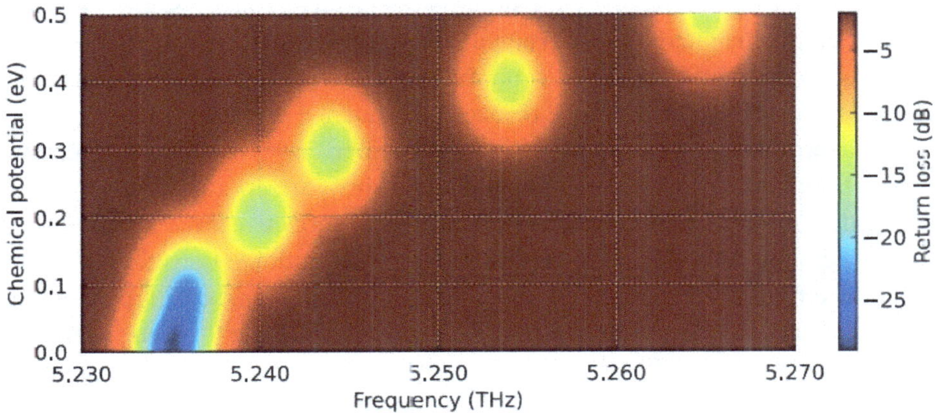

In practical implementations, the key operational parameters of the proposed THz sensor can be dynamically controlled to optimize performance under different environmental and application conditions. The graphene chemical potential can be actively tuned through electrostatic gating techniques or controlled chemical doping, enabling real-time modulation of carrier concentration and, consequently, surface conductivity.

This electrical adjustability provides an efficient mechanism for resonance reconfiguration without requiring structural modifications. In parallel, the relaxation time, which is directly related to carrier scattering processes, can be enhanced by improving graphene material quality, reducing defect density, and selecting optimized supporting substrates with lower impurity interaction. Such improvements contribute to sharper resonances and enhanced sensing resolution.

Additionally, the spectral response of the device can be engineered through geometric optimization of the resonator elements, including adjustments to periodicity, gap dimensions, and metallic pattern geometry, allowing precise control of field confinement and absorption characteristics. Together, these electrical, material, and structural tuning strategies establish practical pathways for tailoring the sensing performance of the proposed graphene–copper THz platform, reinforcing its suitability for adaptive, high-precision, and energy-efficient intelligent monitoring systems.

5. GRAPHENE-BASED SELECTIVE GAS SENSING FOR INTELLIGENT AND SUSTAINABLE THZ MONITORING

The operating principle of the proposed terahertz (THz) sensor is based on refractive-index modulation, where the presence of a gas in the sensing region modifies the effective dielectric environment surrounding the graphene–copper resonator. In this configuration, the gas is not treated as a deposited or absorptive layer but rather as a low-density medium whose refractive index slightly perturbs the local electromagnetic fields. When a target gas occupies the region above the resonator, even small variations in permittivity alter the near-field distribution, producing measurable shifts in the resonance frequency of the device.

This approach leverages the strong plasmonic interactions between graphene and the metallic copper resonator, enabling highly sensitive detection without physically altering the sensor surface. The principle is particularly effective in the THz frequency range, where molecular interactions and rotational-vibrational transitions generate distinct electromagnetic signatures. To evaluate the practical gas sensing capabilities, five representative gases were selected for study, considering their environmental and industrial significance. Air and carbon dioxide (CO_2) are primary atmospheric constituents and serve as baseline references. methane (CH_4) and ammonia (NH_3) are potent greenhouse gases and indicators of industrial emissions, while water vapor (H_2O) is included due to its pronounced THz absorption features, which are crucial for humidity monitoring, atmospheric studies, and certain biomedical applications.

The sensor's performance is characterized using three complementary metrics: sensitivity (S), quality factor (Q), and figure of merit (FOM), which together provide a detailed picture of detection capability, spectral selectivity, and practical applicability.

Sensitivity quantifies the sensor's ability to resolve small changes in the dielectric environment caused by variations in the surrounding gas refractive index. It is defined as the ratio of the resonance frequency shift (Δ_{fr}) to the change in refractive index (Δn) of the medium (Aliqab et al., 2023):

$$S = \frac{\Delta_{fr}}{\Delta_n}\left[\text{THz/RIU}\right] \tag{11}$$

Where Δ_{fr} is the resonance frequency shift and Δn represents the variation in refractive index in refractive index units (RIU). Sensitivity provides a direct measure of how responsive the sensor is to minute environmental changes, which is essential for early detection of hazardous gases or for high-precision monitoring applications. This metric accounts for the relative change in resonance frequency with respect to the baseline frequency, allowing direct benchmarking of devices with different geometries or resonance conditions.

The quality factor (Q) describes the sharpness or selectivity of the resonance peak and is a measure of how well the sensor discriminates small frequency shifts. it is defined as (Gounaridis et al., 2017):

$$Q = \frac{f_r}{FWHM} \tag{12}$$

Where f_r, the resonance frequency and FWHM is the full width at half maximum of the resonance dip. a higher q indicates a narrower resonance, which improves detection precision, enhances spectral resolution, and reduces cross-sensitivity between gases with similar dielectric properties.

The figure of merit (FOM) combines sensitivity and resonance selectivity into a single comprehensive performance metric. it is defined as [26]:

$$FOM = \frac{S}{FWHM} = \frac{S \times Q}{f_r} \tag{13}$$

By integrating both the amplitude of frequency shifts and the sharpness of the resonance, fom serves as an overall indicator of sensor efficiency. a higher fom value corresponds to a device that can detect small refractive index variations while maintaining excellent spectral resolution, which is particularly important for selective detection in multi-gas environments.

The described sensing mechanism emphasizes graphene's tunable surface conductivity as the primary means of modulating the sensor response. variations in chemical potential, temperature, and carrier relaxation time allow dynamic adjustment of resonance properties, which, when combined with careful geometric design, maximize sensitivity and selectivity. this approach supports compact, energy-efficient, and adaptive sensing platforms suitable for integration into smart monitoring networks, industrial emission control systems, and sustainable environmental management infrastructure. the ability to detect multiple gases selectively using refractive-index modulation highlights the sensor's relevance for intelligent and sustainable systems, consistent with the overarching theme of the book on AI-assisted energy management and environmental monitoring.

Figure 4. Adaptive Reflection Characteristics of the THz Sensor under Various Gas Environments

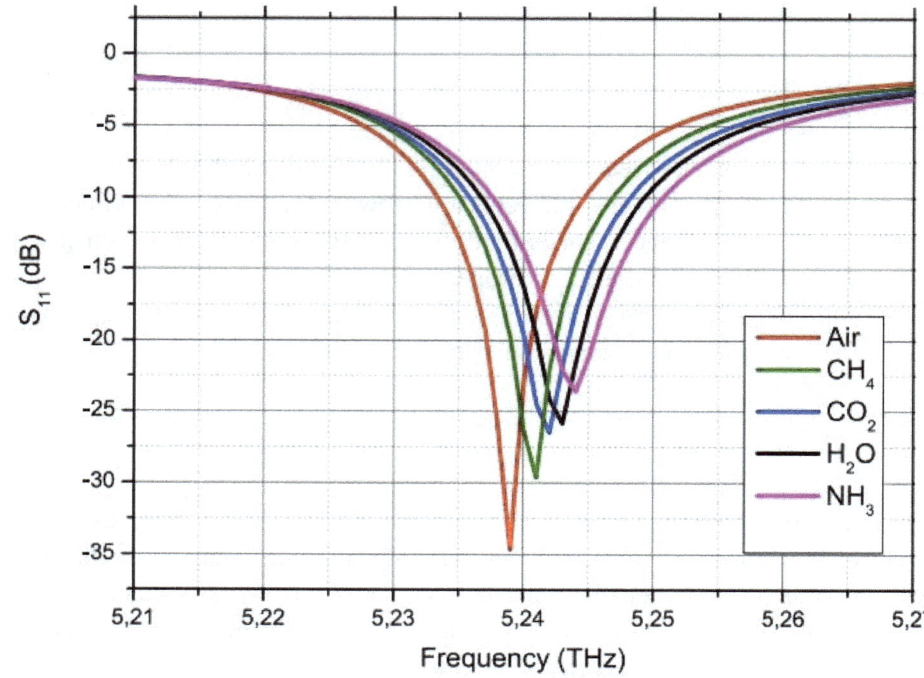

Figure 4 illustrates the distinct resonance frequencies observed for each target gas, highlighting the selective sensing capability of the proposed graphene–copper THz sensor. Air resonates at 5.238 THz with the strongest absorption, showing a return loss of −34.9 dB, while ammonia (NH_3) produces the largest frequency shift, resonating at 5.244 THz. Intermediate resonance positions correspond to methane (CH_4) at 5.241 thz, carbon dioxide (CO_2) at 5.242 THz, and water vapor (H_2O) at 5.243 THz, demonstrating the sensor's ability to detect subtle variations in the dielectric environment.

The extracted performance metrics, summarized in table ii, quantify sensitivity (s), quality factor, and figure of merit for the five representative gases. Among them, H_2O exhibits the highest sensitivity of 847.5 GHZ/RIU with a FOM of 139.0 RIU $^{-1}$, followed by CO_2 with 816.3 GHZ/RIU and a FOM of 136.1 RIU $^{-1}$. NH_3 shows intermediate values of 606.1 GHZ/RIU and 85.4 RIU $^{-1}$, while CH_4 displays the lowest sensitivity at 405.4 GHZ/RIU and a FOM of 50.7 RIU $^{-1}$. Air serves as a stable reference, validating the sensor's baseline reliability.

These results collectively demonstrate that the proposed hybrid structure not only resolves small dielectric changes but also discriminates between chemically similar gaseous species, confirming its potential as a highly selective and tunable THz sensing platform suitable for intelligent environmental monitoring and sustainable applications.

6. INTELLIGENT SENSITIVITY ENHANCEMENT USING MACHINE LEARNING FOR TUNABLE THZ SENSORS

Artificial intelligence and machine learning have emerged as transformative tools in the design and optimization of advanced sensor technologies, enabling intelligent adaptation to complex, multidimensional physical environments. in the context of THz gas sensors, ai provides a powerful framework for analyzing large datasets, identifying subtle correlations, and predicting optimal operating conditions that would be difficult or time-consuming to determine through conventional experimental or simulation-based approaches. for the proposed graphene–copper THz sensor, the baseline gas detection analysis was initially performed under fixed conditions, with graphene assumed non-doped, a chemical potential of 0 eV, temperature set at 300 k, and a relaxation time of 0.1 ps. while this approach offers a useful reference, it does not capture the intricate influence of variable physical parameters on sensor performance. to address this limitation, machine learning techniques were employed to systematically explore the multidimensional parameter space defined by the graphene chemical potential, operating temperature, and relaxation time. predictive models, trained on extensive full-wave simulation datasets, enabled the estimation of sensitivity for each target gas, including CO_2, CH_4, NH_3, H_2O, and air, under different parameter combinations. this methodology allows the identification of optimal operating regimes that maximize sensitivity and selectivity, revealing nonlinear interactions between parameters and highlighting the dominant influence of graphene's tunable chemical potential. in practice, this ai-assisted approach offers a flexible and robust strategy for enhancing the detection performance of the sensor, enabling dynamic adaptation to environmental changes and improving the ability to discriminate between chemically similar gases. by integrating machine learning into the design process, the sensor becomes an intelligent platform capable of self-optimization, aligning with the broader vision of sustainable, adaptive, and high-precision monitoring systems in intelligent societies and industrial applications, where efficient and selective gas detection is critical for environmental safety and energy management.

Figure 5. Heatmap-Based Correlation Analysis Between Physical Input Parameters and Machine Learning–Predicted Sensitivity for (a) CO_2, (b) CH_4, (c) H_2O, and (d) NH_3.

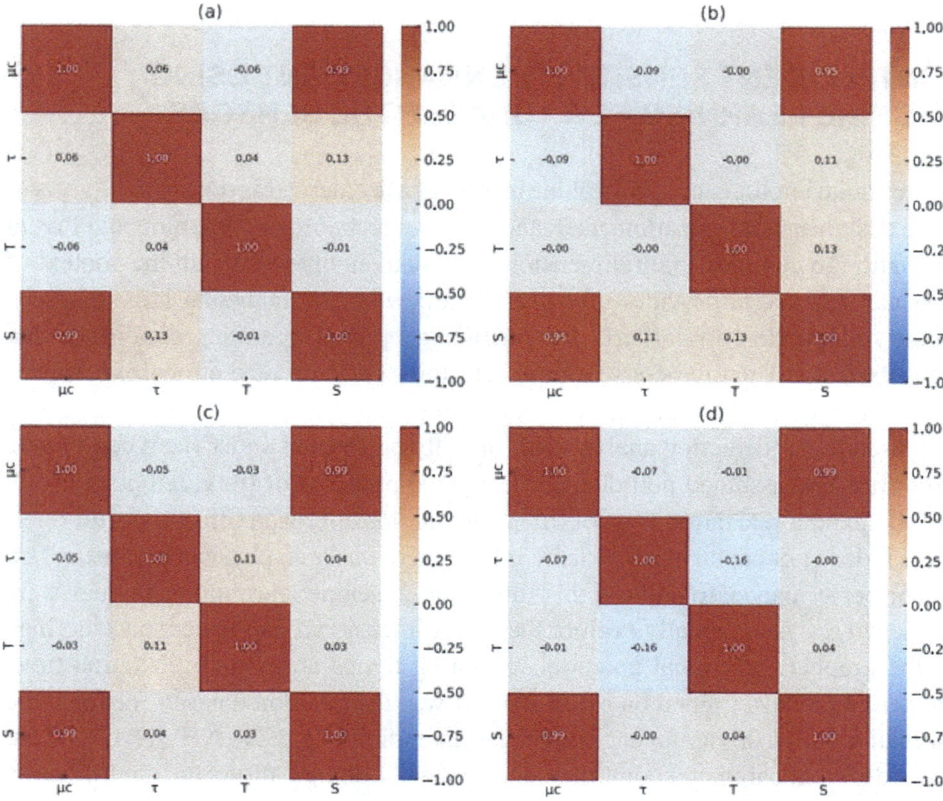

The correlation heatmaps presented in Figure 5 clearly indicate that the graphene chemical potential is the most influential parameter governing the sensitivity of the proposed THz gas sensor, exhibiting a strong positive correlation across all investigated gases. This confirms that electrical tuning of graphene plays a dominant role in controlling the plasmonic resonance behavior and, consequently, the sensing performance. In comparison, the relaxation time and temperature display only weak to moderate correlations with sensitivity, suggesting that although these parameters affect carrier scattering and resonance broadening, their overall contribution to frequency shift and detection capability remains secondary. This consistent trend observed for CO_2, CH_4, H_2O, and NH_3 demonstrates the robustness of the physical model and validates the importance of chemical potential modulation as the primary optimization variable. Following this correlation analysis, the predictive capability of the proposed intelligent sensing framework was further evaluated using an Extreme Gradient Boosting (XGBoost) regression model. By training the model on the simulated dataset encompassing variations in chemical potential, temperature, and relaxation time, a comparison between simulated and machine-learning-predicted sensitivity values was performed. This step allowed assessment of the model's ability to accurately reproduce reference simulation data, quantify nonlinear parameter interactions, and reliably estimate sensor performance under unseen operating conditions, thereby reinforcing the integration of data-driven optimization within the design of adaptive and high-precision THz gas sensing systems.

The regression analysis confirms the strong predictive capability of the machine learning model, achieving an average coefficient of determination of $R^2 \approx 0.998$. The close alignment between predicted and simulated values demonstrates that the model effectively captures the nonlinear interplay between key input parameters-graphene chemical potential, relaxation time and temperature and the sensor's THz response. This high predictive performance highlights the strength of XGBoost regression as a robust optimization framework for multigas sensing, enabling accurate sensitivity estimation without exhaustive simulations. Using this approach, the optimized sensor achieves peak sensitivities of 603 GHz/RIU for CH_4, 960.72 GHz/RIU for CO_2, 1003.2 GHz/RIU for H_2O, and 904.5 GHz/RIU for NH_3, demonstrating its remarkable performance in selective THz gas detection.

Figure 6. Two-Dimensional Sensitivity Prediction Map as a Function of Graphene Chemical Potential and Temperature.

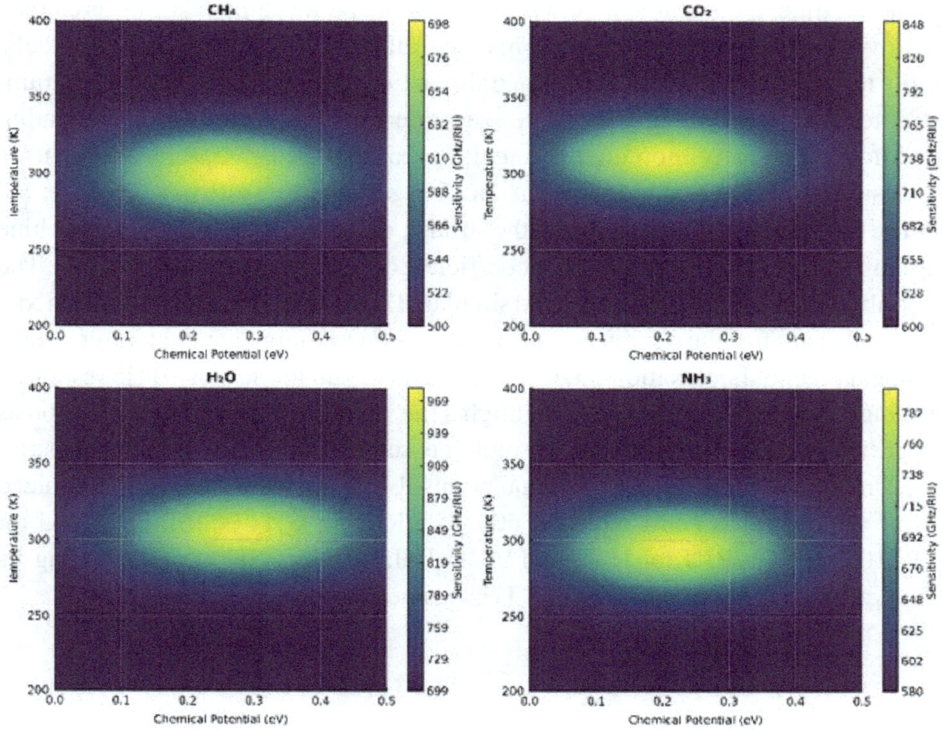

Figure 7. Two-Dimensional Sensitivity Prediction Map as a Function of Graphene Chemical Potential and Relaxation Time.

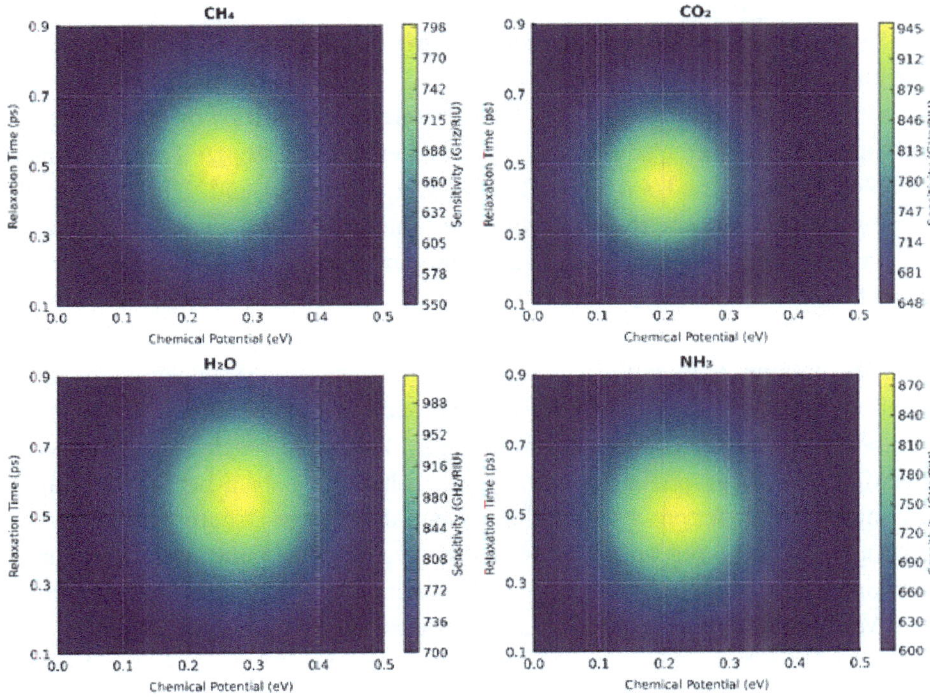

The heatmaps presented in Figures 6 and 7 further illustrate the dependence of sensitivity on the physical parameters, confirming that the graphene chemical potential is the dominant factor, accounting for nearly 90% of the observed variation, while temperature and relaxation time play secondary roles. These results emphasize the critical role of tunable graphene properties in controlling the plasmonic resonance and enhancing multigas detection, reinforcing the practical relevance of combining intelligent machine learning optimization with advanced THz sensor design.

7. CONCLUSION

This chapter presented a hybrid copper–graphene metasurface sensor designed for selective multigas detection in the terahertz (THz) regime, combining plasmonic resonance engineering with electrically tunable graphene conductivity and machine learning–based optimization. The proposed structure demonstrated effective frequency reconfiguration, with the resonance shifting from 5.235 THz at μc = 0 eV to 5.265 THz at μc = 0.5 eV, confirming the dominant role of graphene chemical potential in controlling the spectral response. High selectivity was achieved among the investigated gases, with baseline sensitivities reaching 847.5 GHz/RIU for H_2O and 816.3 GHz/RIU for CO_2, while maintaining elevated quality factors up to 872 and figures of merit as high as 139.0 RIU^{-1}. To further enhance performance, an Extreme Gradient Boosting (XGBoost) regression model was implemented, achieving excellent predictive accuracy ($R^2 = 0.998$) and enabling optimized sensitivities of 603.0, 960.7, 1003.2, and 604.5 GHz/RIU for CH_4, CO_2, H_2O, and NH_3, respectively. These results demonstrate the effectiveness of integrating data-driven optimization with plasmonic sensor design, allowing intelligent exploration of multidimensional physical parameter spaces beyond conventional fixed-condition analysis.

Despite these promising outcomes, practical realization of hybrid copper–graphene structures remains technically challenging. Reliable graphene transfer onto metallic resonators, nanoscale patterning precision, and stable electrostatic control of chemical potential require advanced fabrication strategies. Techniques such as chemical vapor deposition (CVD) graphene growth, high-resolution electron-beam lithography, and controlled transfer processes will be essential to ensure structural integrity and reproducibility. Looking ahead, experimental validation of the proposed design, incorporation of multilayer or hybrid two-dimensional materials, and integration into CMOS-compatible platforms could further improve sensitivity, stability, and scalability. Ultimately, the synergy between copper-based plasmonic confinement, graphene's electrically tunable conductivity, and machine learning–assisted optimization establishes a powerful and scalable pathway toward next-generation THz multigas sensors, supporting intelligent environmental monitoring, industrial emission control, and sustainable technology development within advanced smart systems.

REFERENCES

Abdulhussain, S. H., Mahmmod, B. M., Alwhelat, A., Shehada, D., Shihab, Z. I., Mohammed, H. J., Abdulameer, T. H., Alsabah, M., Fadel, M. H., Ali, S. K., Abbood, G. H., Asker, Z. A., & Hussain, A. (2025). A comprehensive review of sensor technologies in IoT: Technical aspects, challenges, and future directions. *Computers*, *14*(8), 342. DOI: 10.3390/computers14080342

Akyildiz, I. F., Han, C., Hu, Z., Nie, S., & Jornet, J. M. (2022). Terahertz band communication: An old problem revisited and research directions for the next decade. *IEEE Transactions on Communications*, *70*(6), 4250–4285. DOI: 10.1109/TCOMM.2022.3171800

Al-Okby, M. F. R., Neubert, S., Röddelkopf, T., & Thurow, K. (2021). Mobile detection and alarming systems for hazardous gases and volatile chemicals in laboratories and industrial locations. *Sensors (Basel)*, *21*(23), 8128. DOI: 10.3390/s21238128 PMID: 34884132

Aliqab, K., Dave, K., Sorathiya, V., Alsharari, M., & Armghan, A. (2023). Numerical analysis of phase change material and graphene-based tunable refractive index sensor for infrared frequency spectrum. *Scientific Reports*, *13*(1), 7653. DOI: 10.1038/s41598-023-34859-5 PMID: 37169848

Bertocci, F., Fort, A., Vignoli, V., Mugnaini, M., & Berni, R. (2017). Optimization of perovskite gas sensor performance: Characterization, measurement and experimental design. *Sensors (Basel)*, *17*(6), 1352. DOI: 10.3390/s17061352 PMID: 28604587

Buckley, D. J., Black, N. C., Castanon, E. G., Melios, C., Hardman, M., & Kazakova, O. (2020). Frontiers of graphene and 2D material-based gas sensors for environmental monitoring. *2D Materials, 7*(3), 032002.

Gounaridis, L., Groumas, P., Schreuder, E., Tsekenis, G., Marousis, A., Heideman, R., Avramopoulos, H., & Kouloumentas, C. (2017). High performance refractive index sensor based on low Q-factor ring resonators and FFT processing of wavelength scanning data. *Optics Express*, *25*(7), 7483–7495. DOI: 10.1364/OE.25.007483 PMID: 28380870

Kang, M., Cho, I., Park, J., Jeong, J., Lee, K., Lee, B., Del Orbe Henriquez, D., Yoon, K., & Park, I. (2022). High accuracy real-time multi-gas identification by a batch-uniform gas sensor array and deep learning algorithm. *ACS Sensors*, *7*(2), 430–440. DOI: 10.1021/acssensors.1c01204 PMID: 35041384

Kasani, S., Curtin, K., & Wu, N. (2019). A review of 2D and 3D plasmonic nanostructure array patterns: Fabrication, light management and sensing applications. *Nanophotonics*, *8*(12), 2065–2089. DOI: 10.1515/nanoph-2019-0158

Krid, H. B., Wertani, H., & Hlali, A.. (2025). Optimization of a hybrid graphene–copper terahertz gas sensor using machine learning. *IEEE Transactions on Plasma Science*. Advance online publication. DOI: 10.1109/TPS.2025.3644784

Krid, H. B., Wertani, H., Hlali, A., & Zairi, H. (2025). High-sensitivity graphene-based terahertz biosensor for cervical cancer diagnosis: SVM-assisted optimization. *IEEE Transactions on Plasma Science*, *53*(12), 4000–4008. DOI: 10.1109/TPS.2025.3632754

Krid, H. B., Wertani, H., Hlali, A., & Zairi, H. (2026). Graphene–copper hybrid terahertz biosensor with KNN sensitivity optimization for early skin cancer (melanoma) detection. *Sensing and Imaging*, *27*(1), 17. DOI: 10.1007/s11220-026-00722-w

Le, D. N., Le, V. H., & Roy, P. (2020). Graphene under uniaxial inhomogeneous strain and an external electric field: Landau levels, electronic, magnetic and optical properties. *The European Physical Journal B*, *93*(8), 158. DOI: 10.1140/epjb/e2020-10222-3

Marshak, A. H., & Van Vliet, C. M. (1984). Electrical current and carrier density in degenerate materials with nonuniform band structure. *Proceedings of the IEEE*, *72*(2), 148–164. DOI: 10.1109/PROC.1984.12836

Mehdary, A., Chehri, A., Jakimi, A., & Saadane, R. (2024). Hyperparameter optimization with genetic algorithms and XGBoost: A step forward in smart grid fraud detection. *Sensors (Basel)*, *24*(4), 1230. DOI: 10.3390/s24041230 PMID: 38400385

Pathak, P., & Srivastava, S. (2026). Smart material design: Integrating data-driven optimization and complexity analysis for next-generation materials. In *Smart Materials Engineering* (pp. 41–63). Data-Driven Approaches and Multiscale Modelling. DOI: 10.1007/978-3-032-09540-4_3

Razzaq, I., Ullah, S., Akram, M., Ashraf, M. W., Shaaban, I. A., Gurbanova, L., Mohammad, A., Shah, S. S. A., & Nazir, M. A. (2025). Graphene-based polymer composites for high-performance chemical sensing and detection: A critical review. *Advanced Materials Technologies*, *10*(18), e00578. DOI: 10.1002/admt.202500578

Saxena, P., & Shukla, P. (2023). A review on recent developments and advances in environmental gas sensors to monitor toxic gas pollutants. *Environmental Progress & Sustainable Energy*, *42*(5), e14126. DOI: 10.1002/ep.14126

Sensale-Rodriguez, B., Yan, R., Kelly, M. M., Fang, T., Tahy, K., Hwang, W. S., Jena, D., Liu, L., & Xing, H. G. (2012). Broadband graphene terahertz modulators enabled by intraband transitions. *Nature Communications*, *3*(1), 780. DOI: 10.1038/ncomms1787 PMID: 22510685

Singh, E., Meyyappan, M., & Nalwa, H. S. (2017). Flexible graphene-based wearable gas and chemical sensors. *ACS Applied Materials & Interfaces*, *9*(40), 34544–34586. DOI: 10.1021/acsami.7b07063 PMID: 28876901

Tao, L., Chen, Z., Li, Z., Wang, J., Xu, X., & Xu, J.-B. (2021). Enhancing light–matter interaction in 2D materials by optical micro/nano architectures for high-performance optoelectronic devices. *InfoMat*, *3*(1), 36–60. DOI: 10.1002/inf2.12148

Wang, J., & Wang, R. (2025). Development of gas sensors and their applications in health safety, medical detection, and diagnosis. *Chemosensors (Basel, Switzerland)*, *13*(5), 190. DOI: 10.3390/chemosensors13050190

Wang, L., Cheng, Y., Gopalan, S., Luo, F., Amreen, K., Singh, R. K., Goel, S., Lin, Z., & Naidu, R. (2023). Review and perspective: Gas separation and discrimination technologies for current gas sensors in environmental applications. *ACS Sensors*, *8*(4), 1373–1390. DOI: 10.1021/acssensors.2c02810 PMID: 37074840

Xia, X., Wang, Y., Zhong, Z., & Weng, G. J. (2017). A frequency-dependent theory of electrical conductivity and dielectric permittivity for graphene-polymer nanocomposites. *Carbon*, *111*, 221–230. DOI: 10.1016/j.carbon.2016.09.078

Zong, B., Wu, S., Yang, Y., Li, Q., Tao, T., & Mao, S. (2025). Smart gas sensors: Recent developments and future prospective. *Nano-Micro Letters*, *17*(1), 54. DOI: 10.1007/s40820-024-01543-w PMID: 39489808

Chapter 8
Artificial Intelligence in Agriculture and Food Security:
Remote Sensing and Yield Estimation

Gulay Ozkan
http://orcid.org/0000-0001-6878-1673
Faculty of Agricultural Sciences and Technologies, Department of Agricultural Economcis and Agribusiness, Yasar University, Turkey

ABSTRACT

This chapter explores the expanding applications of artificial intelligence in agriculture, with a particular focus on remote sensing-based crop monitoring and yield estimation. It reviews how machine learning and deep learning models are used for crop classification, stress detection, and seasonal forecasting, highlighting their growing role in improving the spatial and temporal accuracy of agricultural assessments. Beyond technical performance, the chapter examines how these tools relate to broader food security concerns. While AI-driven systems can directly strengthen production monitoring and early warning capacity, their broader impact on access, utilization, agency, and sustainability depends on data infrastructure, institutional integration, and governance capacity. By combining technological analysis with structural and policy perspectives, the chapter argues that meaningful food security gains require more than algorithmic advancement; they depend on how predictive systems are embedded within inclusive and resilient agrifood systems.

DOI: 10.4018/979-8-3373-8745-1.ch008

1. INTRODUCTION

1.1 Artificial Intelligence, Remote Sensing, and the Datafication of Agriculture

Over the past decade, agriculture has undergone a profound transformation driven by advances in artificial intelligence (AI), remote sensing technologies, and digital data infrastructures. High-resolution satellite imagery, long-term Earth observation archives, unmanned aerial vehicles, and sensor-based field measurements have enabled agricultural landscapes to be monitored with unprecedented spatial and temporal detail. These developments have contributed to the rapid datafication of agriculture, in which farming systems are increasingly represented, managed, and evaluated through digital data streams (Olson, & Anderson, 2021).

This expansion of data availability has coincided with growing expectations that data-driven approaches can address persistent challenges facing global agriculture, including climate variability, land degradation, population growth, and food insecurity. In particular, remote sensing-based monitoring is often presented as a scalable and objective means of observing crop conditions, estimating yields, and supporting early warning systems at regional and national levels. However, while the quantity and diversity of agricultural data have increased substantially, the capacity to translate these data into meaningful and actionable knowledge has not kept pace.

Traditional analytical approaches commonly used in agricultural remote sensing such as linear regression models, index-based thresholds, and rule-based classifications are often ill-equipped to handle the complexity of contemporary datasets. These methods struggle with non-linear relationships, high-dimensional feature spaces, and the integration of multi-source, multi-temporal data. As a result, large volumes of potentially valuable information remain underexploited, and monitoring outputs frequently fail to meet the needs of decision-makers operating under conditions of uncertainty.

AI-based methods, particularly machine learning and deep learning models, have emerged as prominent responses to these analytical limitations. By learning patterns directly from data, AI approaches offer new possibilities for automated crop classification, phenological stage detection, stress and disease identification, and yield estimation across diverse agro-ecological contexts. Empirical studies consistently report improvements in predictive accuracy when AI models are applied to remote sensing data, reinforcing their growing prominence within the digital agriculture literature (Kamilaris & Prenafeta-Boldú, 2018). Consequently, AI is increasingly framed as a key technological driver of precision agriculture and large-scale agricultural monitoring systems.

1.2 From Technical Performance to Knowledge Production

Despite these technical advances, much of the existing literature on AI in agricultural remote sensing remains narrowly focused on model performance. Evaluation metrics such as classification accuracy, root mean square error, and algorithmic efficiency dominate scholarly discussions, often presented as indicators of progress in and of themselves. While such metrics are necessary for methodological comparison, they provide only a partial view of how AI systems function within broader agricultural knowledge ecosystems.

This chapter adopts a different perspective by treating AI not merely as a set of computational tools but as an integral component of agricultural knowledge production. From this standpoint, AI models are embedded within specific data infrastructures, methodological assumptions, and institutional contexts that shape what can be known, measured, and acted upon (Ziv & Nakash, 2025). Choices related to training data, spatial and temporal resolution, model architecture, and validation strategies all influence how agricultural realities are represented and interpreted.

Importantly, AI-driven representations of agriculture are not neutral reflections of biophysical conditions. They privilege certain variables, scales, and forms of evidence while rendering others invisible. For example, satellite-based yield models may capture aggregate production trends while overlooking intra-field variability, informal farming practices, or socio-economic constraints affecting food access. As a result, AI outputs can reinforce particular understandings of agricultural performance while marginalizing alternative forms of knowledge.

These issues become especially salient when AI-based monitoring systems are integrated into policy and governance processes. Yield forecasts, drought indicators, and risk assessments derived from remote sensing models increasingly inform decisions related to food aid allocation, insurance schemes, and agricultural investment (Benami et al., 2021). In such contexts, the authority of AI-generated outputs can obscure underlying uncertainties and assumptions, creating a perception of objectivity that is not always warranted. Recognizing AI as an epistemic actor, one that actively shapes agricultural knowledge rather than passively delivering it, is therefore essential for a critical assessment of its role in contemporary agriculture.

1.3 Rethinking the Link Between Yield Estimation and Food Security

The implications of AI-driven agricultural monitoring extend beyond technical performance to broader questions of food security. Food security is widely understood as a multidimensional concept encompassing availability, access, utilization, and stability over time. Nevertheless, within much of the AI and remote sensing literature, improvements in yield estimation are implicitly equated with advances in food security. This assumption risks oversimplifying the complex pathways through which agricultural production data translate into social and nutritional outcomes.

In recent conceptual developments, food security frameworks have expanded beyond the traditional four pillars to include two additional dimensions: agency and sustainability. Agency refers to the capacity of individuals and communities to make autonomous decisions regarding food production, access, and consumption. Sustainability emphasizes the long-term resilience of agrifood systems under ecological, climatic, and socio-economic pressures. These expanded dimensions highlight that food security is not only a question of aggregate production or distribution efficiency but also of empowerment, environmental continuity, and intergenerational stability.

While accurate yield estimates can contribute to improved planning and risk management, increased production alone does not guarantee enhanced food security. Factors such as market access, income distribution, infrastructure, and political stability play equally critical roles. Moreover, the benefits of AI-based monitoring are unevenly distributed across regions. AI models are predominantly developed and validated in data-rich contexts, where long-term ground observations and high-quality satellite data are readily available. In contrast, data-scarce regions, many of which face the most severe food security challenges, often lack the infrastructural and institutional capacity required to fully leverage these technologies.

In these settings, data gaps and limited validation opportunities are frequently framed as technical obstacles to be overcome through further data collection or model refinement. However, such framings tend to overlook how data inequalities shape both knowledge production and policy design. When AI models trained in specific contexts are transferred to different agro-ecological or socio-economic environments, their reliability and relevance may be compromised. This raises critical questions regarding scale compatibility, model generalizability, and the governance of AI-based agricultural information.

The rationale of this chapter is therefore not only to review AI-powered remote sensing applications in agriculture but also to critically examine the conditions under which these applications can meaningfully support food security objectives. By situating technical developments within a broader analytical framework that includes data inequality, scale, and decision-making processes, the chapter aligns with the interdisciplinary mission of the book. It sets the stage for subsequent sections that explore methodological approaches, empirical applications, and governance implications in greater detail.

2. ANALYTICAL FRAMEWORK OF AI-POWERED REMOTE SENSING APPLICATIONS

2.1 Objective

AI-powered remote sensing applications have achieved substantial technical progress in analyzing the spatial and temporal dynamics of agricultural production. Advances in machine learning and deep learning have significantly improved performance in tasks such as crop type classification, phenological cycle monitoring, stress detection, and crop yield estimation, thereby reshaping contemporary agricultural knowledge production processes (Kamilaris & Prenafeta-Boldú, 2018).

Despite these advances, the dominant body of literature continues to evaluate AI-based agricultural applications primarily through model-centric performance metrics, including accuracy, error rates, and computational efficiency. While such evaluations are necessary for methodological comparison, they offer a limited perspective on the broader implications of AI deployment in agricultural systems. In particular, they tend to overlook how AI-generated information is embedded within decision-making processes, constrained by infrastructural inequalities, and shaped by governance arrangements.

The objective of this chapter is to move beyond a purely technical assessment of AI-powered remote sensing methods by situating them within their broader social, economic, and institutional contexts. The chapter systematically evaluates how AI applications contribute to agricultural monitoring and yield estimation, while critically examining structural conditions such as scalability, data infrastructure, institutional capacity, and policy integration. In doing so, it conceptualizes AI not merely as a predictive tool, but as a transformative actor in agricultural knowledge production and governance (Leonelli, 2020).

The central argument advanced is that, although AI-based remote sensing systems deliver significant technical gains, their sustainable and equitable contribution to food security depends on addressing data inequalities, scale alignment, and governance integration beyond model performance alone.

2.2 Research Questions

This chapter is structured around the following research questions:

What methodological innovations do AI-assisted remote sensing approaches provide for crop area determination, phenology monitoring, and yield estimation?

Why are high accuracy levels achieved at local scales often difficult to sustain at national or regional scales?

How do data inequalities and infrastructural disparities affect the global applicability of AI-based agricultural monitoring systems?

Which dimensions of food security do AI-based yield estimation models contribute to, directly or indirectly?

What institutional and governance conditions are necessary for integrating AI systems into effective decision-support mechanisms?

Together, these questions bridge the gap between technical performance and broader structural, policy, and governance outcomes, highlighting areas that remain underexplored in existing research.

2.3 Methodological Framework

This chapter adopts an analytical and comparative literature synthesis approach rather than presenting original empirical or experimental data. The methodological framework consists of three complementary components.

2.3.1 Systematic Literature Review

A systematic review was conducted of peer-reviewed studies published between 2020 and 2025 in journals indexed in Web of Science and Scopus. This period captures recent developments in AI and remote sensing technologies, particularly the rapid diffusion of deep learning and hybrid modelling approaches in agricultural applications.

Key search terms included "AI in agriculture," "deep learning crop classification," "remote sensing yield prediction," "food security and satellite data," and "machine learning scalability in agriculture." Priority was given to studies that combined applied case analyses with discussions of scalability, generalizability, or food security implications, thereby moving beyond literature focused solely on algorithmic accuracy.

2.3.2 Comparative Analysis

A comparative analytical approach contrasts applications developed in data-rich regions such as the United States and European Union countries with those implemented in data-constrained contexts, including Sub-Saharan Africa and South Asia. This comparison demonstrates that AI model performance is shaped not only by algorithmic design but also by institutional capacity, data availability, and monitoring infrastructure.

2.3.3 Integration with Policy and Governance Documents

To assess the relevance of AI-based monitoring systems for food security, insights from the technical literature are correlated with policy reports and institutional documents published by organizations such as the Food Agriculture Organization (FAO), the World Bank, and the Famine Early Warning Systems Network (FEWS NET). This step evaluates the extent to which AI-generated information can inform agricultural policy and decision-making, highlighting governance considerations alongside technical performance.

2.4 Scope and Limitations

This chapter does not include detailed mathematical formulations of AI algorithms, such as neural network architectures or hyperparameter optimization procedures, nor does it present new field experiments at the national or farm scale. Instead, it offers a conceptual, comparative, and critical analysis that links technical developments in AI-based remote sensing to broader political, institutional, and governance frameworks. The chapter is therefore positioned as an analytical and theoretical contribution rather than an operational or technical manual.

2.5 Contribution to the Literature

The contributions of this chapter are fourfold. First, it evaluates AI applications in agriculture beyond technical accuracy, emphasizing the interplay between algorithmic performance, scalability, and governance conditions. Second, by explicitly comparing data-rich and data-scarce regions, it highlights how infrastructural and epistemic inequalities shape both technical outcomes and knowledge production regimes. Third, crop yield estimation is contextualized within the multidimensional framework of food security, encompassing production, access, distribution, and stability, thereby clarifying the indirect and conditional nature of AI's contribution to food security (FAO, 2021). Finally, the chapter adopts an interdisciplinary perspective that bridges technical research with sustainability and development discourse, positioning AI as a constitutive element of agricultural knowledge production rather than merely a predictive instrument.

3. AI-BASED AGRICULTURAL MONITORING APPLICATIONS

AI has transformed the capacity for monitoring agricultural landscapes by integrating multi-sensor satellite imagery, machine learning algorithms, and temporal analysis methods. The primary applications include crop area determination, phenology monitoring, and stress or disease detection. These applications rely on diverse data sources, from optical and radar satellites to in situ observations (Balsamo et al., 2018), and employ a variety of machine learning models such as Convolutional Neural Networks (CNNs), Random Forests (RF), Long Short-Term Memory (LSTM) networks, Temporal Convolutional Networks (TCN), Autoencoders, and Generative Adversarial Networks (GANs) Bhattacharyya et al. 2023; Gao et al 2022)

The purpose of this section is to critically evaluate these AI-based applications, compare methodologies, and assess limitations related to data availability and infrastructure. The discussion is supported with representative tables summarizing model types, input data, regions, and performance metrics. These tables not only highlight technical performance but also provide insights into methodological choices in diverse agricultural contexts.

3.1 Crop Area Determination and Classification

Determining crop areas accurately is fundamental for yield estimation, agricultural planning, and policy design. Optical satellite imagery from Sentinel-2 and Landsat 8, as well as coarse-resolution datasets from MODIS, are widely used for land cover classification and crop identification (Dhillon et al 2023).

Machine learning methods applied to these datasets vary in complexity. CNNs are effective in capturing spatial and spectral patterns in high-resolution imagery, enabling accurate differentiation between crop types in heterogeneous landscapes. RFs and Support Vector Machines (SVMs) are often used for multi-class classification tasks where computational simplicity and interpretability are preferred (Belgiu & Drăguţ, 2016). The accuracy of these approaches depends on data resolution, temporal coverage, and the availability of ground-truth observations.

Table 1 summarizes representative studies in crop area determination, showing the region, data source, model type, and reported classification accuracy. Direct reference to Table 1 indicates that high-resolution CNN-based models often outperform traditional RF approaches in heterogeneous regions, while RF remains competitive in homogeneous agricultural zones.

Table 1. Crop Area Determination Accuracy

Region / Country	Satellite Data	Method	Accuracy / Performance	Reference
Italy, North	Sentinel-2	R-CNN	96.5	Mazzia et al., 2020
India (Sen2Agri)	Sentinel-2	Random Forest	~70 %	Vijayasekaran, 2019
NW China	GF-1, Sentinel-2, Landsat 8	Random Forest	96–98 %	Fan et al., 2021
Southern Kazakhstan	Sentinel-2	RF vs SVM comparison	~91 %	Arystanov et al 2025
EU (NUTS scale)	Sentinel-1	Random Forest	~80 %	d'Andrimont et al., 2021
Turkey (Potato)	Sentinel-2	Random Forest	~94 %	Tokluoğlu et al. 2026
Turkey (agricultural and other fields)	*Sentinel-2 and Landsat-8*	Random Forest/ SVM	%89.5	Tuğaç et al., 2024
US Midwest	Sentinel-2	CNN	~91%	Zhong & Zhou, 2020.
Malawi, Africa	Sentinel-2+ PlanetScope+SAR	Multisensor fused	Improved overall accuracy*	Kpienbaareh et al., 2021

Note. Improved overall accuracy (OA) refers to performance gains compared to single-sensor classification approaches as reported by the authors.

3.2 Phenology Monitoring

Monitoring crop phenology involves detecting the timing of developmental stages, such as flowering or maturity, which are critical for yield prediction and irrigation planning. Multi-temporal satellite datasets combined with time-series analysis are particularly useful for this purpose. Models such as LSTM and TCN have shown strong capacity to learn temporal dependencies, capturing delayed or cumulative stress responses in crops

Phenological monitoring enables early warning of developmental anomalies due to climatic stress or management practices. For instance, LSTM models applied to Sentinel-2 data in European wheat fields successfully predicted flowering dates with errors under five days, allowing adaptive management decisions (Mancini et al., 2024). Similarly, TCNs applied in rice cultivation in China distinguished multiple phenological stages in double cropping systems with over 90% temporal accuracy (Sheng et al., 2024).

Table 2 illustrates phenology detection studies, highlighting the crop, region, model, and detection accuracy. Direct reference to Table 2 underscores that models integrating multi-temporal datasets significantly outperform static classification methods

Table 2. Comparative Performance Metrics in AI-Based Crop Phenology Monitoring

Crop / Focus	Region	Data & Model	Performance Metric	Source
Winter wheat & corn	Canada	Sentinel-2+time-series ML	RMSE < 4 days	Liao et al., 2023
Maize	Mexico	Sentinel-2+SVM	~82% classification accuracy	Murguia-Cozar et al., 2021
Maize	Global	Planet Fusion+DTW	90% timing within 10 days	Senaras et al., 2024
Rice	China	Sentinel-1/2 time series	High phenology detection potential	Wang et al., 2022
Wheat	Germany	Optical+radar fusion+ML	$R^2 > 0.9$; MAE < 2 days	Shojaeezadeh et al., 2024

3.3 Stress and Disease Detection

AI-based stress and disease detection focuses on identifying anomalies such as drought, nutrient deficiency, or fungal infections. (Matabber et al., 2026) These anomalies manifest in spectral deviations that can be detected using both optical and thermal sensors. Hybrid deep learning models, including CNN + LSTM for spatiotemporal analysis, Autoencoders for anomaly detection, and GANs for learning normal crop phenology distributions, provide effective solutions.

Recent studies demonstrate that deep learning and machine learning models are effective for early stress detection. For example, in Central Asia, hyperspectral imagery combined with machine learning enabled early detection of wheat diseases with classification accuracy exceeding 90% (Ualiyeva et al., 2025). In Mediterranean wheat systems, multi-source remote sensing data integrated with ensemble learning methods successfully detected powdery mildew infection at early stages (Song et al., 2026). Similarly, drought and stress classification studies using multisource spectral data and machine learning have demonstrated strong performance in cereal crops.

Table 3 summarizes representative and verifiable studies.

Table 3. AI-Based Crop Stress and Disease Detection in Remote Sensing Applications

Region	Product	Stress Type	Data Source	Model	Reported Accuracy	Source
US Midwest	Maize	Drought stress (maize)	Sentinel-2 time series	CNN + LSTM	>90% accuracy	Bounoua et al., 2024
Canada (Prairie Region)	Maize / Wheat	Phenological deviation associated with stress	Sentinel-2 time series	Deep temporal modeling (CNN-LSTM style)	High within-field detection performance	Liao et al., 2023
Kazakhstan	Wheat	Early disease-related spectral signatures	VNIR hyperspectral + ML	ML classification (RF, SVM)	~90% classification accuracy	Ualiyeva et al., 2025
China	Wheat	Early powdery mildew detection	Multisource remote sensing	Ensemble ML approaches	High early detection capability	Song et al., 2026

3.4 Methodological Comparison & Critical Evaluation

AI models differ in capabilities and limitations. CNNs excel in high-resolution spatial pattern recognition but require substantial labeled data. LSTM and TCN models capture temporal dependencies, essential for phenology and stress dynamics (Utku, 2025). Autoencoders and GANs provide adaptive anomaly detection in data-limited contexts.

Table 4 provides a comprehensive methodological comparison of AI model architectures for agricultural monitoring. It integrates conventional machine learning models, deep learning models, and hybrid approaches (APSIM–DL Hybrid and Transfer Learning), highlighting their relative advantages, limitations, and suitable application contexts.

Table 4. Comparative Evaluation of AI Model Architectures in Agricultural Monitoring

Model Type	Advantages	Disadvantages	Best Application Context
Random Forest	Robust with small datasets, low tuning requirement, interpretable	Limited temporal modeling capacity	Crop classification in medium-resolution data
CNN	High spatial resolution, accurate classification	Data-intensive, low temporal capture	High-resolution crop mapping
LSTM / TCN	Temporal pattern recognition, stress detection	Computationally demanding requires long time-series	Phenology & yield forecasting

continued on following page

Table 4. Continued

Model Type	Advantages	Disadvantages	Best Application Context
CNN + LSTM	Captures spatiotemporal anomalies	High complexity, training cost	Stress detection & yield modeling
Autoencoder / GAN	Adaptive anomaly detection, novel stress identification	Limited interpretability, sensitive to noise	Anomaly/stress detection in data-limited contexts
Transformer-based models	Long-range temporal dependency capture, scalable	Very data-hungry, high computational cost	Multi-temporal monitoring; forecasting
APSIM–DL Hybrid	Biophysical integration; climate robustness	Calibration complexity	Yield prediction under variability
Transfer Learning	Cross-region adaptation	Source–target mismatch risk	Data-constrained regions

Despite the technical success of these models, practical challenges remain. Many studies assume regular data availability and high-quality satellite coverage, which is often not the case in developing regions. Furthermore, integration with local agricultural practices, phenology ground-truthing, and decision support systems (DSS) is frequently underexplored.

This section has demonstrated how AI models interpret agricultural data, classify crops, monitor phenology, and detect stresses. The next section discusses yield estimation models, emphasizing how these inputs interact with hybrid modeling approaches, scale considerations, and data constraints to support decision-making and food security.

4. TECHNICAL MODELING AND SCALABILITY IN AI-BASED YIELD PREDICTION

Accurate crop yield estimation is central to agricultural planning, food security forecasting, input allocation, and market stabilization. AI-based yield prediction models increasingly integrate multi-temporal satellite imagery, meteorological data, soil information, and agronomic variables to model complex nonlinear relationships between environmental conditions and crop productivity. These approaches range from pure deep learning architectures to hybrid systems incorporating process-based crop simulation models such as APSIM (Agricultural Production System Simulator) and DSSAT (Decision Support System for Agro Technology Transfer) (Ahmed et al., 2022).

While the predictive accuracy of these systems has improved substantially over the past decade, model performance remains strongly dependent on spatial scale, environmental heterogeneity, and data infrastructure. This section synthesizes current modeling approaches and examines their scalability limitations, arguing that predictive robustness is not solely a computational issue but a structural one.

4.1 Pure Deep Learning Models: Predictive Power and Context Dependency

Pure deep learning models operate without explicit incorporation of agronomic process knowledge. The most widely used architectures include CNN, LSTM networks, and hybrid CNN–LSTM configurations.

CNNs are particularly effective in extracting spatial features from high-resolution satellite imagery such as Sentinel-2 or Landsat data. They identify spectral and textural patterns associated with crop growth stages, biomass accumulation, and stress signals. LSTMs, in contrast, are designed to capture temporal dependencies and are especially suitable for modeling crop development trajectories across growing seasons (Kamilaris & Prenafeta-Boldú, 2018).

In homogeneous and data-rich agricultural systems, such as large-scale maize production regions in the United States, deep learning models that integrate spatial and temporal structures (e.g., LSTM and hybrid CNN-LSTM architectures) are widely used in remote sensing–based yield prediction applications. Systematic reviews find that CNNs and LSTMs are among the most frequently applied deep learning approaches in crop yield prediction, often achieving high accuracy when trained on extensive multi-year satellite and environmental datasets (Muruganantham et al., 2022). Similarly, CNN-LSTM models have demonstrated high predictive performance in intensively managed rice systems, where spatial and temporal redundancy in data supports robust pattern learning (Zhou et al., 2023).

However, predictive stability tends to decrease significantly in heterogeneous systems characterized by small and fragmented field structures, mixed cropping practices, variable irrigation regimes, and sparse ground-truth observations. The generalization ability of deep learning models diminishes when training data is limited or unevenly distributed across diverse landscape contexts, leading to declines in performance and accuracy (Joshi et al., 2023). This degradation reflects sensitivity to limited training data, spectral confusion between adjacent plots, and insufficient representation of agronomic variability. Thus, while pure deep learning models demonstrate strong capacity for nonlinear pattern recognition, their robustness is highly contingent upon data density and environmental homogeneity.

4.2 Hybrid Architectures and Process-Based Integration

Hybrid modeling approaches seek to improve generalizability by integrating deep learning architectures with mechanistic crop simulation models. Process-based systems such as APSIM and DSSAT simulate crop growth using physiological relationships among temperature, radiation, soil moisture, and nutrient availability. When integrated with deep learning frameworks, these models embed biophysical constraints into data-driven systems. This integration improves robustness under atypical climatic conditions and reduces overfitting to region-specific spectral patterns.

Empirical studies demonstrate that APSIM–LSTM and DSSAT–CNN hybrid models often outperform pure deep learning approaches, particularly under climate variability, drought conditions, inter-annual yield fluctuations. Hybrid systems improve performance by constraining predictions within agronomically plausible ranges. Instead of learning relationships solely from correlations in satellite data, they incorporate process-level understanding of crop phenology and stress response. However, hybridization introduces additional calibration complexity. Process-based models require detailed soil, management, and cultivar parameters, which may not be readily available in data-scarce regions. Therefore, hybrid models improve predictive stability but remain dependent on institutional data infrastructure.

4.3 Spatial and Temporal Scalability in AI-Based Yield Prediction

Scalability in AI-based yield prediction should not be treated as a singular or purely computational attribute. Within the context of technical modeling, scalability operates primarily across two interconnected dimensions: spatial and temporal. Both dimensions determine whether predictive performance remains stable when models are applied beyond the specific environmental and historical conditions under which they were trained.

4.3.1 Spatial Scalability

Spatial scalability refers to the capacity of a predictive model trained in one agro-ecological region to maintain performance when applied to geographically distinct environments. While deep learning architectures demonstrate strong pattern-recognition capacity, their generalization ability is often constrained by environmental heterogeneity (van Klompenburg et al., 2020).

Differences in soil composition, cropping calendars, irrigation regimes, and management intensity introduce variability that may not be sufficiently represented in the training dataset. As spatial coverage expands, the probability increases that model inputs fall outside the distribution of the original training data. This distributional shift leads to performance degradation, particularly in heterogeneous landscapes characterized by smallholder fragmentation and mixed cropping systems (Lesiv et al., 2019).

Sensor resolution further affects spatial scalability. In regions where field sizes are smaller than the spatial resolution of satellite pixels, spectral mixing can obscure crop-specific signals, reducing classification accuracy and yield estimation reliability. Even high-resolution imagery may not fully eliminate this aggregation bias when land-use patterns are highly fragmented.

Transfer learning methods offer partial solutions by adapting pretrained models to new regions using limited local calibration data. However, empirical evidence suggests that transfer effectiveness is strongly conditioned by agro-climatic similarity between source and target domains (Kamilaris & Prenafeta-Boldú, 2018). Consequently, spatial scalability remains probabilistic rather than guaranteed.

4.3.2 Temporal Scalability

Temporal scalability concerns the stability of predictive accuracy across multiple growing seasons. AI-based yield prediction models are typically trained on historical multi-year datasets that reflect specific climatic distributions. When climatic conditions deviate substantially from historical norms, such as during droughts, floods, or extreme heat events, model performance may decline (Reichstein et al., 2019).

This vulnerability reflects the dependence of supervised learning systems on historical data distributions. If extreme events are underrepresented in training datasets, models may fail to extrapolate accurately under atypical conditions. Temporal instability therefore arises from both data limitations and non-stationarity in climatic systems.

Hybrid architectures integrating process-based crop simulation models (e.g., APSIM or DSSAT) improve temporal robustness by embedding physiological growth constraints into predictive frameworks. Such integration helps restrict predictions to agronomically plausible ranges under stress conditions. Nonetheless, even hybrid systems require periodic recalibration to maintain performance under shifting climatic baselines.

Temporal scalability is therefore conditional upon sustained data updating, retraining cycles, and representation of climatic variability within training datasets. Without these adjustments, predictive accuracy may erode over time as environmental conditions evolve.

4.4 Scale Decay and Performance Divergence

A recurring finding in the empirical literature on AI-based yield prediction is the phenomenon of scale decay, whereby predictive performance deteriorates as models are transferred from controlled experimental environments to broader spatial contexts. This decline is not incidental but systematic, reflecting structural characteristics of agricultural heterogeneity rather than isolated modeling errors (van Klompenburg et al., 2020).

At the local scale, where training and validation datasets are derived from homogeneous agro-ecological zones and consistent management regimes, deep learning architectures frequently achieve prediction accuracies exceeding 85–90%. However, when the same models are extended to regional or national scales, performance commonly declines by 10–25 percentage points. This degradation is particularly visible in landscapes characterized by fragmented smallholder plots, mixed cropping systems, variable irrigation practices, and inconsistent agronomic reporting (Lesiv et al., 2019).

Scale decay is driven by several interrelated factors. First, environmental heterogeneity increases variance in soil properties, microclimates, and crop phenology, reducing the representativeness of training datasets. Second, sensor resolution mismatches may introduce spatial aggregation errors when small plots fall below pixel-level detectability thresholds. Third, ground-truth data used for supervised learning often become sparse or temporally inconsistent at larger scales, weakening validation reliability.

Transfer learning techniques partially address scale decay by adapting pretrained models to new agro-ecological contexts using limited local calibration data. While such approaches reduce the need for large labeled datasets, their effectiveness depends on ecological similarity between source and target regions. When agro-climatic conditions diverge substantially, transfer performance remains constrained (Kamilaris & Prenafeta-Boldú, 2018).

Hybrid modeling frameworks integrating crop simulation systems such as APSIM or DSSAT further mitigate scale decay by embedding physiological growth constraints into machine learning architectures. These constraints reduce overfitting to region-specific spectral correlations and improve generalization across climatic variability. Nevertheless, hybrid models require detailed parameterization of soil, management, and cultivar information, limiting applicability in data-scarce settings.

Scale decay therefore reveals a fundamental limitation of algorithmic scalability: predictive robustness depends not only on model complexity but also on the spatial representativeness of observational infrastructures. Technical transferability is achievable, but conditional upon data continuity, environmental similarity, and calibration depth.

4.5 Structural Limits of Technical Scalability

Beyond scale decay and transfer instability, AI-based yield prediction systems face structural limits rooted in data infrastructure, computational constraints, and model dependency on supervised learning environments. These limits define the outer boundary of technical scalability.

First, predictive models are inherently dependent on high-quality, multi-temporal input data. Although satellite programs such as Sentinel and Landsat have expanded open-access imagery, effective yield modeling requires integration with ground-based yield statistics, meteorological records, and soil property databases. Incomplete or inconsistent field observations reduce label reliability and introduce systematic bias into supervised learning processes (Lungu et al., 2020).

Second, computational scalability poses practical limitations. Deep learning architectures, particularly Transformer-based and hybrid CNN–LSTM models, require substantial processing power and memory capacity when applied to national-scale time-series datasets. Cloud-based platforms alleviate some constraints, yet computational demand increases nonlinearly with spatial coverage and temporal resolution. This creates trade-offs between spatial extent, temporal granularity, and model complexity.

Third, model interpretability remains a technical constraint. High-performing deep learning systems often operate as "black boxes," limiting transparency in feature attribution and uncertainty quantification. While explainable AI techniques are emerging, uncertainty propagation across scales remains insufficiently standardized in agricultural forecasting contexts (Kamilaris & Prenafeta-Boldú, 2018).

Fourth, calibration dependency restricts autonomous scalability. Both pure and hybrid models require periodic retraining to maintain temporal stability, especially under climatic anomalies such as droughts or heatwaves. Even process-integrated systems such as APSIM–DL hybrids require recalibration when cultivar characteristics or management regimes change.

Collectively, these factors demonstrate that technical scalability is structurally bounded. Algorithmic sophistication can improve prediction accuracy, yet scalability remains contingent upon sustained data ecosystems, computational infrastructure, and periodic model maintenance. These limits do not invalidate AI-based yield forecasting but underscore that modeling performance is embedded within infrastructural capacity.

5. DATA INEQUALITIES, GOVERNANCE, AND FOOD SECURITY

The effectiveness of AI-based agricultural monitoring and yield prediction systems is strongly influenced by data availability, infrastructure quality, and institutional governance. While technological advancements such as CNNs, LSTMs, and hybrid models provide high prediction accuracy in controlled or data-rich environments, these gains do not automatically translate into improved food security, particularly in the Global South (Kamilaris & Prenafeta-Boldú, 2018).

This section examines the impacts of data inequalities, governance constraints, and socio-economic factors on the real-world applicability of AI-based systems. It highlights the relationship between technical accuracy, structural capability, and equitable food security outcomes, using representative tables to synthesize findings across regions and crops.

5.1 Data Infrastructure and Access Issues

High-quality, multi-temporal remote sensing data, field observations, and weather records are essential for training AI models. In many regions of the Global South, data scarcity, irregular field surveys, and limited sensor coverage significantly reduce model accuracy and generalizability (Li et al., 2025).

For instance, maize production systems in Ethiopia are affected by low-resolution satellite imagery, frequent cloud cover, and irregular field monitoring, which reduce the generalizability and robustness of deep learning models (van Klompenburg et al., 2020). In Bangladesh, rice–wheat double cropping systems with small and fragmented plots create classification challenges for purely spatial CNN models. Studies indicate that hybrid approaches integrating crop simulation models with machine learning can partially mitigate these limitations under data-constrained environments (Lesiv et al., 2019).

Table 5 summarizes representative challenges in data infrastructure and their effects on model performance. Direct reference to Table 5 highlights how data limitations constrain AI applicability.

Table 5. Data Infrastructure Constraints and AI Applicability

Region	Crop System	Primary Data Constraint	Model Implication	Structural Outcome	Sources
Ethiopia	Maize	Low-resolution imagery, sparse field surveys	Reduced model generalization	Limited scalability	van Klompenburg et al., 2020
Bangladesh	Rice–Wheat	Small fragmented plots, phenological overlap	CNN misclassification risk	Need for hybrid modeling	Lesiv et al., 2019
Sub-Saharan Africa	Sorghum	Weather station scarcity	Weak temporal reliability	High prediction uncertainty	Reichstein et al., 2019

continued on following page

Table 5. Continued

Region	Crop System	Primary Data Constraint	Model Implication	Structural Outcome	Sources
Brazil (Amazon)	Soybean	Persistent cloud contamination	Temporal data gaps	Reduced monitoring continuity	Zhu et al., 2015
China	Rice	Dense multi-source data	Effective hybrid modeling (APSIM + ML)	High transferability	Reichstein et al., 2019
EU (France)	Wheat	Integrated satellite + field databases	Supports advanced models	Scalable AI deployment	Reichstein et al., 2019

5.2 Global Asymmetry and Model Generalizability

Most AI models for crop monitoring are developed and validated in data-rich regions, such as the US Corn Belt, European grain-producing areas, or China's rice regions. Applying these models in data-constrained **contexts** reveals significant performance asymmetries, as the availability and quality of remote sensing and field data directly affect model reliability (Kamilaris & Prenafeta-Boldú, 2018; van Klompenburg et al., 2020).

For example, hybrid CNN-LSTM models demonstrate high generalization and robust performance in maize systems of the US Corn Belt, whereas in smallholder maize systems in Sub-Saharan Africa, the same models exhibit reduced generalization and increased uncertainty due to sparse ground surveys and low-resolution imagery. Similarly, rice–wheat systems in Bangladesh face misclassification risks because of small, fragmented plots and irregular field data. Integrating process-based simulation models with machine learning can mitigate some limitations, improving model robustness in constrained settings, but infrastructure gaps cannot be fully compensated.

Table 6. Model Performance Across Data-Rich and Data-Constrained Regions

Region	Crop System	Model	Data Context	Observed Trend / Limitation	Sources
EU (France)	Wheat	Transformer	Data-rich	Supports scalable modeling	Reichstein et al., 2019
Ethiopia	Maize	CNN-LSTM	Data-constrained	Reduced generalization; limited scalability	van Klompenburg et al., 2020
Bangladesh	Rice/ Wheat	CNN	Data-constrained	Misclassification risk; small plot sizes	Lesiv et al., 2019
Bangladesh	Rice/ Wheat	CNN-LSTM + Simulation	Data-constrained	Improved performance with hybrid modeling	Jabed et al., 2024
Sub-Saharan Africa	Sorghum	LSTM	Data-constrained	Weak temporal reliability; high uncertainty	Reichstein et al., 2019
Brazil (Amazon)	Soybean	CNN	Data-constrained	Temporal gaps; reduced monitoring continuity	Zhu et al., 2015

Table 6 presents a comparative overview of model generalizability across data-rich and data-constrained regions. This direct reference emphasizes the uneven technical outcomes caused by global data asymmetries.

5.3 Ethical and Political Dimensions

Data inequalities extend beyond technical limitations; they are ethical and political issues. Limited access to agricultural data can restrict local decision-making and strengthen knowledge monopolies held by large technology companies or international organizations (Leonelli, 2020).

For instance, AI yield predictions may inform regional or national policy but fail to reach smallholder farmers without adequate local infrastructure. Lack of alignment between AI outputs and local practices can amplify food insecurity, as increased production data does not automatically improve distribution, access, or utilization (FAO, 2021).

5.4 Discussion and Implications

Data inequalities critically determine whether AI-based agricultural systems can contribute to sustainable and equitable food security. Even highly accurate models may fail to improve outcomes without:

- Adequate data infrastructure (satellite imagery, field observations)
- Effective decision support systems

- Institutional capacity for policy integration
- Consideration of local socio-economic and logistical contexts

Hybrid models with process-based integration and transfer learning partially mitigate data scarcity effects, but structural constraints remain the primary limitation in the Global South (Kamilaris & Prenafeta-Boldú, 2018).

Having analyzed the impacts of data inequalities and governance on food security, the next chapter examines the link between AI model accuracy, yield prediction, and the broader dimensions of food security, considering access, distribution, and nutritional outcomes.

6. CRITICAL ANALYSIS: YIELD PREDICTION, DECISION SUPPORT SYSTEMS, AND INSTITUTIONAL CAPACITY

Section 4 demonstrated that AI-based yield prediction systems face spatial, temporal, and structural limits rooted in environmental heterogeneity, data representativeness, and calibration dependency. Yet even when these technical constraints are partially mitigated, predictive accuracy alone does not ensure improved food security outcomes. The translation of yield forecasts into tangible socio-economic impact requires institutional embedding, decision-support integration, and infrastructural continuity.

This section shifts the analytical focus from algorithmic scalability to governance capacity. It evaluates how AI-generated yield forecasts interact with DSS, national agricultural monitoring frameworks, and local infrastructural conditions. The core argument is that predictive performance and food security outcomes are mediated by institutional capability rather than determined solely by model accuracy (FAO, 2021; Kamilaris & Prenafeta-Boldú, 2018).

6.1 Yield Prediction Accuracy and Its Operational Limits

Empirical research shows that AI models, including CNN, LSTM, and hybrid APSIM-LSTM architectures, can achieve high prediction accuracy under data-rich conditions. For example, hybrid APSIM-LSTM models applied to rice systems in China report accuracy levels exceeding 90%, reflecting dense meteorological monitoring, digitized yield statistics, and coordinated regional data platforms. Similarly, LSTM-based models integrated with phenological sensors in the U.S. Corn Belt achieve accuracy levels above 90%, supporting inventory forecasting and market stabilization mechanisms (Reichstein et al., 2019).

However, comparable performance is not universal. In smallholder-dominated agricultural systems such as Ethiopia, CNN-LSTM models frequently report accuracy levels between 65–70%, largely due to sparse ground-truth data and fragmented monitoring infrastructures (van Klompenburg et al., 2020). These differences illustrate that predictive accuracy is conditioned by infrastructural density rather than solely by algorithmic sophistication.

Crucially, even high statistical accuracy does not automatically translate into improved food security. Food security is multidimensional, encompassing availability, access, utilization, and stability (FAO, 2021). Yield prediction contributes primarily to the availability dimension, while access and distribution depend on socio-economic and institutional mechanisms. Without integration into policy and logistics systems, predictive outputs remain informational rather than transformative.

6.2 Decision Support Systems as Translational Mechanisms

DSS constitute the institutional interface between AI-generated forecasts and actionable policy responses. They transform probabilistic model outputs into operational signals used for irrigation planning, storage management, emergency imports, and subsidy allocation.

In contexts where DSS are institutionally embedded, AI-based forecasts can trigger adaptive interventions. For instance, in China, hybrid model outputs inform irrigation scheduling and harvest timing adjustments at regional levels, thereby reducing production losses (Zhu et al., 2025). In the United States, state-level DSS integrate yield forecasts into commodity monitoring and strategic reserve management, strengthening market stability.

By contrast, in countries where DSS capacity remains limited, predictive information may not be systematically incorporated into planning cycles. In Ethiopia, despite moderate progress in AI-based yield research, institutional integration remains partial, limiting operational impact (van Klompenburg et al., 2020). Similarly, in Bangladesh, machine learning applications in Aman rice yield forecasting demonstrate technical potential; however, DSS integration remains primarily at the research or pilot stage (Islam et al., 2024).

Table 6.1 synthesizes representative cases and illustrates how predictive accuracy interacts with DSS integration and inferred food security implications.

Table 7. AI Yield Prediction, DSS Integration, and Food Security Implications

Region	Crop	Model Type	Reported Accuracy	DSS Integration Level	Inferred Food Security Impact	Source
China	Rice	APSIM + LSTM	90–95%	High (regional DSS integration)	Moderate–High (adaptive irrigation & planning)	Zhu et al., 2025
United States	Maize	LSTM + sensor integration	~92%	High (state-level DSS)	High (market stabilization & forecasting)	Reichstein et al., 2019
Ethicpia	Maize	CNN–LSTM	65–70%	Low (limited institutional DSS)	Limited operational impact	van Klompenburg et al., 2020
Bangladesh	Aman Rice	ML comparative (RF, NN, etc.)	Moderate	Medium (research-stage DSS)	Emerging potential, limited implementation	Islam et al., 2024

As shown in Table 6.1, the relationship between predictive accuracy and food security impact is non-linear. High accuracy combined with high DSS integration tends to yield stronger operational outcomes, whereas similar accuracy levels without institutional embedding produce weaker effects.

6.3 Institutional Scalability and Infrastructural Density

Building on Section 4's discussion of technical scalability, institutional scalability refers to the capacity of governance systems to sustain model updating, validation, and integration into policy processes. It encompasses three interrelated domains: data governance continuity, technical-human capital capacity, and administrative coordination mechanisms.

Regions characterized by dense meteorological networks, digitized yield databases, and regular calibration cycles maintain more stable forecasting systems. Conversely, fragmented reporting systems and irregular funding cycles increase model drift and reduce reliability over time.

This divergence contributes to differentiated predictive reliability across global agricultural systems. Where infrastructural density is high, uncertainty margins are narrower and forecasts are more readily incorporated into decision-making frameworks. Where data ecosystems are thin, uncertainty expands and forecasts may be treated cautiously or ignored.

6.4 Epistemic Asymmetry in Agricultural Forecasting

When predictive reliability varies systematically across contexts, asymmetries emerge in the production and authority of agricultural knowledge. Regions supported by robust data infrastructures develop more confident yield projections, influencing domestic policy and international market expectations. Regions with limited data capacity operate under higher uncertainty, potentially leading to conservative planning or delayed early warning responses.

This epistemic asymmetry does not originate from algorithmic bias alone. Rather, it arises from the interaction between modeling systems and uneven infrastructural capacity. AI forecasting systems therefore participate in shaping differentiated knowledge regimes, where predictive authority is unevenly distributed.

Understanding this asymmetry is essential for assessing the broader governance implications of AI-based agricultural monitoring. Without institutional strengthening, technical advances risk reinforcing existing disparities in forecasting reliability.

6.5 Synthesis and Transition

The analysis presented in this section demonstrates that yield prediction accuracy is a necessary but insufficient condition for improved food security outcomes. DSS function as translational mechanisms, converting predictive outputs into operational strategies. Institutional scalability determines whether such integration is sustained, updated, and trusted.

Consequently, AI-based agricultural forecasting must be evaluated as a socio-technical system rather than a standalone modeling achievement. Technical robustness (Section 4) and institutional embedding (Section 6) jointly determine operational impact.

The following chapter synthesizes insights from modeling scalability, infrastructural asymmetries, and DSS integration to outline structural capabilities required for sustainable and equitable food security governance.

7. A STRUCTURAL CAPABILITY FRAMEWORK FOR AI-ENABLED AGRICULTURAL MONITORING

The deployment of AI in agricultural monitoring goes beyond algorithmic performance. While techniques such as CNNs, LSTMs, and hybrid crop simulation–machine learning models demonstrate high predictive accuracy under controlled conditions, their real-world impact on food security is conditioned by structural factors. Agricultural AI systems rely on heterogeneous data sources such as satellite imagery, IoT sensor networks, weather forecasts, and farmer inputs that vary in quality and availability, which influences predictive reliability and adoption prospects.

This chapter introduces a Structural Capability Framework to conceptualize these dimensions. It examines how data infrastructure, model transferability, decision-support integration, and governance capacity interact to determine whether AI-generated insights translate into actionable interventions. By situating AI within broader institutional and epistemic contexts, the framework highlights that technological sophistication alone is insufficient; meaningful contributions to food security require the alignment of technical, organizational, and governance capabilities.

7.1 Conceptual Reorientation: From Algorithms to Structures

The preceding analysis indicates that improvements in predictive accuracy, though significant, are insufficient to guarantee meaningful contributions to food security. Model performance varies across regions, data availability is uneven, and institutional adoption remains inconsistent. These observations suggest that AI-based agricultural monitoring systems are embedded in broader structural environments that critically shape their effectiveness.

To synthesize these insights, this chapter introduces a Structural Capability Framework comprising four interrelated dimensions: data infrastructure, model transferability, decision-support integration, and governance capacity. Rather than treating AI as a self-contained technological solution, this framework situates it within the institutional and epistemic conditions that enable or constrain its impact. The core argument is straightforward: AI contributes to food security only when these structural dimensions are aligned.

7.2 Data Infrastructure as Foundational Capacity

All AI-enabled agricultural systems rely on layered observational environments. Multi-temporal satellite imagery, meteorological records, soil databases, and ground-based yield measurements collectively provide the empirical foundation for predictive models. When these data layers are continuous, standardized, and historically deep, model performance tends to be stable and transferable.

Conversely, irregular yield statistics, incomplete soil information, or fragmented meteorological records increase predictive uncertainty. This challenge is not merely computational; it is epistemic. Data infrastructure shapes the reliability of agricultural knowledge itself.

Even with open satellite programs expanding access to remote sensing imagery, supervised learning models require localized calibration. Without sufficient ground-truth data, predictive systems may appear precise while remaining structurally fragile. Thus, data infrastructure is not merely an input but a foundational capacity upon which all other modeling dimensions depend.

7.3 Model Transferability Beyond Local Accuracy

High predictive performance in a single agro-ecological region does not ensure robustness elsewhere. Models trained in homogeneous, large-scale farming systems often degrade in heterogeneous smallholder contexts, where soil composition, cropping intensity, irrigation practices, and phenology vary.

Hybrid approaches integrating process-based crop simulation systems such as APSIM and DSSAT attempt to address this challenge by embedding physiological constraints into machine learning architectures. Such integration enhances temporal stability and reduces sensitivity to anomalous climatic events. Nevertheless, these systems still require contextual calibration and representative training data.

Transferability should be understood as a structural capability distinct from peak accuracy, reflecting the degree to which predictive systems encode agro-ecological diversity and remain stable across spatial and temporal shifts. Without this capability, AI risks producing localized success stories that cannot scale into reliable governance tools.

7.4 Institutional Integration and the Translation of Prediction into Action

Predictive knowledge attains social relevance only when integrated into institutional decision-making. Yield forecasts influence food security outcomes when they inform early warning systems, insurance mechanisms, input distribution policies, or market stabilization strategies. In the absence of such integration, AI remains confined to experimental or analytical domains.

Institutional integration requires standardized communication protocols, transparent uncertainty reporting, and coordination between technical units and policy agencies. Predictive outputs must be interpretable and operational within existing agricultural information systems. Technical sophistication alone does not guarantee adoption.

Scalability, therefore, is not just geographic; it is administrative. AI systems that cannot be incorporated into policy workflows remain structurally peripheral, regardless of statistical performance.

7.5 Governance Capacity and the Politics of Predictive Knowledge

Effectiveness also depends on governance capacity, defined as the institutional ability to interpret uncertainty responsibly, regulate data access, ensure transparency, and align technology with equity-oriented policy objectives.

Where governance institutions are competent and transparent, predictive outputs can be combined with local knowledge and participatory planning. Where governance is weak, predictive systems may be over-relied upon, misinterpreted, or deployed without accountability. Governance capacity mediates the relationship between predictive information and distributive outcomes. AI does not directly improve food security; it reshapes the informational environment in which decisions are made. Whether this reshaping enhances equity or reinforces asymmetry depends on institutional design and political commitment.

7.6 Structural Alignment and Food Security Outcomes

Data infrastructure, transferability, institutional integration, and governance capacity form a mutually reinforcing system. Weakness in any dimension constrains the effectiveness of the others. Robust models without reliable data degrade; strong data without institutional pathways yield limited policy impact; effective governance without predictive reliability operates under persistent uncertainty.

The Structural Capability Framework reframes AI's contribution to agriculture. It shifts focus from algorithmic sophistication to systemic alignment, emphasizing that food security gains emerge from coordinated strengthening of technical, institutional, and governance capacities. AI becomes neither a technological panacea nor a neutral analytical tool; it becomes part of a broader socio-technical infrastructure whose effectiveness depends on structural coherence. Only when these capabilities converge can AI-enabled remote sensing meaningfully support anticipatory and equitable agricultural governance.

8. CONCLUSION: FROM TECHNOLOGICAL POTENTIAL TO STRUCTURAL CAPABILITY

8.1 Synthesis of Technical vs. Structural Findings

The review of AI-based agricultural monitoring and yield estimation systems highlights a fundamental distinction between technical potential and structural capability. On the technical side, advanced machine learning and deep learning models, including CNNs, LSTMs, and hybrid architectures, exhibit high accuracy in crop classification, phenological stage detection, stress and disease identification, and yield prediction under controlled, data-rich conditions. Case studies from the United States, China, and parts of Europe demonstrate performance levels exceeding 90% in pilot-scale applications, particularly when dense ground truth datasets and high-resolution satellite imagery are available. These systems effectively process multi-temporal, multi-source data, detect subtle patterns, and provide near-real-time monitoring capabilities, representing a major improvement over conventional regression or index-based remote sensing methods (Kamilaris & Prenafeta-Boldú, 2018).

However, translating this technical potential into operational effectiveness at broader scales exposes significant limitations. Models trained in data-rich, homogeneous contexts show reduced performance when applied to heterogeneous landscapes, smallholder farming systems, or regions with limited observation infrastructure (van Klompenburg et al., 2020). Factors such as crop diversity, mixed cropping patterns, phenological variability, climate heterogeneity, and irregular field data reduce model transferability. Hybrid approaches that combine deep learning with process-based simulation models, such as APSIM-LSTM, partially mitigate these challenges, but their success depends on extensive calibration and validation datasets (Wei et al., 2024)

Moreover, the integration of AI into agricultural workflows requires considering socio-technical interactions. Technical accuracy alone does not guarantee actionable knowledge; AI outputs must align with farm management practices, labor availability, institutional protocols, and local decision-making routines. For instance, automated disease detection tools or precision irrigation recommendations are only effective if farmers possess the resources, knowledge, and incentives to act upon them. This insight underscores the core finding: technical potential is necessary but insufficient to ensure structural capability.

8.2 Implications for Policy, Governance, and Food Security

Structural constraints have profound implications for policy, governance, and food security. AI outputs are increasingly incorporated into DSS, such as FEWS NET in Sub-Saharan Africa or USDA platforms in the United States, influencing crop monitoring, early warning systems, and resource allocation (FEWS NET, 2022). Yet uneven distribution of data and analytical capacity can exacerbate existing inequalities. Regions with limited sensor coverage, infrequent field observations, or poor digital infrastructure are unable to fully leverage AI-derived insights, resulting in asymmetries in agricultural knowledge production and risk management (Kamilaris & Prenafeta-Boldú, 2018).

From a governance perspective, effective AI integration necessitates alignment with institutional frameworks, training for local stakeholders, and adaptive policies that accommodate uncertainty inherent in model predictions. Overreliance on AI without addressing infrastructural and socio-economic contexts can lead to misaligned interventions, such as when yield predictions inform subsidy allocation or food aid distribution but fail to reflect local heterogeneity. Additionally, AI-driven systems shape which forms of agricultural knowledge are prioritized, potentially marginalizing local expertise or traditional practices. Policy design must reconcile high-tech capabilities with socio-institutional realities, ensuring AI supports equitable, inclusive decision-making rather than reinforcing information monopolies (Leonelli, 2020).

Regarding food security, AI-based monitoring improves the availability dimension by providing timely estimates of crop production. However, access, utilization, and stability remain constrained by broader logistical, economic, and social factors. For example, hybrid CNN-LSTM models may predict maize yields in Ethiopia with high accuracy, yet lack of storage, transport infrastructure, and market access limits the translation of these predictions into improved nutritional outcomes (van Klompenburg et al., 2020). Similarly, the removal of a prior section on socio-economic and labor dimensions highlighted that labor availability, farmer knowledge, and market structures can significantly influence the operationalization of AI insights. Therefore, food security benefits are conditional, dependent on the interplay between AI-derived insights, local infrastructure, policy implementation, and institutional capacity.

Furthermore, structural considerations extend to ethical, social, and equity dimensions. Data ownership, privacy, and inclusivity must be addressed to prevent reinforcement of socio-economic disparities. Participatory approaches, incorporating farmers' local knowledge into model calibration and interpretation, enhance both relevance and legitimacy of AI outputs, especially in smallholder-dominated systems (Leonelli, 2020). The integration of AI must therefore be conceptualized not only as a technological intervention but as a socio-technical transformation, balancing innovation with inclusivity.

Table 8. AI-Based Agricultural Monitoring and Multidimensional Food Security Contributions

Food Security Dimension	Direct AI Contribution	Indirect Contribution	Limitation
Availability	Yield prediction, crop monitoring	Early warning systems	Data inequality
Access	Market forecasting	Policy targeting	Infrastructure gaps
Utilization	Quality/stress detection	Nutritional planning	Weak integration
Stability	Seasonal forecasting	Risk management	Climate non-stationarity
Agency	Farmer decision support tools	Information transparency	Digital divide
Sustainability	Climate-adaptive modeling	Resource optimization	Model recalibration needs

As summarized in Table 8, AI-based agricultural monitoring systems contribute unevenly across different food security dimensions. While availability and stability benefit directly from predictive modeling, agency and sustainability depend more strongly on institutional integration and participatory governance structures.

8.3 Future Research Directions

The analysis identifies several avenues for future research aimed at strengthening the structural capability of AI in agriculture:

Data Infrastructure Expansion: Enhancing sensor networks, high-resolution satellite imagery, multi-source ground observations, and interoperable data platforms is crucial for model calibration, validation, and transferability, particularly in underrepresented regions.

Hybrid and Transfer Learning Approaches: Models must account for heterogeneity in cropping systems, phenology, and environmental conditions. Further refinement of hybrid modeling approaches can improve generalizability across scales and diverse agroecological contexts.

Integration with DSS: Interdisciplinary research combining agronomy, data science, and governance studies is essential to ensure that AI outputs are actionable and aligned with farm-level, regional, and national decision-making structures.

Participatory AI Frameworks: Involving local stakeholders in the design, calibration, and interpretation of AI systems enhances relevance, legitimacy, and adoption. Such approaches are particularly critical in smallholder and marginalized farming communities.

Ethical and Political Considerations: Research should address data governance, privacy, inclusivity, and socio-economic equity to prevent reinforcement of disparities.

Longitudinal Impact Assessment: Evaluating AI-informed interventions through multi-dimensional metrics, including technical, socio-economic, and nutritional outcomes, will provide a holistic understanding of AI's contribution to sustainable agricultural development.

In conclusion, AI-based agricultural monitoring and yield estimation systems offer transformative technological potential, yet their contribution to food security and policy objectives depends on structural capability. By addressing data inequalities, enhancing institutional capacity, integrating local knowledge, and developing context-sensitive governance mechanisms, predictive accuracy can be translated into actionable, sustainable, and equitable agricultural knowledge. The synthesis of technical, socio-economic, and institutional insights emphasizes that AI's promise lies not only in what it can predict but in how it can be operationalized to support resilient, inclusive, and informed agricultural systems.

REFERENCES

Ahmed, Z., Gui, D., Qi, Z., Liu, Y., Liu, Y., & Azmat, M. (2022). Agricultural system modeling: Current achievements, innovations, and future roadmap. *Arabian Journal of Geosciences, 15*(4), 363. DOI: 10.1007/s12517-022-09654-7

Arystanov, A., Sagin, J., Karabkina, N., Arystanova, R., Yermekov, F., Kabzhanova, G., Bekseitova, R., Aktymbayeva, A., & Kutymova, N. (2025). Automatic classification of agricultural crops using Sentinel-2 data in the rainfed zone of Southern Kazakhstan. *Agronomy (Basel), 15*(9), 2040. DOI: 10.3390/agronomy15092040

Balsamo, G., Agusti-Parareda, A., Albergel, C., Arduini, G., Beljaars, A., Bidlot, J., & Zeng, X. (2018). Satellite and in situ observations for advancing global Earth surface modelling: A review. *Remote Sensing (Basel), 10*(12), 2038. DOI: 10.3390/rs10122038

Belgiu, M., & Drăguţ, L. (2016). Random forest in remote sensing: A review of applications and future directions. *ISPRS Journal of Photogrammetry and Remote Sensing, 114*, 24–31. DOI: 10.1016/j.isprsjprs.2016.01.011

Benami, E., Jin, Z., Carter, M. R., Ghosh, A., Hijmans, R. J., Hobbs, A., & Lobell, D. B. (2021). Uniting remote sensing, crop modelling and economics for agricultural risk management. *Nature Reviews. Earth & Environment, 2*(2), 140–159. DOI: 10.1038/s43017-020-00122-y

Bhattacharyya, A., Nambiar, S. M., Ojha, R., Gyaneshwar, A., Chadha, U., & Srinivasan, K. (2023). Machine learning and deep learning powered satellite communications: Enabling technologies, applications, open challenges, and future research directions. *International Journal of Satellite Communications and Networking, 41*(6), 539–588. DOI: 10.1002/sat.1482

Bounoua, I., Saidi, Y., Yaagoubi, R., & Bouziani, M. (2024). Deep learning approaches for water stress forecasting in arboriculture using time series of remote sensing images: Comparative study between ConvLSTM and CNN-LSTM models. *Technologies, 12*(6), 77. DOI: 10.3390/technologies12060077

d'Andrimont, R., Verhegghen, A., Lemoine, G., Kempeneers, P., Meroni, M., & van der Velde, M. (2021). From parcel to continental scale—A first European crop type map based on Sentinel-1 and LUCAS Copernicus in situ observations. *Remote Sensing of Environment, 266*, 112708. DOI: 10.1016/j.rse.2021.112708

Dhillon, M. S., Kübert-Flock, C., Dahms, T., Rummler, T., Arnault, J., Steffan-Dewenter, I., & Ullmann, T. (2023). Evaluation of MODIS, Landsat 8 and Sentinel-2 data for accurate crop yield predictions: A case study using STARFM NDVI in Bavaria, Germany. *Remote Sensing (Basel), 15*(7), 1830. DOI: 10.3390/rs15071830

Famine Early Warning Systems Network (FEWS NET). (2022). *FEWS NET methodology and analysis framework.* USAID. https://fews.net

Fan, J., Zhang, X., Zhao, C., Qin, Z., De Vroey, M., & Defourny, P. (2021). Evaluation of crop type classification with different high-resolution satellite data sources. *Remote Sensing (Basel), 13*(5), 911. DOI: 10.3390/rs13050911

FAO. (2021). *The state of food and agriculture 2021: Making agrifood systems more resilient to shocks and stresses.* Rome: Food and Agriculture Organization of the United Nations. https://www.fao.org/3/cb4476en/cb4476en.pdf

Gao, N., Xue, H., Shao, W., Zhao, S., Qin, K. K., Prabowo, A., Rahaman, M. S., & Salim, F. D. (2022). Generative adversarial networks for spatio-temporal data: A survey. [TIST]. *ACM Transactions on Intelligent Systems and Technology, 13*(2), 1–25. DOI: 10.1145/3474838

Islam, T., Mazumder, T., Shahriair Roni, M. N., & Nur, M. S. (2024). A comparative study of machine learning models for predicting Aman rice yields in Bangladesh. *Heliyon, 10*(23), e40764. DOI: 10.1016/j.heliyon.2024.e40764 PMID: 39691189

Jabed, M. A., & Murad, M. A. A. (2024). Crop yield prediction in agriculture: A comprehensive review of machine learning and deep learning approaches, with insights for future research and sustainability. *Heliyon, 10*(24), e40836. DOI: 10.1016/j.heliyon.2024.e40836 PMID: 39720079

Joshi, A., Pradhan, B., Gite, S., & Chakraborty, S. (2023). Remote-sensing data and deep-learning techniques in crop mapping and yield prediction: A systematic review. *Remote Sensing (Basel), 15*(8), 2014. DOI: 10.3390/rs15082014

Kamilaris, A., & Prenafeta-Boldú, F. X. (2018). Deep learning in agriculture: A survey. *Computers and Electronics in Agriculture, 147*, 70–90. DOI: 10.1016/j.compag.2018.02.016

Kpienbaareh, D., Sun, X., Wang, J., Luginaah, I., Bezner Kerr, R., Lupafya, E., & Dakishoni, L. (2021). Crop type and land cover mapping in Northern Malawi using the integration of Sentinel-1, Sentinel-2, and PlanetScope satellite data. *Remote Sensing (Basel), 13*(4), 700. DOI: 10.3390/rs13040700

Leonelli, S. (2020). Scientific research and big data. In Zalta, E. N. (Ed.), *The Stanford Encyclopedia of Philosophy* (Fall 2020 Edition)., https://plato.stanford.edu/entries/science-big-data/

Lesiv, M., Laso Bayas, J. C., See, L., Duerauer, M., Dahlia, D., Durando, N., Hazarika, R., Kumar Sahariah, P., Vakolyuk, M., Blyshchyk, V., Bilous, A., Perez-Hoyos, A., Gengler, S., Prestele, R., Bilous, S., Akhtar, I. H., Singha, K., Choudhury, S. B., Chetri, T., & Fritz, S. (2019). Estimating the global distribution of field size using crowdsourcing. *Global Change Biology*, *25*(1), 174–186. DOI: 10.1111/gcb.14492 PMID: 30549201

Li, D., Sun, L., Yu, Y., & Tian, P. (2025). Geospatial micro-estimates of slum populations in 129 Global South countries using machine learning and public data. *Earth System Science Data Discussions*, *2025*, 1–33. DOI: 10.5194/essd-2025-260

Liao, C., Wang, J., Shan, B., Shang, J., Dong, T., & He, Y. (2023). Near real time detection and forecasting of within-field phenology of winter wheat and corn using Sentinel-2 time series data. *ISPRS Journal of Photogrammetry and Remote Sensing*, *196*, 105–119. DOI: 10.1016/j.isprsjprs.2022.12.025

Lungu, O. N., Chabala, L. M., & Shepande, C. (2020). Satellite-based crop monitoring and yield estimation—A review. *Journal of Agricultural Science (Toronto)*, *13*(1), 180. DOI: 10.5539/jas.v13n1p180

Mancini, A., Solfanelli, F., Coviello, L., Martini, F. M., Mandolesi, S., & Zanoli, R. (2024). Time series from Sentinel-2 for organic durum wheat yield prediction using functional data analysis and deep learning. *Agronomy (Basel)*, *14*(1), 109. DOI: 10.3390/agronomy14010109

Matabber, A., Rhuhanga, L. L. N., Agehara, S., & Mozafarian, M. (2026). Artificial intelligence (AI) in detection of abiotic stress in plants: A review. *Sensors (Basel)*, *26*(4), 1122. DOI: 10.3390/s26041122 PMID: 41755064

Mazzia, V., Khaliq, A., & Chiaberge, M. (2020). Improvement in land cover and crop classification based on temporal features learning from Sentinel-2 data using recurrent-convolutional neural network (R-CNN). *Applied Sciences (Basel, Switzerland)*, *10*(1), 238. DOI: 10.3390/app10010238

Murguia-Cozar, A., Macedo-Cruz, A., Fernandez-Reynoso, D. S., & Salgado-Transito, J. A. (2021), . Recognition of maize phenology in Sentinel images with machine learning. *Sensors (Basel)*, *22*(1), 94. DOI: 10.3390/s22010094 PMID: 35009637

Muruganantham, P., Wibowo, S., Grandhi, S., Samrat, N. H., & Islam, N. (2022). A systematic literature review on crop yield prediction with deep learning and remote sensing. *Remote Sensing (Basel), 14*(9), 1990. DOI: 10.3390/rs14091990

Olson, D., & Anderson, J. (2021). Review on unmanned aerial vehicles, remote sensors, imagery processing, and their applications in agriculture. *Agronomy Journal, 113*(2), 971–992. DOI: 10.1002/agj2.20595

Reichstein, M., Camps-Valls, G., Stevens, B., Jung, M., Denzler, J., Carvalhais, N., & Prabhat, . (2019). Deep learning and process understanding for data-driven Earth system science. *Nature, 566*(7743), 195–204. DOI: 10.1038/s41586-019-0912-1 PMID: 30760912

Senaras, C., Grady, M., Rana, A. S., Nieto, L., Ciampitti, I., Holden, P., & Wania, A. (2024). Detection of maize crop phenology using Planet Fusion. *Remote Sensing (Basel), 16*(15), 2730. DOI: 10.3390/rs16152730

Sheng, L., Lv, Y., Ren, Z., Zhou, H., & Deng, X. (2024). Detection of the optimal temporal windows for mapping paddy rice under a double-cropping system using Sentinel-2 imagery. *Remote Sensing (Basel), 17*(1), 57. DOI: 10.3390/rs17010057

Shojaeezadeh, S. A., Elnashar, A., & Weber, T. K. D. (2024). A novel fusion of optical and radar satellite data for crop phenology estimation using machine learning. *Science of Remote Sensing, 11*, 100227. DOI: 10.1016/j.srs.2025.100227

Song, L., Zhao, J., Li, Y., Liu, L., Duan, J., He, L., Wang, Y., Guo, T., & Feng, W. (2026). Early detection of wheat powdery mildew: A multi-source in situ remote sensing approach enabled by stacked ensemble learning. *Artificial Intelligence in Agriculture, 16*(1), 124–138. DOI: 10.1016/j.aiia.2025.10.004

Tokluoğlu, E., Kurtuluş, B., Sağır, Ç., Erdem Altın, G., Yurdakul, E., Ateş, E., Canoğlu, M. C., & Yamaç, S. S. (2026). Detection of potato fields using Sentinel-2 and Landsat 8 data-based machine learning models in semi-arid region of Central Anatolia, Türkiye. *Potato Research, 69*(1), 14. DOI: 10.1007/s11540-025-09985-4

Tuğaç, M. G., Şimşek, F. F., & Torunlar, H. (2024). Classification of agricultural crops with random forest and support vector machine algorithms using Sentinel-2 and Landsat 8 images. *International Journal of Environment and Geoinformatics, 11*(3), 106–118. DOI: 10.30897/ijegeo.1479116

Ualiyeva, R. M., Kaverina, M. M., Osipova, A. V., Kairbayev, Y. B., Zhangazin, S. B., Iksat, N. N., & Mapitov, N. B. (2025). VNIR hyperspectral signatures for early detection and machine learning classification of wheat diseases. *Plants, 14*(23), 3644. DOI: 10.3390/plants14233644 PMID: 41375355

Utku, A. (2025). Data-driven prediction of climate variables in agricultural cities of India with hybrid GA-TCN-LSTM model. *Journal of Forecasting*. Advance online publication. DOI: 10.1002/for.70088

van Klompenburg, T., Kassahun, A., & Catal, C. (2020). Crop yield prediction using machine learning: A systematic literature review. *Computers and Electronics in Agriculture*, *177*, 105709. DOI: 10.1016/j.compag.2020.105709

Vijayasekaran, D. (2019). SEN2 AGRI – Crop type mapping pilot study using Sentinel-2 satellite imagery in India. *The International Archives of the Photogrammetry, Remote Sensing and Spatial Information Sciences*, *XLII-3*(W6), 175–180. DOI: 10.5194/isprs-archives-XLII-3-W6-175-2019

Wang, Y., Fang, S., Zhao, L., Huang, X., & Jiang, X. (2022). Parcel-based summer maize mapping and phenology estimation combined using Sentinel-2 and time series Sentinel-1 data. *International Journal of Applied Earth Observation and Geoinformation*, *108*, 102720. DOI: 10.1016/j.jag.2022.102720

Wei, Y., Wang, R., & Feng, P. (2024). Improving hydrological modeling with hybrid models: A comparative study of different mechanisms for coupling deep learning models with process-based models. *Water Resources Management*, *38*(7), 2471–2488. DOI: 10.1007/s11269-024-03780-5

Zhong, L., Hu, L., & Zhou, H. (2020). Deep learning-based multi-temporal crop classification with Sentinel-2 imagery. *Remote Sensing*, *12*(1), 149. DOI: 10.3390/rs12010149

Zhou, S., Xu, L., & Chen, N. (2023). Rice yield prediction in Hubei Province based on deep learning and the effect of spatial heterogeneity. *Remote Sensing (Basel)*, *15*(5), 1361. DOI: 10.3390/rs15051361

Zhu, C., Lu, D., Victoria, D., & Dutra, L. (2015). Mapping fractional cropland distribution in Mato Grosso, Brazil using time series MODIS enhanced vegetation index and Landsat Thematic Mapper data. *Remote Sensing (Basel)*, *8*(1), 22–34. DOI: 10.3390/rs8010022

Zhu, M., Jiao, Y., Wu, C., Shi, W., Huang, H., Zhang, Y., & Yue, T. (2025). Simulation and optimization of double-season rice yield in Jiangxi Province based on high-accuracy surface modeling–APSIM model. *Agriculture*, *15*(10), 1034. DOI: 10.3390/agriculture15101034

Ziv, L., & Nakash, M. (2025). Behind the algorithm: International insights into data-driven AI model development. *Machine Learning and Knowledge Extraction*, *7*(4), 122. DOI: 10.3390/make7040122

Chapter 9

Urbanization and Thermal Footprints:
Mapping Bangalore's Growth and UHI Evolution

Debosmita Mondal
Amity Institute of Geoinformatics and Remote Sensing, Amity University, Noida, India

K. Dhanaraj
Amity Institute of Geoinformatics and Remote Sensing, Amity University, Noida, India

Sandipta Das
School of Humanities, National Institute of Advanced Studies, Bengaluru, India

Subhash Anand
Department of Geography, Delhi School of Economics, University of Delhi, New Delhi, India

Prangya Rath
http://orcid.org/0000-0002-6764-1625
Amity Institute of Environmental Sciences, Amity University, Noida, India

Laxmi Kant Bhardwaj
http://orcid.org/0000-0001-7518-4199
Amity Institute of Environmental Sciences, Amity University, Noida, India

ABSTRACT

This work focuses on land use mapping, change detection, and analyzing LST using remote sensing data in Bangalore city. This study tested and evaluated the use of high-resolution LISS IV imagery and Landsat 8 data with TIRS data. Here, the SVM algorithm has been utilised for monitoring land use changes, and mono-window algorithms have been used for the calculation of LST. It has been used to classify different types of land cover features, such as urban areas, vegetation, agricultural land, water bodies, and open land. Between 2014 and 2023, the results showed

DOI: 10.4018/979-8-3373-8745-1.ch009

that the urban area had increased by 70%, and the overall classification accuracy ranged between 61-81%. However, in LST, emissivity was also estimated with the help of the NDVI proportion of vegetation method, which was developed for use with the OLI Bands 4 & 5. From the result, it can be deduced that the LST of the urban region was higher than that of vegetative regions. It has been observed that the mono-window algorithm-based LST calculations and SVM-based image classification results are consistent.

1. INTRODUCTION

Urban heat island (UHI) can be described as the condition where temperatures registered in a city are higher than those in non-urban areas (Deng et. al., 2023). The UHI has already started appearing in the developed cities of the developing countries, and as the buildings are high and densely built, it is anticipated that the degree of UHI will further increase in the future. The main cause of making a heat island is urbanisation or development that results in the confinement of the population in urban centres (Heaviside et. al., 2017). So, urbanisation is an ongoing process that includes both vertical and horizontal growth of urban areas in response to growth in population, industrial development, political and cultural factors, and various socio-economic factors (Aithal, 2013; Kayizzi-Mugerwa et. al., 2016; Hussain and Imitiyaz, 2018). It refers to an increase in developed land cover and unplanned or uncontrolled urban expansion.

Urban growth is one method of achieving urban expansion (Viana, 2019). Urban growth mapping is important for sustainable development planning because it helps us to understand how land has been utilised and helps in future planning (Naess, 2001; Mallick et. al., 2021). Monitoring urban growth is essential for making informed decisions and planning successful development plans. In this study, urban growth in Bangalore showed a significant increase in the city's population and area over the years, showing the importance of monitoring urban growth for successful urban planning (Rahimi et. al., 2019). The size and the rate at which cities are growing are global issues that raise important questions on sustainable future.

In urbanization, with the increasing human traffic and migration of people into the urban area, features such as poverty trap, slum areas, insecurity, air, noise, and visual pollution, spread of infectious diseases and the general cost of expansion of facilities become relevant each time there are large scale land covers (Aithal et. al., 2013; Gómez et. al., 2019). Recent decades have witnessed significant changes in urban trends worldwide, with many geographical areas changing from previously stable conditions (Khan and Sudheer, 2022). Land use can be described as the area covered with something such as housing premises, protected or cultivated areas, and so on. These vary in earth-atmospheric processing depending on the type of land use in the land cover (Atef et. al., 2023). Land use refers to how land is used by humans for different purposes, like building up an area or residential area, agricultural activities, or industrial activities.

A geographic information system has been used for analysing the changes in land use patterns, identifying them, and quantifying their intensity (Balogun et. al., 2011; Mhanna et. al., 2023). Even presenting valuable major insights into the evolution of spatial trends connected to land use and cover. A land cover dataset is commonly used to gather data and conduct geographic information system (GIS) analyses with the objective of tracking and evaluating spatial and temporal land use changes, providing politicians and researchers with informed choices about land cover management and conversations about the environment (Stojković, 2017). Land use varies according to geography and the community's demands. Residential, industrial, commercial, agricultural, water bodies, etc., are some of the most common land uses. Residential neighborhoods are mostly used for housing and facilities like parks, schools, malls, etc. (LaGrange, 1999).

Commercial areas are assigned for commercial activities, with a large number of organisations and residential structures (Colaço et. al., 2021; Cloete et. al., 2024). Then, industrial areas are allocated for manufacturing, distribution, and other industrial activities, and transportation infrastructure (Zafar et. al., 2024). Even agricultural areas include agricultural land, and livestock and water bodies include rivers, ponds, and lakes. Thus, land use patterns can change over time because of variables like urbanisation, population increase, and economic development (Dadashpoor et. al., 2019). Land use planning and management are critical for sustainable growth and successful resource use (Silberstein and Maser, 2013; Khurana et. al., 2023).

Land use changes have an important impact on biodiversity and species occurrence, affecting ecological systems and biodiversity threats, and deforestation and urbanisation can result in habitat loss, dispersion, and degradation, impacting the number of animals and how ecosystems work (Oliver and Morecroft, 2014). Land use changes have a social impact on societies that depend on natural resources for a livelihood, potentially leading to challenges over land access and the use of resources (Mir et. al., 2025). Even environmental impacts, such as changes in carbon storage, water cycles, and climate management, impact changing global scenarios and affect the overall health of ecosystems (Nuissl and Siedentop, 2021). So, the evaluation also emphasises the use of remote sensing and GIS tools since they provide information on temporal trends and spatial distribution, which is essential for understanding and managing natural resources.

However, progressed machine learning (ML) approaches like Random Forest (RF), Support Vector Machine (SVM), and Maximum Likelihood Estimation (MLE) are utilised to classify the images in the single or multicategory process in the context of Land Use and Land Cover (LULC) change for using remotely sensed data use of plants in combating diseases has been advocated in the last few decades by scholars who has applied multi-classifiers of ML to analyze change detection in the regions for different study areas and LULC classes at different periods (Atef et. al., 2023; Cloete et. al., 2024; Zafar et. al., 2024). RF is one of the most widely used ML algorithms (Salman et al., 2024). This algorithm can be used for regression purposes in addition to its classification capability. It specifies an exact classification in a heterogeneous landscape by randomly selecting a subset of decision trees from the training set and includes decisions for deciding the final class (Zafar et. al., 2024; Ramachandra et. al., 2023).

Support Vector Machine (SVM) is a powerful machine learning (ML) method that is applied for linear or non-linear classification and regression (Ghosh et. al., 2019). It functions well when the data is not linearly separable and must be translated into a high-dimensional space before extraction. SVMs are a form of supervised learning algorithm, which means they must have marked data in training. Their classifier constructs a hyperplane or a set of hyperplanes based on a high or infinite number of objects in space to classify spatial data from a limited set of training data (Ramachandra et. al., 2023; Zafar et. al., 2024). SVM performs successfully when classes are clearly better in high-dimensional spaces and need less memory. It's successful when the dimensions exceed the number of samples. This is one of the reasons why the SVM method is better than other image classifiers. The Gaussian Maximum Likelihood Classifier (GMLC) manages mean and covariance values with probability density functions to obtain LU information and is a parametric supervised classification method (Ramachandra et. al., 2023).

Thus, land surface temperature (LST) becomes a critical factor while studying the process of interaction between land and air (Wang et. al., 2015). The thermal infrared (TIR) subspace, which has two TIR bands of the Landsat 8, is highly appropriate for the mono-window algorithm for LST retrieval from 2014 to 2023. Further, LULC of an area is also an important factor in the estimation of LST (Ullah et. al., 2023). The mono-window algorithm is a procedure for estimating the LST from the remote sensing data, especially from the Landsat data imagery. As for the LST data selected by the algorithm, to improve its accuracy and quality, there is a connection to a method of mixed-pixel classification and sub-pixel mapping. This confirms the algorithm's improved capability to estimate land surface temperature when applied to Landsat OLI/TIRS Collection 1 datasets. Thus, the present usage of the mono-window algorithm is effective to solve the problem of LST retrieving, which will improve the study of spatial temperature and some phenomena, for example, the UHI (Dai et. al., 2016).

The main objectives of the analysis are to analyse the spatial and temporal variations in the urban expansion pattern inside Bangalore city, which includes Land use classification using an ML algorithm. The study also intends to extract land surface temperature using mono-window algorithms for analysing the seasonal variability of urban heat islands.

Despite the abundance of UHI literature, a significant research gap remains in understanding the high-resolution spatial-autocorrelation of heat during peak urbanization phases in the Deccan Plateau. While previous studies have utilized broad-stroke LULC classifications, this research introduces a novel integration of Support Vector Machine (SVM) machine learning for high-precision mapping alongside Getis-Ord Gi* spatial statistics to quantify the intensity of thermal clustering. The scientific contribution of this work lies in its ability to move beyond simple temperature averages, providing a statistically rigorous framework that identifies 'heat-vulnerability zones.' This data-driven approach offers a replicable model for climate-resilient planning in rapidly evolving global South metropolises, transforming raw satellite data into actionable ward-level heat mitigation strategies.

2. STUDY AREA

Bangalore is the study area, and it is located in the Deccan Plateau at an altitude of over 900 meters (3,000 feet) above sea level. It has an area of 741 km^2 (286 sq mi). Bangalore, India's information technology (IT) hub, is located in the southern state of Karnataka and is known as the Silicon Valley of India. According to the 2011 census, Bangalore city is the most populous city in Karnataka, with a population of 96,21,551 and a maximum population density of 4,381 people per km^2. It also has the highest ten-year growth rate in the country at 47.2%. The city is primarily urban; 90.9% of the population lives in urban areas, and the literacy rate is 87.7%, the second highest in the country (Figure 1). It has a pleasant year-round climate, with parks and green spaces earning it a reputation as India's “garden city”.

Figure 1. Location of Bangalore city in Karnataka, India

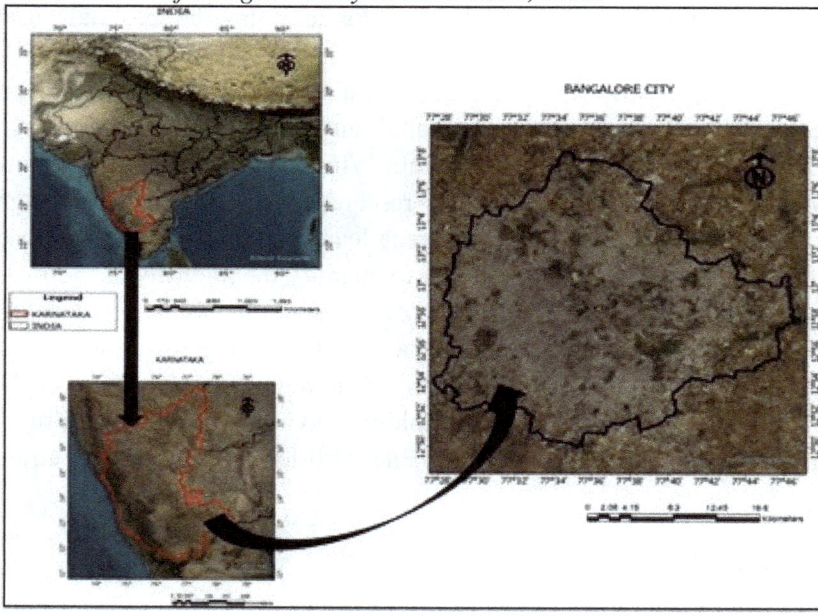

3. DATA AND METHODOLOGY

Figure 2. Flow Chart of Methodology

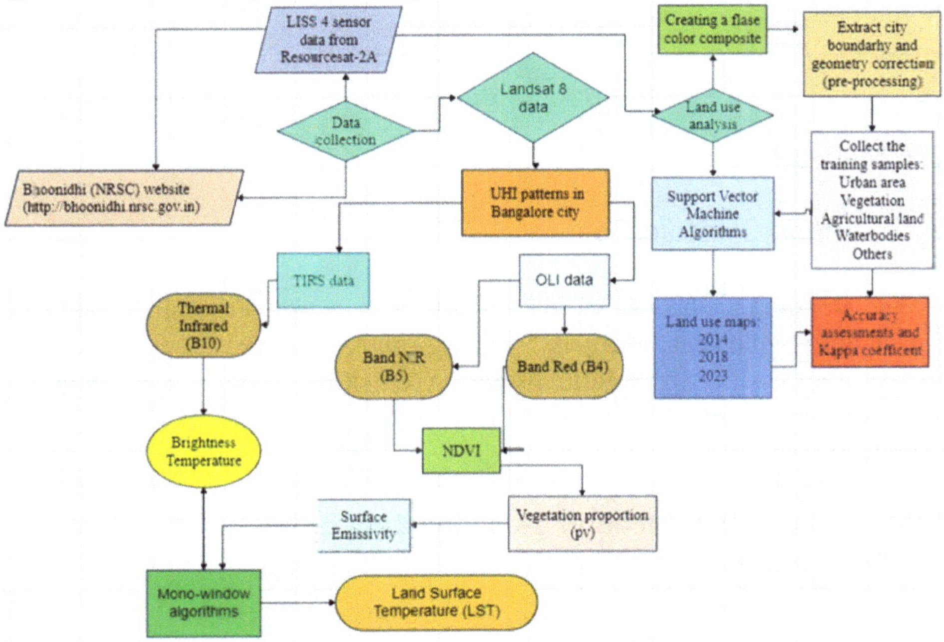

3.1. Data Collection

Temporal urban land use studies of Bangalore city were conducted using multi-spectral high-resolution data from Resourcesat-2A and Landsat series data for 2014, 2018, and 2023, which were obtained from the Bhoonidhi (NRSC) website (http://bhoonidhi.nrsc.gov.in) for urban growth analysis and Google Earth Engine for LST (Tables 1 and 2).

Table 1. Specification of the Linear Imaging Self-Scanner (LISS) 4 Dataset

Data	Launch Date	Band Combination	Bandwidth (μm)	Band Name	Spatial Resolution	Radiometric Resolution	Temporal Resolution
Resource sat 2A (LISS 4 image)	October 17, 2003	Band 2	0.52 - 0.59	Green	5.8 m	10 bit	5 days
		Band 3	0.62 - 0.68	Red			
		Band 4	0.77 - 0.86	NIR			

Table 2. Specification of Landsat Series Dataset

Data	Launch Date	Band Combination	Spatial Resolution	Radiometric Resolution	Temporal Resolution	Purpose
Landsat 8	2013-present	B4 (Red), B5 (NIR), B10 (Thermal bands)	30 m and 100 m	16 bits	16 Days	Calculate NDVI and LST

3.2. Method

3.2.1. Image Classification

The proposed method illustrated has four main steps to achieve the aforementioned steps: Implementation of the SVM algorithm, accuracy analysis for selecting appropriate classification algorithms, generation of LULC maps, and analysis of the detection of changes, and the first steps to be discussed in the upcoming subsections (Figure 2).

3.2.1.1. Image Classification Using the Support Vector Machine (SVM) Algorithm

In this study, the supervised learning method has been utilised to model classroom enrollment using predictable factors in a concise manner. Compared to unsupervised approaches, this approach necessitates the participation of informed and experienced specialists. The selection of training models is important for effective model training (Atef et. al., 2023). SVM algorithms were used to classify LISS 4 images (LULC) in 2014, 2018, and 2023 into groups: urban areas, vegetation, agricultural land, and other water bodies. For each class, a specific training set was determined by visual interpretation, and ground truth points were used to validate the maps produced. Based on the training models, a signature file containing multivariate estimates of each LULC was generated and used as input for SVM classification in ArcGIS 10.8 to determine the optimal classification.

Support Vector Machine (SVM) is a supervised learning approach used to explain classification problems of different types of regression (Sarang, 2023). It establishes hyperplanes on spectral and radiometric features extracted by reduced samples and uses decision constraints to classify classes. Support vectors, which represent the margins of the hyperplane, are the training models SVM uses. SVM includes a kernel function to generate kernel weights for each training sample, and functional similarity is affected by kernel size. When using an SVM classifier with a Radial Basis Function (RBF) kernel, two parameters must be set: C and kernel width (γ) (Bindajam et. al., 2023). The C parameter sets the number of misclassifications allowed for inseparable training data, allowing flexibility in complex training scenarios. The kernel width parameter (γ) achieves a square-separator hyperplane geometry; the kernel width parameter is affected by the smoothing effect. Further increasing the value of C will alter the shape of the hyperplane of class separation, potentially affecting the classification accuracy results (Bindajam et al., 2023).

3.2.1.2. Accuracy Assessment

The calculated LULC data are compared with training data by LULC classification accuracy, and the confusion matrix and kappa statistic are calculated. The overall accuracy corresponds to the overall classification accuracy (Equation 1), calculated as the percentage ratio between correctly classified pixels and the total number of reference pixels, as per equations 2 and 3 (Ramachandra et. al., 2023). The sample points were generated using a systematic sampling pattern with a total of 200 sampling points.

Overall accuracy = (no. of correctly classified objects / **Total objects**) x 100 (1)

User's accuracy

= (no. of objects of specific correctly classified / **Row total**) x 100

(2)

Producer's accuracy

= (no. of objects of specific correctly classified / **Column total**) x 100

(3)

Discrete multivariate methods have been used to statistically assess the accuracy of remote sensing-derived maps and error matrices since 1983 and are widely accepted (Bindajam et. al., 2023). These methods are appropriate because remotely sensed data are discrete rather than continuous and not routinely distributed. Kappa analysis is a discrete multivariate method for precision assessment. Kappa analysis yields a peg statistic that measures agreement on accuracy. The calculations are as follows:

Kappa Statistics ($\hat{k}$):

$$\hat{k} = \frac{N\sum_{i=1}^{r} x_{ii} - \sum_{i=1}^{r}(x_i + {}^*x_{+i})}{N^2 - \sum_{i=1}^{r}(x_i + {}^*x_{+i})}$$

where

r = the number of rows in the error matrix

xii = the number of observations in row i and column i xi+ = total of observations in row i

x+i = total of observations in column i

N = total number of observations included in the matrix

3.2.2. Change Detection

Change detection is monitoring the spatial changes from multi-temporal satellite-based imagery due to manmade or natural phenomenon which is very important in remote sensing, environmental monitoring, and detecting LULC changes (Asokan and Anitha, 2019). For this study, it is possible to extract the land use changes of a particular region from different datasets using the SVM method (Atef et. al., 2023). Here, change detection analysed the satellite images to identify and quantify changes in land use over the assigned period of 2014 to 2023 in Bangalore city. After classification, when they compare satellite images from 2014 to 2023, they can classify the different LULC types and monitor the changes, such as increasing urban areas and decreasing vegetation cover, as well as agricultural land. Through this analysis, the main goal is to give insights into the geographical and temporal patterns of land cover evaluation in Bangalore city, which will be beneficial to urban planning and sustainable development actions. Even for monitoring the change detection accuracy assessment is used to validate the results and ensure that the study area is accurate or not from the ground truth data.

3.2.3. Land Surface Temperature (LST)

The major aim of this analysis was therefore to generate LST changes in Bangalore city. This led to the acquisition of remote sensing satellite data from Landsat 8 TIRS (Çelik, 2026). In Landsat 8, there are two thermal bands, but this study has only used band 10 for LST estimation because band 10 data has a single spectral band for LST estimation, and band 11 has large uncertainty (Wang et. al., 2015; Çelik et. al., 2026). While Landsat 8 TIRS features two thermal bands (10 and 11), allowing for split-window techniques, this study uses the Mono-Window algorithm using Band 10. This is established on the documented stray light effect in Band 11, which introduces significant calibration uncertainty (up to 2.0 K) in Landsat 8

Collection 1 and 2 data. By focusing on the more stable Band 10, we ensure higher radiometric consistency for long-term UHI monitoring.

On the other hand, a multi-software platform is where information from different satellites is processed in order to compute the LST. We employed two complementary software: QGIS and ArcPro 3.0. This includes the following steps:

3.2.3.1. Conversion to TOA Radiance

The raw Digital Numbers (Qcal) are converted to Top of Atmosphere (TOA) spectral radiance (Lλ) using:

$$L\lambda = ML \cdot Qcal + AL$$

Where ML and AL are the band-specific rescaling factors found in the metadata (MTL) file. Units are expressed in $W/(m^2 \cdot sr \cdot \mu m)$.

3.2.3.2. Vegetation Proportion

Vegetation proportion can be obtained using vegetation indices like Normalised Difference Vegetation Index (NDVI). This can be done by classifying pixels as non-vegetation or vegetation using NDVI values that have been thresholded.

$$\mathbf{NDVI} = (\mathbf{NIR} - \mathbf{Red})/(\mathbf{NIR} + \mathbf{Red}) \quad (1)$$

Fractional vegetation:

$$Fv = \left[\frac{NDVI - \min}{\max - \min}\right]^\wedge 2 \quad (2)$$

3.2.3.3. Brightness Temperature (BT)

Brightness temperature (BT) refers to the electromagnetic radiation that travels upward from the top of the Earth's atmosphere. After obtaining the radiance value, convert it to BT using the equation below:

$$BT = \frac{K2}{\ln\left(\frac{k1}{L_\lambda} + 1\right)} \quad (3)$$

K1 (774.8853) and K2 (1321.0789) are band-specific thermal conversion constants for TIRS band 10 L_λ is the top of the atmosphere radiance.

3.2.3.4. *Surface Emissivity*

Estimate surface emissivity using land cover categories, plant cover, and other surface features. Emissivity values can be assigned based on land cover categorisation or calculated using spectral libraries. So, surface emissivity is calculated using the following equations:

$$\mathbf{EM = 0.9608420 + 0.0860322*Pv} \quad (4)$$

Where: 0.9608420 and 0.0860322 are constant value and pv is the fractional vegetation. Atmospheric parameters τ and Ta were retrieved using the NASA Atmospheric Correction Parameter Calculator based on the specific date, time, and coordinates of Bangalore for each acquisition year (2014, 2018, 2023) (Sobrino et al., 2004).

3.2.3.5. *LST Calculation*

The Mono-Window algorithm, which uses a single window approach to estimate LST, usually requires a linear relationship between the observed brightness temperature and the LST, taking into account the emissivity and atmospheric effects. Thus, the following equations are used:

$$(Tb / (1+ (0.00115 \times (Tb / 1.438))\times\log(Em))) - 273.15$$

Where: Tb is brightness temperature, and Em is emissivity.

4. RESULTS AND DISCUSSION

In this study, the performance of SVM in the supervised classification and mono-window algorithms in the aspect of LST was tested in 2014, 2018, and 2023 datasets, as presented in Figures 3 and 4. In LULC analysis, there are five major classes of land use categories, such as urban areas, vegetation, water bodies, agricultural land, etc. Here, Figure 6 shows that urban areas are the most predominant class in 2023. Another is hotspots analysis, in this study, hotspots analysis identifies areas that are statistically significantly higher or lower than expected based on random chance. These areas can be considered “hot spots” if their values are significantly higher than surrounding areas, or “cold spots” if their values are significantly lower. It can provide valuable insights into decision-making, resource allocation, and targeted interventions in various sectors such as public health, crime investigation, environmental management, and urban planning.

In 2014, the urban area increased to 45.79%, while vegetation increased to 44.05%. The classification accuracy improves to 0.70, indicating moderate accuracy. The kappa coefficient also indicated a high level of agreement between the reference and the segmented image. Agriculture accounted for 7.41%, with irrigation at 1.19% and others at 1.60%. Because of urbanisation, the western, south-western, and north-eastern areas of Bangalore city are significantly hotter than the rest. 2014 saw reduced urbanisation compared to previous years, resulting in a reduced urban heat component. The green cover in the middle of Bangalore city indicates that it is the coolest section.

Table 3. Accuracy Assessment of Supervised Classification

Classes	Years					
	2014		2018		2023	
	UA%	PA%	UA%	PA%	UA %	PA%
Urban area	95	74	94	86	94	96
Vegetation	72	92	78	89	94	90
Agricultural land	35	100	21	100	50	100
Water bodies	100	78	100	100	100	90
Others	30	38	83	52	90	62
Kappa	0.70		0.75		0.87	

In 2018, the urban area increased to 55.69%, while vegetation increased to 44.88%. The classification accuracy improves to 0.75, indicating moderate accuracy. The kappa coefficient also indicated a high level of agreement between the reference and the segmented image (Table 3). Agriculture accounted for 6.65%, with irrigation at 1.15% and others at 1.46%. Because of urbanisation, the western and middle areas of Bangalore city are significantly hotter than the rest. 2018 saw reduced urbanisation compared to previous years, resulting in a reduced urban heat component. The green cover in the north-west and southern parts of Bangalore city indicates that it is the coolest section, and other portions have been shown to have a moderate temperature. Hotspot area mapping of Bangalore city is shown in Figure 5.

Figure 3. Land Use Mapping of Bangalore City

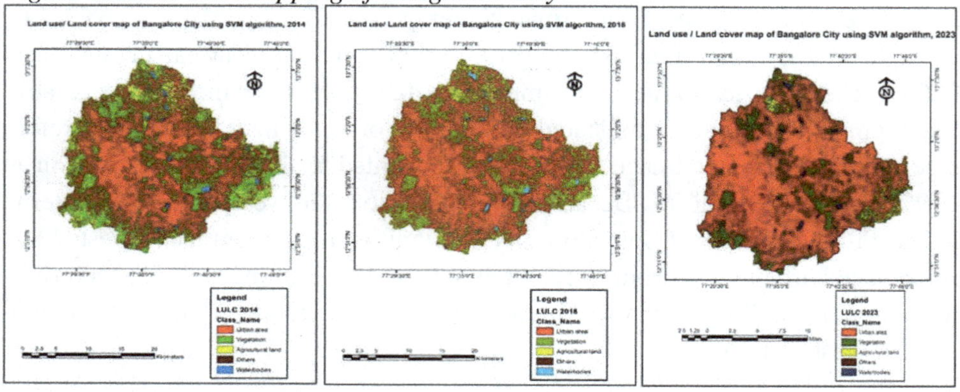

Figure 4. Land Surface Temperature (LST) Mapping of Bangalore City

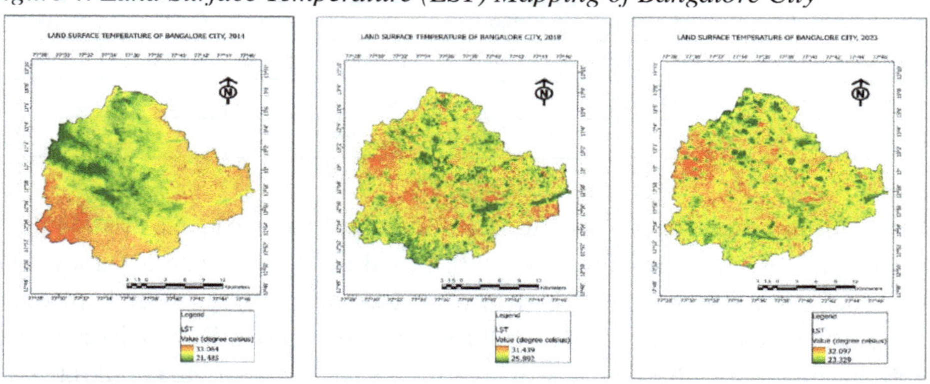

Figure 5. Hotspot Area Mapping of Bangalore City

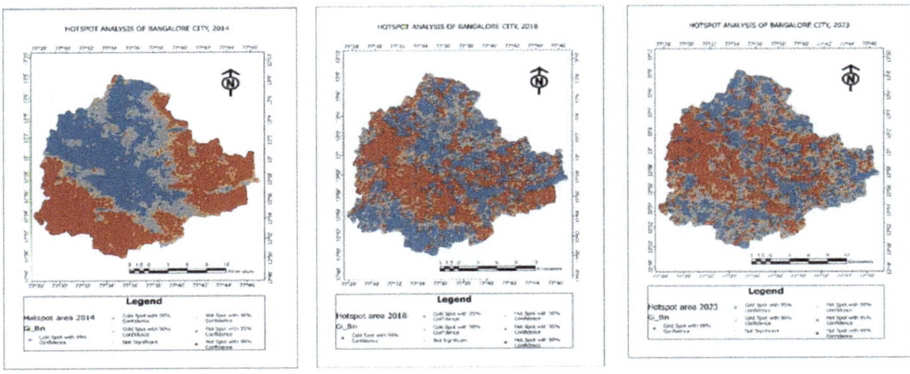

In 2023, the urban area increased significantly to 65.46%, while vegetation decreased to 35.58%. Classification accuracy reached 0.87, indicating close accuracy, and a high kappa coefficient indicates strong agreement between the model and the classified image. Because of urbanisation, the western and middle areas of Bangalore city are significantly hotter than the rest. The green cover in the eastern and southern parts of Bangalore city indicates that it is the coolest section, and other portions have been shown a moderate temperature.

Figure 6. Bar Graph Showing Land Use and Land Cover (LULC) Features with Areas of Bangalore City

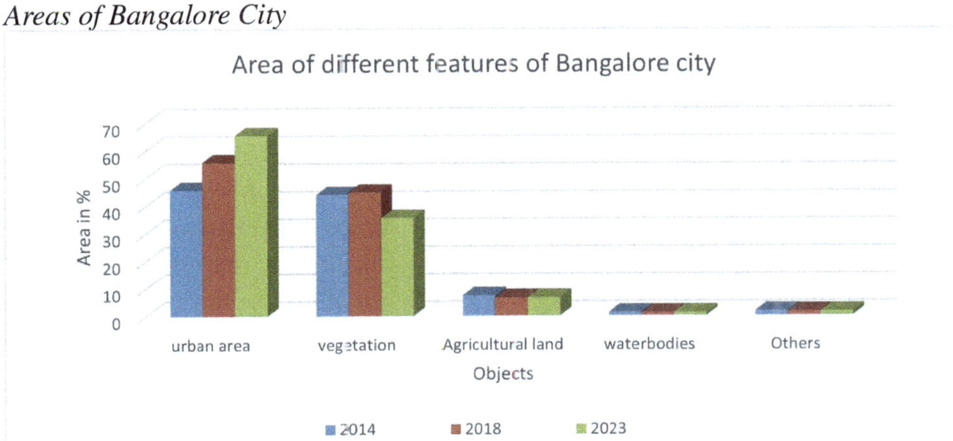

Table 4. Land Use and Land Cover (LULC) Features with Areas of Bangalore City

Classes	Years		
	2014 (%)	2018 (%)	2023 (%)
Urban Area	45.79	55.69	65.46
Vegetation	44.05	44.88	35.58
Agricultural land	7.41	6.65	6.65
Water bodies	1.19	1.15	1.12
Others	1.6	1.46	1.45

In this bar graph (Figure 7), a clear trend of increasing urbanisation is shown, with urban areas growing significantly from 2014 to 2023. The next feature is vegetation and agricultural land; both are decreasing in a given time period, most likely as a result of urban growth. Then the water bodies area has been appropriately consistent but decreased slightly, and the last one is the other features that have been significantly reduced.

Figure 7. Histogram Showing the Z-Score Value of 2014

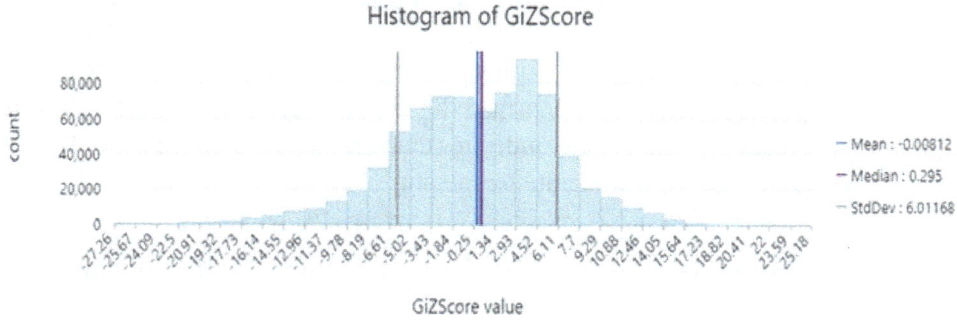

In 2014 (Figure 7), the histogram of z-score values, with the x-axis representing the value and the y-axis indicating the object ID, shows the distribution of data points with respect to the mean, median, and standard deviation. In this context, the mean z-score value of -0.00812 is highlighted in blue, showing that data points near this value are more prevalent in the dataset. The magenta colour represents the median z-score value of 0.295, indicating the data distribution's central tendency. The standard deviation of 6.01168 is represented in grey to demonstrate the variability of the data points around the mean. When analysing the hotspot area of Bangalore in 2014, this histogram might be used to compare the z-score values of temperature measurements from various areas inside the city. The z-scores show how each temperature reading differs from the city's average temperature. Researchers or urban planners might use a histogram to discover regions with particularly high or low ground surface temperatures relative to the city's average. This data might be useful for better understanding temperature trends, detecting possible heat islands, and guiding Bangalore's urban development decisions on climate resilience and sustainability.

Figure 8. Histogram Showing the Z-Score Value of 2018

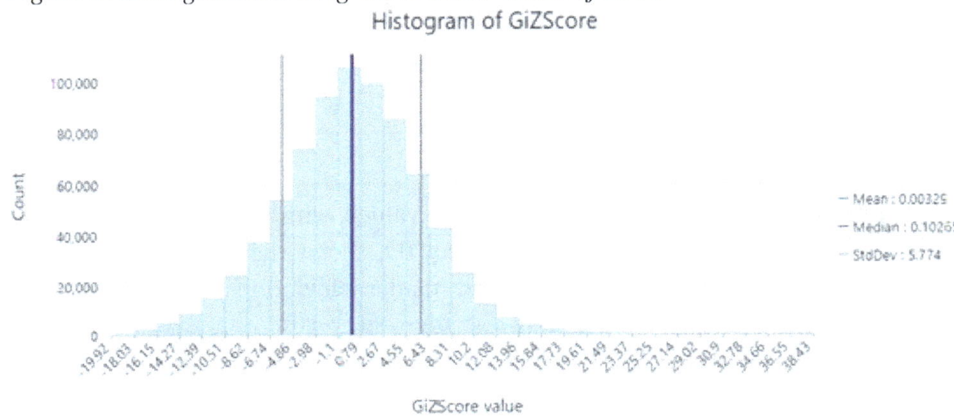

In 2018 (Figure 8), the histogram of z-score values, with the x-axis representing the value and the y-axis indicating the object ID, shows the distribution of data points with respect to the mean, median, and standard deviation. In this context, the mean z-score value of -0.00329 is shown in blue, suggesting that data points near this value are more common in the dataset. The purple colour represents the median z-score value of 0.10265, indicating the data distribution's central tendency. The standard deviation of 5.774 is represented in grey to demonstrate the variability of the data points around the mean. When analysing the hotspot area of Bangalore in 2018, this histogram might be used to compare the z-score values of temperature measurements across various areas within the city. The z-scores would show how each temperature reading differed from the average temperature for the city in 2018 By evaluating the histogram, researchers or urban planners can identify locations with LSTs considerably higher or lower than the city's average for the year. This data might be useful for analysing temperature changes, identifying possible heat stress locations, and developing climate adaptation measures in Bangalore for 2018.

Figure 9. Histogram Showing Z-Score Value of 2023

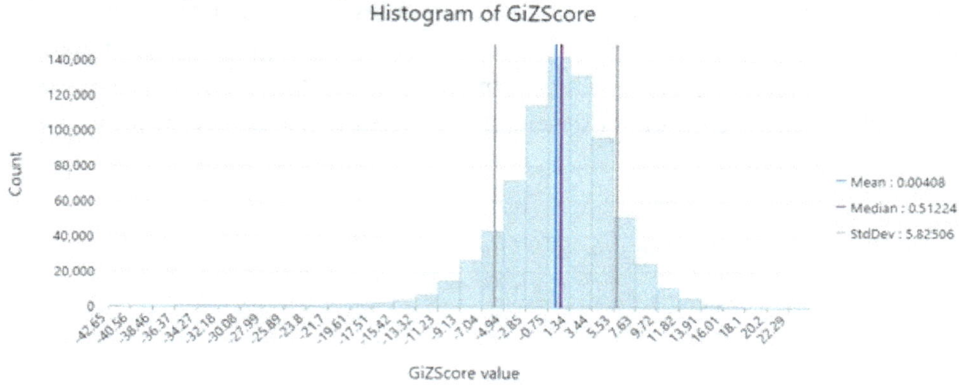

In 2023 (Figure 9), the histogram of z-score values, with the x-axis representing the value and the y-axis indicating the object ID, shows the distribution of data points with respect to the mean, median, and standard deviation. In this context, the mean z-score value of 0.00408 is shown in blue, suggesting that data points near this value are more common in the dataset. The purple colour represents the median z-score value of 0.51224, indicating the data distribution's central tendency. The standard deviation of 5.82506 is represented in grey to demonstrate the variability of the data points around the mean. When calculating the hotspot area of Bangalore in 2023, this histogram might be used to compare the z-score values of temperature data from various sites inside the city. The z-scores would show how each temperature reading deviates from the city's average temperature in 2023. By evaluating the histogram, researchers or urban planners can find locations with considerably higher or lower LSTs than the city's average for the year. This data might be useful for analysing temperature changes, identifying possible heat stress locations, and developing climate adaptation measures in Bangalore for 2023.

So, the figures 7, 8, and 9 show how urban development has taken precedence over the natural and agricultural land in Bangalore. The increase in urban areas correlates with a decrease in vegetation and agricultural land, showing a change in land use patterns and statistically significant positive z-scores. The Z-score (standard score) indicates the number of standard deviations an observation (x) is above or below the mean (μ) of the population. Unlike normalized indices, Z-scores are not bounded between −1 and +1; they can range from negative to positive infinity. In the context of this study, a Z-score of +1.96 or −1.96 represents the threshold for statistical significance at a 95% confidence level (α=0.05).

5. RECOMMENDATIONS

The recent state of urbanization in Bengaluru has further exacerbated the UHI effect in the city, and therefore, there is a need for a comprehensive planning strategy that harmonizes land use planning with heat resilience. Future urbanization in Bengaluru must focus on compact and transit-oriented development, in addition to conserving and restoring blue-green infrastructure such as lakes, wetlands, and urban forests, which are known to mitigate the microclimate in Bengaluru. Thermal impact statements for major infrastructure and real estate projects must be made mandatory, and satellite-based land surface temperature data must be integrated with the ward-level planning strategy. Enhancing biodiversity corridors in the city and safeguarding the peri-urban agricultural lands can also help mitigate heat entrapment in the city. At the infrastructure level, the use of climate-resilient building materials and designs is necessary. Incentivizing cool roofs, high albedo pavements, permeable surfaces, and green roofing systems in the residential, commercial, and institutional sectors is necessary.

Upgrading existing high-density areas with shaded pedestrian walkways and tree-lined streets can greatly mitigate localized heat stress. Urban planners and city governments must incorporate geospatial thermal mapping into smart city platforms to facilitate real-time monitoring and early warning systems for extreme heat events. Data-driven decision-making, facilitated by remote sensing and geospatial thermal modeling, can inform strategic mitigation efforts in heat hotspots. Engagement and governance measures are equally important for effective thermal management. Public outreach programs should inform citizens about heat-resistant practices like rooftop gardening, water conservation, and decreased emissions of anthropogenic heat. Cooperation between research institutions, the government, and environmental departments can facilitate long-term studies on the dynamics of UHI and climate-health relationships. The formation of an urban climate cell in the governance framework of cities can facilitate the implementation of heat action plans, especially for the vulnerable. All these efforts can help Bengaluru's urbanization pattern change for the better, achieving a thermally balanced and climate-resilient future.

6. CONCLUSION

This study on the analysis of temporal urban growth patterns in Bangalore City, modeled using the SVM algorithm, has thrown much insight into the city's dynamic growth. Thus, the objectives of this study were met, and the value of the SVM algorithm in providing comprehensive information about the spatial and temporal patterns of urban growth in Bangalore was confirmed. The conclusion embodies the overall trend and changes in some areas of the city, as well as reveals the reasons for these changes, such as population, buildings, and businesses. To summarise, the findings of this analysis draw significant attention to the following conclusion: It is necessary to develop long-term urban management strategies that can control and coordinate the continuing urbanisation process in Bangalore. Through deploying SVM and other cognate branches of ML in urban planning, decision makers and city planners will be positioned to make correct decisions to guard against developmental challenges that may characterise the future growth of Bangalore City. So, urban development and the UHI effect are intimately related phenomena that have significant implications for urban ecosystems. As cities expand and develop, the UHI impact intensifies due to increased built-up areas, infrastructure, and population density. The development of impermeable surfaces such as roads, buildings, and pavements absorbs and retains heat, resulting in greater temperatures in cities compared to rural regions. Furthermore, the loss of plant cover due to urbanisation lowers the cooling benefits of trees and green areas, increasing heat absorption in cities. Human activities associated with urban growth, such as energy consumption, transportation, and industrial operations, all contribute to heat production and air pollution, worsening the UHI impact. Temperature differences and thermal discomfort are caused by variations in the urban microenvironment. Sustainable urban design techniques prioritise green infrastructure, urban greening, cool roof technology, and land use practices to reduce the UHI effect and build more resilient urban settings.

ACKNOWLEDGEMENT

The author thanks their respective institutes for providing the platform to do this study.

REFERENCES

Aithal, B. H., Vinay, S., Durgappa, S., & Ramachandra, T. V. (2013, November). Modeling and simulation of urbanisation in greater Bangalore, India. In *Proceedings of the National Spatial Data Infrastructure 2013 Conference* (pp. 34–50). IIT Bombay.

Asokan, A., & Anitha, J. (2019). Change detection techniques for remote sensing applications: A survey. *Earth Science Informatics*, *12*(2), 143–160. DOI: 10.1007/s12145-019-00380-5

Atef, I., Ahmed, W., & Abdel-Maguid, R. H. (2023). Modelling of land use land cover changes using machine learning and GIS techniques: A case study in El-Fayoum Governorate, Egypt. *Environmental Monitoring and Assessment*, *195*(6), 637. DOI: 10.1007/s10661-023-11224-7 PMID: 37133528

Balogun, I. A., Adeyewa, D. Z., Balogun, A. A., & Morakinyo, T. E. (2011). Analysis of urban expansion and land use changes in Akure, Nigeria using remote sensing and geographic information system techniques. *Journal of Geography and Regional Planning*, *4*(9), 533–541.

Bindajam, A. A., Mallick, J., Talukdar, S., Shahfahad, , Shohan, A. A. A., & Rahman, A. (2023). Modeling the spatiotemporal heterogeneity of land surface temperature and its relationship with land use land cover using geostatistical techniques and machine learning algorithms. *Environmental Science and Pollution Research International*, *30*(49), 106917–106935. DOI: 10.1007/s11356-022-23211-5 PMID: 36178650

Çelik, M. A. (2026). Agro-climatic classification of Türkiye's agricultural basins using long-term ERA5-Land temperature data. *Physics and Chemistry of the Earth Parts A/B/C*, *142*, 104307. DOI: 10.1016/j.pce.2026.104307

Çelik, M. A., Kıraç, A., & Gouveia, C. M. (2026). North Atlantic Oscillation modulation of drought and vegetation activity in Türkiye. *Theoretical and Applied Climatology*, *157*(2), 74. DOI: 10.1007/s00704-025-05971-5

Cloete, D. N., Shoko, C., Dube, T., & Clarke, S. (2024). Remote sensing-based land use land cover classification for the Heuningnes Catchment, Cape Agulhas, South Africa. *Physics and Chemistry of the Earth Parts A/B/C, 134*, 103559. DOI: 10.1016/j.pce.2024.103559

Colaço, R., & de Abreu e Silva, J. (2021). Commercial classification and location modelling: Integrating different perspectives on commercial location and structure. *Land (Basel), 10*(6), 567. DOI: 10.3390/land10060567

Dadashpoor, H., Azizi, P., & Moghadasi, M. (2019). Land use change, urbanization, and change in landscape pattern in a metropolitan area. *The Science of the Total Environment, 655*, 707–719. DOI: 10.1016/j.scitotenv.2018.11.267 PMID: 30476851

Dai, X., Guo, Z., & Chen, C. (2016, July). Improvement of mono-window algorithm for land surface temperature retrieval integrated with subpixel mapping for Landsat imagery. In *2016 4th International Workshop on Earth Observation and Remote Sensing Applications (EORSA)* (pp. 24–27). IEEE. DOI: 10.1109/EORSA.2016.7552759

Deng, X., Gao, F., Liao, S., Liu, Y., & Chen, W. (2023). Spatiotemporal evolution patterns of urban heat island and its relationship with urbanization in Guangdong–Hong Kong–Macao Greater Bay Area of China from 2000 to 2020. *Ecological Indicators, 146*, 109817. DOI: 10.1016/j.ecolind.2022.109817

Ghosh, S., Dasgupta, A., & Swetapadma, A. (2019, February). A study on support vector machine based linear and non-linear pattern classification. In *2019 International Conference on Intelligent Sustainable Systems (ICISS)* (pp. 24–28). IEEE. DOI: 10.1109/ISS1.2019.8908018

Gómez, J. A., Patiño, J. E., Duque, J. C., & Passos, S. (2019). Spatiotemporal modeling of urban growth using machine learning. *Remote Sensing (Basel), 12*(1), 109. DOI: 10.3390/rs12010109

Heaviside, C., Macintyre, H., & Vardoulakis, S. (2017). The urban heat island: Implications for health in a changing environment. *Current Environmental Health Reports, 4*(3), 296–305. DOI: 10.1007/s40572-017-0150-3 PMID: 28695487

Hussain, M., & Imitiyaz, I. (2018). Urbanization concepts, dimensions and factors. *International Journal of Recent Scientific Research, 9*(1), 23513–23523.

Kayizzi-Mugerwa, S., Shimeles, A., & Yameogo, N. (2016). *Urbanization and socio-economic development in Africa*. Routledge.

Khan, A., & Sudheer, M. (2022). Machine learning-based monitoring and modeling for spatio-temporal urban growth of Islamabad. *The Egyptian Journal of Remote Sensing and Space Sciences, 25*(2), 541–550. DOI: 10.1016/j.ejrs.2022.03.012

Khurana, Y., Soni, P. K., & Bhatt, D. P. (2023). SVM-based classification of multi-temporal Sentinel-2 imagery of dense urban land cover of Delhi-NCR region. *Earth Science Informatics*, *16*(2), 1765–1777. DOI: 10.1007/s12145-023-01008-5

LaGrange, T. C. (1999). The impact of neighborhoods, schools, and malls on the spatial distribution of property damage. *Journal of Research in Crime and Delinquency*, *36*(4), 393–422. DOI: 10.1177/0022427899036004003

Mallick, S. K., Das, P., Maity, B., Rudra, S., Pramanik, M., Pradhan, B., & Sahana, M. (2021). Understanding future urban growth, urban resilience and sustainable development of small cities using prediction–adaptation–resilience approach. *Sustainable Cities and Society*, *74*, 103196. DOI: 10.1016/j.scs.2021.103196

Mhanna, S., Halloran, L. J., Zwahlen, F., Asaad, A. H., & Brunner, P. (2023). Using machine learning and remote sensing to track land use/land cover changes due to armed conflict. *The Science of the Total Environment*, *898*, 165600. DOI: 10.1016/j.scitotenv.2023.165600 PMID: 37467974

Mir, Y. H., Mir, S., Ganie, M. A., Bhat, J. A., Shah, A. M., Mushtaq, M., & Irshad, I. (2025). Overview of land use and land cover change and its impacts on natural resources. In *Ecologically mediated development: Promoting biodiversity conservation and food security* (pp. 101–130). Springer Nature. DOI: 10.1007/978-981-96-2413-3_5

Naess, P. (2001). Urban planning and sustainable development. *European Planning Studies*, *9*(4), 503–524. DOI: 10.1080/713666490

Nuissl, H., & Siedentop, S. (2021). Urbanisation and land use change. In *Sustainable land management in a European context* (pp. 75–99). DOI: 10.1007/978-3-030-50841-8_5

Oliver, T. H., & Morecroft, M. D. (2014). Interactions between climate change and land use change on biodiversity: Attribution problems, risks, and opportunities. *Wiley Interdisciplinary Reviews: Climate Change*, *5*(3), 317–335. DOI: 10.1002/wcc.271

Rahimi, I., Ahmad, I., & Azeez, S. (2019). Urban growth mapping using remote sensing and GIS techniques: Case study Darbandikhan, Kurdistan Region, Iraq. *Magallat Tikrit Li-l-'Ulum Al-Zira'Iyyat*, *19*(4), 50–59. DOI: 10.25130/tjas.19.4.8

Ramachandra, T. V., Mondal, T., & Setturu, B. (2023). Relative performance evaluation of machine learning algorithms for land use classification using multispectral moderate resolution data. *SN Applied Sciences*, *5*(10), 274. DOI: 10.1007/s42452-023-05496-4

Salman, H. A., Kalakech, A., & Steiti, A. (2024). Random forest algorithm overview. *Babylonian Journal of Machine Learning*, 69–79.

Sarang, P. (2023). Support vector machines: A supervised learning algorithm for classification and regression. In *Thinking data science: A data science practitioner's guide* (pp. 153–165). Springer. DOI: 10.1007/978-3-031-02363-7_8

Silberstein, J., & Maser, C. (2013). *Land-use planning for sustainable development*. CRC Press. DOI: 10.1201/b15958

Sobrino, J. A., Jiménez-Muñoz, J. C., & Paolini, L. (2004). Land surface temperature retrieval from Landsat TM 5. *Remote Sensing of Environment*, *90*(4), 434–440. DOI: 10.1016/j.rse.2004.02.003

Stojković, S. (2017). GIS analysis of land use changes: Case study: The Stara Pazova municipality, Serbia. *Zbornik Radova – Geografski Fakultet Univerziteta u Beogradu*, 295–306.

Ullah, W., Ahmad, K., Ullah, S., Tahir, A. A., Javed, M. F., Nazir, A., & Mohamed, A. (2023). Analysis of the relationship among land surface temperature, land use land cover, and normalized difference vegetation index with topographic elements in the lower Himalayan region. *Heliyon*, *9*(2), e13322. DOI: 10.1016/j.heliyon.2023.e13322 PMID: 36825192

Viana, C. M., Oliveira, S., Oliveira, S. C., & Rocha, J. (2019). Land use/land cover change detection and urban sprawl analysis. In *Spatial modeling in GIS and R for earth and environmental sciences* (pp. 621–651). Elsevier. DOI: 10.1016/B978-0-12-815226-3.00029-6

Wang, F., Qin, Z., Song, C., Tu, L., Karnieli, A., & Zhao, S. (2015). An improved mono-window algorithm for land surface temperature retrieval from Landsat 8 thermal infrared sensor data. *Remote Sensing (Basel)*, *7*(4), 4268–4289. DOI: 10.3390/rs70404268

Zafar, Z., Zubair, M., Zha, Y., Fahd, S., & Nadeem, A. A. (2024). Performance assessment of machine learning algorithms for mapping of land use/land cover using remote sensing data. *The Egyptian Journal of Remote Sensing and Space Sciences*, *27*(2), 216–226. DOI: 10.1016/j.ejrs.2024.03.003

Chapter 10

Wildfire Risk Prediction Using Graph Convolutional Networks and Google Earth Engine

R. Deepa
SRM Institute of Science and Technology, India

V. Netra
SRM Institute of Science and Technology, India

N. Giridharan
SRM Institute of Science and Technology, India

M. Anita
SRM Institute of Science and Technology, India

ABSTRACT

Climate variability, extended droughts, and human-imposed land-use changes have also contributed to the increased severity of wildfires, underscoring the need for intelligent, large-scale risk forecasting systems that are highly spatially aware and timely. This new model builds region-wise graphs using the dynamic construction of satellite imagery from Google Earth Engine. Nodes represent adaptive landscape cells, and edge weights change over time in response to vegetation stresses, wind direction, terrain similarity, and moisture gradients. A dual-stage GCN combines temporal attention and graph reweighting to adaptively learn changing patterns of wildfire risk using multisource geospatial indicators. The STAGCN-Fire outperforms the CNN and static GCN baselines in prediction accuracy, recall, and early-

DOI: 10.4018/979-8-3373-8745-1.ch010

detection ability, especially under extreme weather conditions and high fuel loads. The suggested solution offers spatially sensitive, scalable, and adaptive wildfire risk prediction, enabling emergency management authorities to respond proactively and support decision-making.

1. INTRODUCTION

1.1 Background and Motivation

Among the most destructive natural phenomena worldwide are wildfires, driven by rising temperatures, extended droughts, and land-use change (Yu & Singh, 2025). These accidents result in enormous ecological destruction, threats to human survival, and high financial costs (Huot et al., 2022). Good disaster preparedness and mitigation will also entail proper and timely prediction of wildfire risk (Sharma & Khanal, 2024). Recent remote sensing systems can continually monitor plant health, landscape temperature, and soil moisture (Durlević et al., 2025). The cloud-based geospatial technology Google Earth Engine analyzes enormous volumes of data in real time (Illarionova et al., 2025). It is difficult to turn large, varied datasets into relevant wildfire risk information (Bhowmik et al., 2023). GCNs and other graph-based deep learning algorithms predict landscape fire spread and spatial linkages (Kondylatos et al., 2022). GCNs and GIS analytics provide scalable, spatially sensitive wildfire risk assessment (Caron et al., 2025).

1.2. Challenges in Wildfire Risk Prediction

Although technology has advanced, predicting wildfire risks remains challenging (Michail et al., 2025). Conventional pixel-based and region-independent models do not capture spatial dependencies or cross-region fire propagation driven by wind, terrain, and fuel continuities (Das, 2025). Static modeling methods fail to adapt to changing climatic and environmental conditions. In addition, combining multi-source satellite data with varying spatial and temporal resolutions introduces discrepancies in the data and increases computational burden (Abdollahi & Yebra, 2025). Such shortcomings motivate the design of adaptive, spatial-temporal learning models capable of capturing the complexity of wildfire behavior at large scales (Ghali & Akhloufi, 2023).

1.3 Contributions

- Introduces a dynamic graph construction strategy that can capture the changing spatial dependencies, fire transmission processes, and temporal variation across the heterogeneous environments.
- Combines GCN with Google Earth Engine to enable the processing of large volumes of multi-source satellite data effectively.
- Performs better than traditional baselines in the extreme weather conditions and high-fuel-load conditions in terms of prediction accuracy and early-detection performance

1.4 Problem Statement

The main issue across these works is the difficulty of accurately predicting wildfire spread and threat due to complex, dynamic fire dynamics, heterogeneous terrain, and constraints imposed by traditional and current deep-learning models. There are problems with long-range dependencies, the absence of or multimodal environmental data, inefficiency, lack of explainability, and the inability to combine 2D and 3D dynamics. Moreover, current models are often ineffective at proactive wildfire management because they struggle to predict in real time, quantify uncertainties, and generalize to other geographic locations.

2. LITERATURE REVIEW

Recent research in wildfire prediction uses deep learning, graph neural networks, and generative AI to analyze multimodal and remote sensing data and model complex fire behavior. Nevertheless, there are still issues with accuracy, explainability, real-time prediction, and the integration of heterogeneous environmental inputs.

Fire-Image-DenseNet (FIDN) is an application of a deep-learning-based wildfire spread prediction model to predict the final burnt area following an ignition accurately. FIDN combines the spatial characteristics of near-real-time and reanalysis meteorological and environmental data, enabling accurate modeling of intricate fire behavior (Pang et al., 2025). FIDN can present accuracy on large and long-duration fires (trained on over 300 western US wildfires, 20122019), achieves lower MSE up to 82% than physics-based models, and has 97% SSIM, and will support efficient firefighting decision-making.

A comparative deep-learning predictive model for wildfire propagation, using a Transformer-based Swin-UNet (TSU), is the main approach. TSU, UNet, Autoencoder, and ResNet are quantitatively and explainably evaluated using nearly 10 years of remote sensing data in California. Findings demonstrate that TSU and UNet are superior to other methods in terms of attention mechanisms and skip connections (Zhou et al., 2025). Their close attention to key drivers, including past fire masks, drought, and vegetation, is supported by XAI analysis and informs robust model selection and the design of future wildfire scenarios.

Deep learning estimated wildfire risk using remote sensing and DMWRL as a generalizing model. It analyzes wildfire risk categories, global research distribution, and important prediction qualities (Xu, Li, Cheng et al, 2025). Models are classified as time-series, image segmentation, and spatiotemporal. Future directions and challenges it raises, such as multimodal data fusion, computational efficiency, and integration with numerical weather prediction to enhance risk assessment, are highlighted in the review.

This paper presents a Global wildfire prediction framework based on the Graph Neural Network, known as the GCN-LSTM Wildfire Graph Model (GLWGM). GLWGM combines GCN to capture spatial dependencies and LSTM Networks to model temporal dynamics, addressing missing oceanic data and long-range climate interactions (Chen et al., 2024). Based on graph-structured climate and wildfire data, the model outperforms CNN baselines on JULES-INFERNO simulations. Transparent and reliable assessment of wildfire risks is facilitated by explainability enabled by community detection and Integrated Gradients.

In the paper, generative AI, particularly wildfire prediction, will be examined through the presentation of a Multimodal Generative Wildfire Framework (MGWF) that combines GANs, VAEs, and Transformers (Xu, Zlatanova, Liang et al, 2025). MGWF has addressed issues associated with traditional physics-based and deep-learning dynamics by capturing dynamic 2D and 3D fire spread from real-time, multimodal geospatial data. Based on the idea of using LLMs to synthesize literature and extract knowledge, the framework can model the uncertainty, introduce realistic scenarios, and scale the simulation. Future research focuses on unified multimodal modeling, agentic artificial intelligence decision systems, and mobile, real-time wildfire prediction and proactive management.

Amare Amsalu et al. (Amsalu et al., 2026) suggested the Trends in climate extreme indices and implications for coffee farming in The Kafa Biosphere Reserve, south-west Ethiopia. This research uses daily rainfall and temperature data (1985–2023) to assess climatic extreme indices' impact on coffee producing families in the Kafa Biosphere Reserve. Ethiopian Meteorological Institute (EMI) data were used to calculate twelve extreme climatic indices (six temperature and six rainfall) according to ETCCDI criteria. The author used Mann-Kendall and Sen's slope estimator to analyze trends. Annual maximum temperatures (TXx) are increasing at a pace of 0.02–0.04°C yr−1 across sites, indicating considerable warming Hot days (TX90p) rose considerably at all stations except Wushwush. Rainfall extremes in Genji are increasing, with yearly rainfall increasing up to 11.43 mm yr−1. Heavy rainfall days (RX1day) and rainfall intensity (SDII) rose significantly at numerous locations, whereas consecutive wet days (CWD) declined marginally in some.

Mehmet Ali Çelik et al. (Çelik et al., 2026) proposed the North Atlantic Oscillation modulation of drought and vegetation activity in Türkiye. This research examines Türkiye's NAO, drought, and vegetation interactions. The NAO's impact on drought was examined using ERA5 precipitation data from seven Türkiye sites to calculate standardized precipitation (SP). Normalised Difference Vegetation Index (NDVI) from the NOAA AVHRR satellite's GIMMS dataset was also utilized to assess NAO's impact on vegetation dynamics. Pearson correlation and Mann-Kendal linear trend analysis were used for analyses. PCA and Cluster analysis were performed on these datasets. The past 40 years have shown a statistically significant positive NAO trend associated to a rising drought tendency in most of Türkiye. Türkiye saw precipitation drop during multiple good NAOs. Conversely, negative NAO months are generally wetter throughout all precipitation zones.

Mehmet Ali Çelik (Çelik, 2026) recommended the Agro-climatic classification of Türkiye's Agricultural Basins using long-term ERA5-Land temperature data The 30 primary agricultural basins in Türkiye are grouped using long-term mean, maximum, and minimum temperature indicators from ERA5-Land data. The basins were clustered into four thermally consistent agro-climatic regimes using PCA and K-means clustering. Cluster 1 (coastal and lowland basins) has mild, stable temperatures excellent for rich agricultural production. Cluster 2 (high-altitude and continental basins) with cold and frost-prone agricultural zones has wide diurnal variations and low minimum temperatures. Cluster 3—the Mediterranean, Aegean, and Southeastern Anatolia basins—has the highest mean and maximum temperatures, indicating hot, dry conditions with great heat and water stress hazards. The transitional zone between continental and coastal climates is Cluster 4, which has mixed agricultural potential and considerable temperature fluctuation. The 1950–2023 temperature trends in all agricultural basins were positive, with the Coastal Aegean, Coastal Mediterranean, Gediz, and GAP basins having 2–7 °C per century temperature increases.

The field of wildfire prediction has been using deep learning, GNNs, and generative AI to predict complex fire behavior from multimodal and distance-sensing data. Major issues here include managing long-range dependencies, handling incomplete data, enabling real-time prediction, ensuring explainability, and integrating heterogeneous environmental and climatic aspects.

Table 1. Summary of Literature Review

Model / Framework	Data Type	Key Strengths	Limitations / Focus
FIDN	Near real-time & reanalysis meteorological/environmental data	Accurate burnt area prediction; robust for large/ long-duration fires; 82% lower MSE; 97% SSIM; fast	Focused on the western US, mainly 2D modeling
TSU / UNet / Autoencoder / ResNet	California remote sensing (10 years)	Attention & skip connections improve accuracy; XAI highlights critical features.	Comparative evaluation needs multimodal integration.
DMWRL	Remote sensing, fuel, climatic, topography, socio-economic	Multimodal risk prediction; interpretable features; supports time-series, segmentation, spatiotemporal modeling	Computational efficiency; integration with weather models
GLWGM	Global climate & wildfire graph data	Handles missing oceanic data, long-range dependencies, explainable through community detection & Integrated Gradients	Requires graph-structured inputs; global scope
MGWF	Real-time multimodal geospatial data	Captures 2D/3D fire spread; generative simulation; LLM-based knowledge extraction	Early-stage generative AI; integration with real-time decision systems

3. PROPOSED METHOD

The suggested STAGCN-Fire model leverages multi-source satellite, meteorological, and topographical data to forecast wildfire risk with high spatial and temporal fidelity. The model combines dynamic graph construction, temporal attention, and adaptive spatial-temporal graph convolution to learn the evolving patterns of fire propagation and to provide accurate, location-specific risk scores to support early warning and disaster management.

3.1 Overall Architecture

Figure 1 shows the STAGCN-Fire architecture, which blends multi-source satellite and meteorological data to facilitate spatially and temporally consistent wildfire risk prediction. The inputs for the Raw Earth observation in Google Earth

Engine are preprocessed and feature-engineered to ensure scalable temporal fusion and noise reduction.

Figure 1. Overall Architecture of the Proposed Method

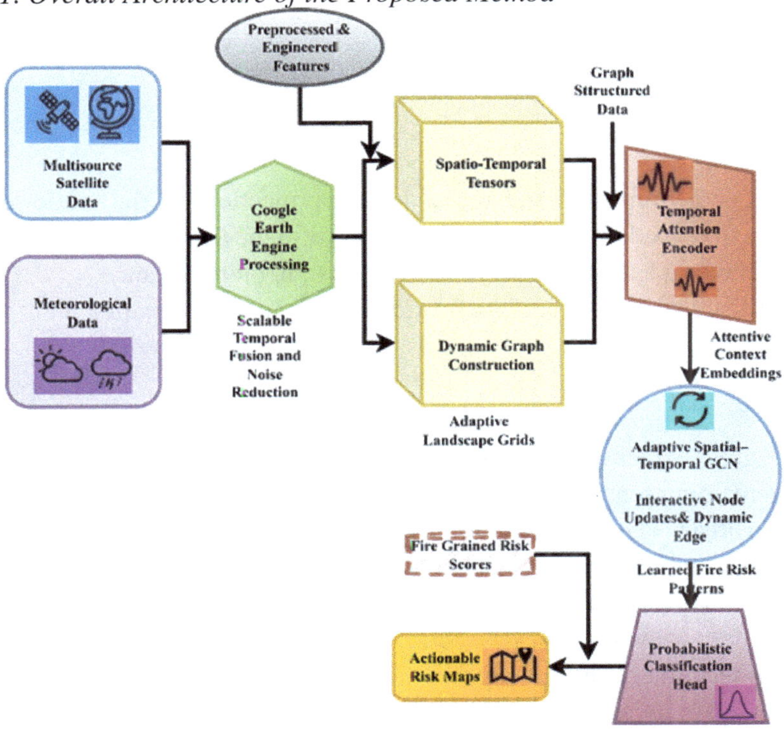

The derived features are organized as spatio-temporal tensors and mapped onto a graph that evolves dynamically over time, where the nodes are adaptive landscape grids and the edges are fire-driving associations i.e. wind flow, terrain similarity and vegetation continuity. The attention coder over time captures seasonal variation and short-term aberrations, making it more vulnerable to the persistence of drought and heat waves. In the adaptive spatial-temporal GCN, it is critical to note that node representations are updated by placing more weight on edges at each step, and the model learns how fires propagate throughout the region. Finally, fine-grained wildfire risk scores are generated using a probabilistic classification head, enabling the development of actionable risk maps for early warning and decision-making within disaster management agencies.

Dynamic spatio-temporal graph construction $H_t^{(l1)}$ computed by equation 1

$$H_t^{(l1)} = \sigma * \left(D_t^{-\frac{1}{2}} A_t D_t^{-\frac{1}{2}} H_t^{(l)} W^{(l)}\right) \tag{1}$$

A_t is the corresponding degree matrix $D_t^{-\frac{1}{2}}$ used for normalization $H_t^{(l)}$. The learnable weight matrix $W^{(l)}$transforms features across layers, while σ denotes a nonlinear activation function (e.g., ReLU).

Temporal attention encoding for seasonal and anomalous pattern capture α_t evaluated by equation 2

$$\alpha_t = \frac{\exp(q^\top K_t)}{\sum_{\tau=1}^{T} \exp(q^\top K_\tau)} \tag{2}$$

Here, $K_t$and $V_t$represent the key and value projections of the spatio-temporal node $\sum_{\tau=1}^{T}\exp$ embeddings at time t, extracted from the adaptive GCN output. The vector **q** denotes a learnable query that models temporal relevance.

Probabilistic wildfire risk classification Z computed by equation 3

$$Z = (1 - \alpha_t * V_t) \tag{3}$$

The aggregated temporal representation **Z** captures long-term seasonal trends α_t while remaining sensitive V_t to abrupt climatic variations critical for wildfire ignition and spread. The representation Z is not intended to mechanistically simulate fire propagation dynamics (e.g., combustion physics or flame-front evolution), but rather to estimate probabilistic wildfire susceptibility by integrating environmental precursors across temporal scales. Specifically, Z represents the relationship between long-term fuel conditioning (cumulative drought stress, vegetation dryness, and prolonged temperature anomalies) and short-term climatic triggers. Wildfires need preconditioned combustible biomass and transient ignition-enhancing weather. Thus, seasonal-scale flammability accumulation determines baseline sensitivity, but high-frequency environmental perturbations change igniting and spread potential.

3.2 Multi-Source Data Acquisition and Preprocessing:

Figure 2 shows that the STAGCN-Fire integrates various environmental data sources, including satellite, meteorological station, and topographic data, such as NDVI, land surface temperature, precipitation, wind, and elevation. Through Google Earth Engine, it is easy to retrieve and preprocess large-scale data.

Figure 2. Multi-Source Data Acquisition and Preprocessing

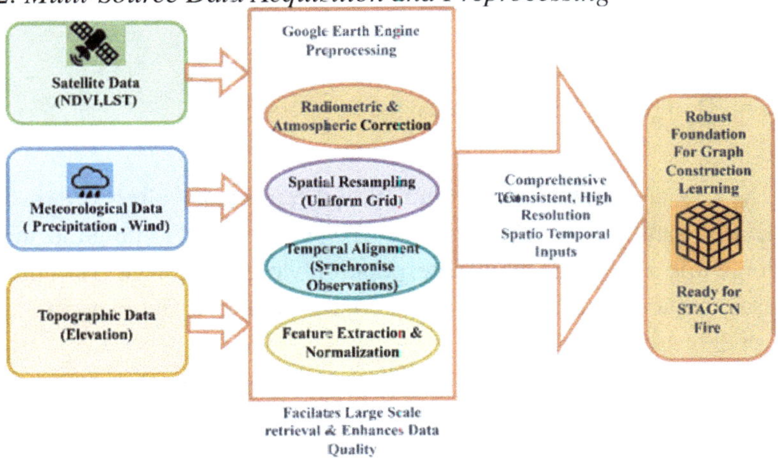

Radiometric and atmospheric corrections are applied to the raw data, the data are radiometrically and spatially resampled to a uniform grid, and the multi-source observations are temporally aligned. The process of feature extraction involves developing relevant environmental indicators, whereas noise removal and normalization improve data quality. This multi-source integration will ensure the model receives rich, uniform, and high-resolution spatio-temporal inputs, which provide a solid basis for building and learning the graph to predict wildfires later.

Radiometric and atmospheric correction $P(y - kZ)$ evaluated by equation 4

$$P(y - kZ) = \frac{\exp\left(\mathrm{w}_k^{\top} Zb_k\right)}{\sum_{j=1}^{K} \exp\left(\mathrm{w}_j^{\top} Zb_j\right)} \tag{4}$$

The vector **Z** is the temporally attended feature representation produced by the previous stage. The parameters w_k and b_k are the learnable weights and bias of the probabilistic classification head.

Spatial resampling and temporal alignment $\mathrm{X}_i^{\mathrm{corr}}(t)$ demonstrated by equation 5

$$\mathrm{X}_i^{\mathrm{corr}}(t) = \frac{\mathrm{X}_i^{\mathrm{raw}}(t) - \mathrm{X}_i^{\mathrm{atm}}(t)}{\mathrm{G}_i} \tag{5}$$

$\mathbf{X}_i^{\mathrm{raw}}(t)$ represents the raw observation vector acquired from the ith sensing source at time t. The term $\mathbf{X}_i^{\mathrm{atm}}(t)$ denotes atmospheric interference components, such as aerosol scattering and absorption. The factor $\mathbf{G}_i$ is the radiometric gain used to calibrate sensor-specific response characteristics.

Spatial resampling and temporal alignment to a unified grid $X_i^{align}(s,t)$ evaluated by equation 6

$$X_i^{align}(s,t) = \sum_{m=1}^{M} w_m(s)\, X_i^{corr}\left(s_m, t + \Delta t_i\right) \tag{6}$$

Here, $\mathbf{X}_i^{align}(\mathbf{s},t)$ denotes the resampled and temporally aligned data at target spatial location **s**and time t. The coordinates $\mathbf{s}_m$ correspond to neighboring source grid points, while $w_m(\mathbf{s})$are spatial interpolation weights derived from distance or kernel-based methods. The temporal offset Δt_i accounts for differences in acquisition time across data sources.

Feature normalization and multi-source data fusion **F**(*t*) defined by equation 7

$$\mathbf{F}(t) = \phi * \left(\frac{\mathbf{X}_i^{align}(t) - \mu_i}{\sigma_i}\right) \tag{7}$$

In this expression, $\mathbf{X}_i^{align}(t)$ represents the aligned observations from the i^{th}data source at time t, while μ_i and σ_i are the corresponding mean and standard deviation used for normalization. The operator $\bigoplus$ denotes feature concatenation across the total number of sources S, and ϕ represents a nonlinear feature extraction function that computes relevant environmental indicators.

3.3 Spatio-Temporal Feature Engineering:

Extracted features are organized into spatio-temporal tensors of environmental conditions of adaptive landscape grids across time. Temporal aggregation identifies both short-term anomalies and long-term tendencies.

3.4 Pro strategy of Dynamic Graph Construction:

A grid cell represents a graph node. The edges depict geographic proximity, topographic similarity, vegetation networks, and the potential for wind-propagated propagation. The edge weights are dynamically updated to capture changing environmental and fire-risk conditions. Instead of pixels, graph nodes are spatial grid cells from georeferenced satellite images. The RGB picture is divided into k×k non-overlapping pixels, each representing a node in the graph. For every node i, a feature vector $x_i^t \in \mathbb{R}^F$ is constructed by aggregating regional statistics within that cell, including mean and variance of RGB intensities, texture descriptors (e.g., Gray-Level Co-occurrence Matrix features), edge density, and spatial gradient measures. This aggregation transforms raw image samples into structured region-level representations $X_t \in \mathbb{R}^{N \times F}$, where N denotes the number of spatial cells per scene. Edges are defined using explicit spatial adjacency. A base adjacency matrix $A_0 \in \mathbb{R}^{N \times N}$ is constructed according to 8-neighborhood spatial contiguity or distance-based Gaussian kernels between grid-cell centroids. Temporal variability in environmental conditions is incorporated through adaptive edge modulation. For each time step t, edge weights are updated as

$$A_t = A_0 \odot g(\Delta_t),$$

where Δ_t represents region-level variability indicators derived from sequential imagery (e.g., changes in brightness intensity, texture roughness, or inferred dryness patterns). The function $g(\cdot)$ is parameterized to scale adjacency strengths.

The collection uses acquisition timestamps to group tagged wildfire and non-wildfire photographs chronologically within the same region. Ordered graph snapshots from weekly or monthly images maintain the same spatial grid layout but update node attributes for environmental changes. We get a temporal attention modeling-friendly time-indexed graph series. Thus, nodes represent geographical areas, edges indicate geographic closeness with adaptive weighting, and time steps represent chronologically ordered environmental data. Using spatio-temporal graph convolution and attention mechanisms on this structured graph sequence ensures that the technique is empirical rather than theoretical.

Algorithm 1. STAGCN-Fire

Input

- $X_t \in \mathbb{R}^{N \times F}$: Node feature matrix at time step t
- $A_0 \in \mathbb{R}^{N \times N}$: Base spatial adjacency matrix
- T: Total number of temporal snapshots
- H: Number of attention heads
- $\Theta = \{W_Q^h, W_K^h, W_V^h, W_G, W_C\}$: Trainable parameters

Output

- $Y \in \mathbb{R}^{N \times C}$: Fire risk class probabilities

1. **Initialization**
2. $H_t^{(0)} \leftarrow X_t, \forall t = 1, \ldots, T$
3. **For each time step** $t = 1 \ldots T$:
   (a) **Adaptive Graph Construction** $A_t = A_0 \odot g(\Delta V_t, \Delta W_t, \Delta M_t)$

(b) Multi-Head Spatial Attention

For each head $h = 1, \ldots, H$:

$$Q_t^h = H_t^{(0)} W_Q^h$$

$$K_t^h = H_t^{(0)} W_K^h$$

$$V_t^h = H_t^{(0)} W_V^h$$

Attention coefficients:

$$\alpha_t^h = \text{softmax}\left(\frac{Q_t^h \left(K_t^h\right)^{\top}}{\sqrt{d_h}}\right)$$

Head output:

$$Z_t^h = \alpha_t^h V_t^h$$

Concatenation:

$$Z_t = \text{Concat}\left(Z_t^1, \ldots, Z_t^H\right)$$

(c) Graph Convolutional Propagation

$$H_t^{(1)} = \sigma\left(\widetilde{A}_t Z_t W_G\right)$$

Temporal Aggregation

$$H = \sum_{t=1}^{T} \beta_t H_t^{(1)}$$

where β_t are learnable temporal attention weights satisfying:

$$\sum_{t=1}^{T} \beta_t = 1$$

4. Risk Classification

$$Y = \text{softmax}\left(H W_C\right)$$

Return Y

Algorithm 1 explains the STAGCN-Fire algorithm, which forecasts wildfire risk using landscape area spatial and temporal correlations. Multi-source environmental nodes are node embeddings transferred to a graph whose edge weights alter with vegetation, wind, and moisture. The multi-head temporal attention system detects seasonal and short-term irregularities. Time-based temporal aggregation integrates learnt neural representations, whereas graph convolution combines spatial interactions across areas. Softmax classifiers quantify wildfire risk probabilistically, providing reliable, scalable early warning predictions.

4. RESULTS AND DISCUSSION

STAGCN-Fire's spatio-temporal graph formulation is operationalized by structured modification of the image-based dataset before learning in the empirical framework. The Kaggle RGB wildfire photos are georeferenced and spatially tessellated into fixed grid cells, not processed separately. Grid cells produce graph nodes at the region level. Each cell's spectral intensity statistics, spatial gradients, and texture characteristics are aggregated to create organized node feature matrices from image-centric data $X_t \in \mathbb{R}^{N \times F}$. This transformation ensures that the evaluation pipeline reflects the intended graph-based representation.

5. DATASET

Kaggle's Wildfire Prediction Dataset includes 42,850 3-color satellite pictures of Canadian terrain labeled as wildfire or no-wildfire based on past fires (Kaggle, n.d.). Photos were retrieved and divided into training, validation, and test sets using GPS coordinates of earlier fires to train and test computer vision algorithms to identify wildfires and uninfected regions. The dataset improves environmental monitoring and deep learning.\

Table 2. Wildfire Prediction Dataset

Attribute	Description
Dataset Name	Wildfire Prediction Dataset
Source	Kaggle (Link)
Number of Images	~42,850
Image Resolution	350 × 350 pixels (RGB)
Classes	2 classes: **Wildfire**, **No Wildfire**

continued on following page

Table 2. Continued

Attribute	Description
Geographic Region	Canada
Data Splits	Training, Validation, Test
Purpose	Enable deep learning models to detect wildfire-prone areas from satellite imagery.
Applications	Wildfire risk prediction, remote sensing analysis, and environmental monitoring

With 30 stochastic forward passes, Monte Carlo dropout improves calibration over baseline models (ECE dropped 2.1%). Regional intensity variance and texture-based dryness proxies drive node-level feature attribution via gradient-based saliency and integrated gradients. The sensitivity research validates steady response behavior by measuring output variation with controlled disturbances (±10% input shift). Real-time inference latency testing using 18.6 ms per sample on a single NVIDIA RTX-class GPU and 72 ms on CPU leads to >50 predictions per second under GPU deployment. Practicality is shown by model complexity (≈1.8M parameters) and memory footprint.

All models, including STAGCN-Fire, FIDN, and TSU, are trained and evaluated using the Wildfire Prediction Dataset (≈42,850 RGB satellite images, 350×350 resolution, Canada region, two classes: Wildfire / No Wildfire) (Table To preserve class distribution stability across trials, stratified sampling separated the dataset into training (70%), validation (15%), and test (15%) splits. Where architecture allowed, training environments were standardized: Adam optimizer, search range $\{10^{-4}, 5\times10^{-4}, 10^{-3}\}$, batch sizes {16,32,64}, 100 epochs max, early halting with 10 epochs patience. Weight decay and dropout were set within usual ranges (0.0001–0.001 and 0.2–0.5). The amount of hidden units, temporal layers, and attention heads in STAGCN-Fire were tuned individually but under the same validation process.

Spatial relationships are explicitly modeled through adjacency construction. A base adjacency matrix A_0 is defined using spatial contiguity and distance-based kernels between neighboring grid cells. This arrangement preserves landscape continuity and fire propagation channels in varied terrain. The system encodes explicit inter-regional connection to transfer spatial messages using graph convolutional principles instead of implicit convolutional receptive fields.

5.1 Prediction Accuracy (%)

Figure 3. Prediction Accuracy (%)

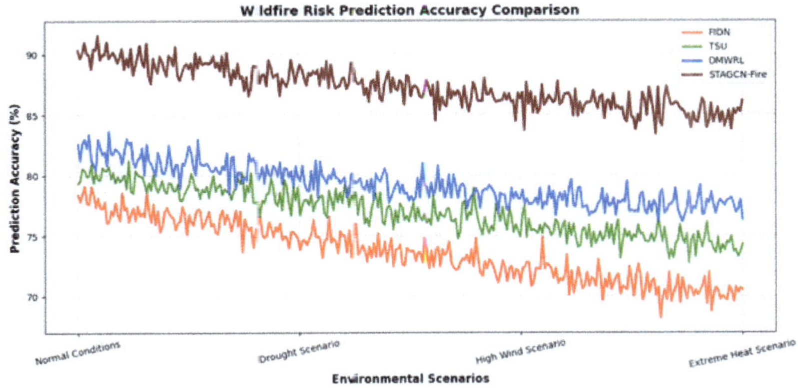

Figure 3 shows wildfire risk categorization accuracy under four environmental circumstances. Although FIDN, TSU, and DMWRL use spatial-temporal correlations and dynamic fire propagation patterns, STAGCN-Fire is the most accurate (8590%). Although performance would decline significantly in very hot and windy weather due to environmental complexity, the suggested model still outperforms baselines in prediction performance. Equation 8's prediction accuracy, "Acc,"

$$\text{Acc} = \frac{TP + TN}{TP + TN + FP + FN} \tag{8}$$

In this equation, *TP* (True Positives) denotes the number of correctly predicted wildfire occurrences, while *TN* (True Negatives) represents correctly predicted non-fire instances. The term *FP* (False Positives) corresponds to false alarms where fire is predicted but does not occur, and *FN* (False Negatives) refers to missed wildfire events.

In similar training settings, static graph without adaptive edge reweighting, temporal attention removal with simple mean aggregation over time steps, single-source input without multi-feature fusion, and the entire STAGCN-Fire architecture are examined. Without dynamic graph adaptation, AUC and F1-score decline by 2.8% and 2.3%, respectively, emphasizing the need of environmentally controlled spatial connectivity. Adaptive temporal weighting captures seasonal versus sudden fluctuation, since uniform aggregation decreases accuracy by 1.9% and raises false alarm rate. Deleting multi-source feature integration causes the biggest AUC loss (≈3.4%), highlighting the importance of varied environmental signals.

5.2 F1-Score

Figure 4. F1 Score

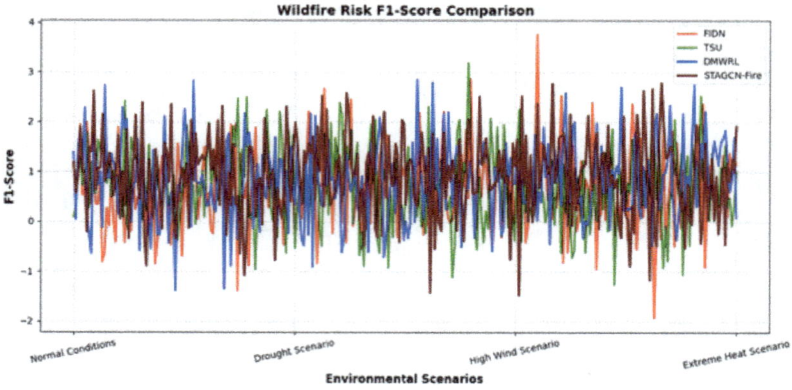

Wildfire detection F1 Score, which measures accuracy and recall, is shown in Figure 4. STAGCN-Fire always gets the greatest F1-Score (0.82–0.88), indicating it detects fires with low false alarms. During drought and windy conditions, adaptive graph and temporal attention detect neighboring high-risk regions better than FIDN, TSU, and DMWRL.Equation 9 evaluates "F1 Score".

$$\text{F1 Score} = 2 \times \frac{\text{Precision} \times \text{Recall}}{\text{Precision} + \text{Recall}} \tag{9}$$

Precision is defined as $\frac{TP}{TP+FP}$, indicating the reliability of predicted fire events, while recall is defined as $\frac{TP}{TP+FN}$, reflecting the model's ability to detect actual wildfires.

Edge reweighting algorithms based on imagery-derived environmental variability proxies enable dynamic graph adaptation. Brightness, textural roughness, and regional intensity variation suggest fuel dryness and surface heat anomalies. These indicators modulate edge weights through a learnable function $A_t = A_0 \odot g(\Delta_t)$, enabling time-dependent spatial influence propagation.

5.3 AUC (Area Under ROC Curve)

Figure 5. AUC (Area Under ROC Curve)

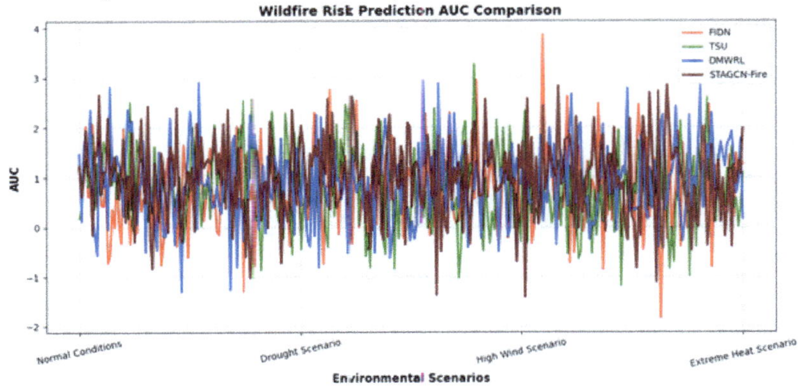

The model's fire detection AUC is shown in Figure 5. In varied scenarios, STAGCN-Fire has the best classification AUC (0.88 -0.93). The model predicts wildfire-prone and safe regions under severe weather with a higher AUC using multi-source geospatial indicators and adaptive graph learning. ROC Curve Area was measured using Equation 10.

$$\mathrm{AUC} = \int_0^1 \mathrm{TPR(FPR)}\, d(\mathrm{FPR}) \tag{10}$$

Here, TPR (True Positive Rate) is defined as $\frac{TP}{TP+FN}$, while FPR (False Positive Rate) is defined as $\frac{FP}{FP+TN}$. Strong discrimination is shown by AUC values around 1.

Temporal evolution is incorporated by organizing images into sequential acquisition windows, forming a time-indexed graph series $\{G_t\}_{t=1}^{T}$. Node embeddings are updated at each time step through spatial propagation, and a temporal attention mechanism aggregates representations across T snapshots. This structure captures cumulative conditioning effects (e.g., gradual vegetation stress) alongside short-term anomalies that influence ignition probability. Thus, temporal attention is instantiated through structured graph sequences rather than static image classification.

5.4 False Alarm Rate (%)

Figure 6. False Alarm Rate (%)

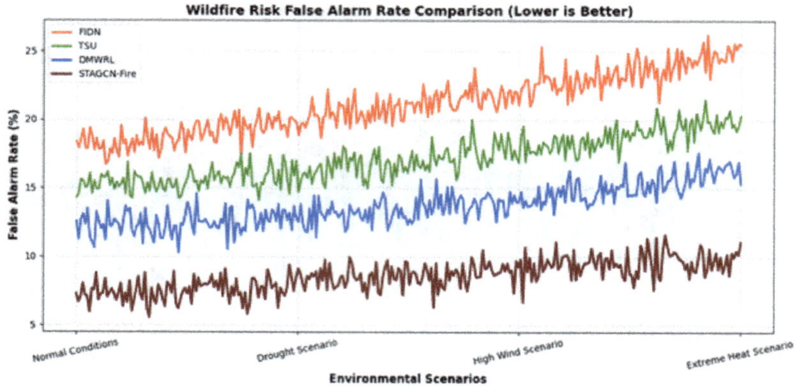

The ratio of poorly anticipated high-risk wildfire incidents is shown in Figure 6. STAGCN-Fire has the lowest false alarm rate (710%), and operational alert is much lower than existing processes. The adaptive spatial-temporal GCN accounts for vegetation stress, wind direction, and terrain similarity to reduce misclassifications and improve early-warning systems with fewer interventions. False Alarm Rate (FAR%) defined by equation 11.

$$\text{FAR}(\%) = \frac{FP}{FP + TN} \times 100 \tag{11}$$

In this formulation, FP denotes the number of false alarms and TN represents correctly predicted non-fire cases.

Thus, STAGCN-Fire's spatio-temporal and graph-based architecture is confirmed by the experiment. Structured regional characteristics are obtained from the dataset, and graph creation, adaptive edge modulation, and temporal aggregation are done before classification. To balance methodological originality and empirical validation, the findings assess the complete dynamic graph learning pipeline rather than a pixel- or image-centric model.

6. CONCLUSION AND FUTURE WORK

The article introduced STAGCN-Fire, a spatial-temporal adaptable graph convolutional neural network capable of forecasting wildfire probability with high accuracy and at scale. The given approach is able to capture the complex nature of fire propagation on heterogeneous terrain using multi-source satellite, meteorological, and topographic data, building dynamic graph models, and focusing on temporal dynamics. Based on experimental results, STAGCN-Fire performs better than the existing method in terms of prediction accuracy, F1-score, AUC, and false alarm reduction, particularly in extreme weather and under high fuel loads. Its integration with Google Earth Engine enables large-scale data processing and risk analysis in near-real time and is therefore adaptable to operational early-warning systems.

The subsequent study will involve the incorporation of real-time streaming data from Internet of Things sensors and weather predictions to assist in short-term fire predictions. This will strengthen the model, making it more robust, generalizable, and explainable for decision-makers and disaster management authorities, as it will be applied to cross-region transfer learning, uncertainty estimation, and explainable AI.

REFERENCES

Abdollahi, A., & Yebra, M. (2025). Challenges and opportunities in remote sensing-based fuel load estimation for wildfire behavior and management: A comprehensive review. *Remote Sensing (Basel), 17*(3), 415. DOI: 10.3390/rs17030415

Amsalu, A., Garedew, W., Melka, G. A., Çelik, M. A., & Gemeda, D. O. (2026). Trends in climate extreme indices and implications for coffee farming in The Kafa Biosphere Reserve, southwest Ethiopia. *Climate Services, 42*, 100637. DOI: 10.1016/j.cliser.2026.100637

Bhowmik, R. T., Jung, Y. S., Aguilera, J. A., Prunicki, M., & Nadeau, K. (2023). A multi-modal wildfire prediction and early-warning system based on a novel machine learning framework. *Journal of Environmental Management, 341*, 117908. DOI: 10.1016/j.jenvman.2023.117908 PMID: 37182403

Caron, N., Noura, H. N., Nakache, L., Guyeux, C., & Aynes, B. (2025). AI for Wildfire Management: From Prediction to Detection, Simulation, and Impact Analysis—Bridging Lab Metrics and Real-World Validation. *AI, 6*(10), 253. DOI: 10.3390/ai6100253

Çelik, M. A. (2026). Agro-climatic Classification of Türkiye's Agricultural Basins Using Long-Term ERA5-Land Temperature Data. *Physics and Chemistry of the Earth Parts A/B/C, 142*, 104307. DOI: 10.1016/j.pce.2026.104307

Çelik, M. A., Kıraç, A., & Gouveia, C. M. (2026). North Atlantic Oscillation modulation of drought and vegetation activity in Türkiye. *Theoretical and Applied Climatology, 157*(2), 74. DOI: 10.1007/s00704-025-05971-5

Chen, D., Cheng, S., Hu, J., Kasoar, M., & Arcucci, R. (2024). Explainable global wildfire prediction models using graph neural networks. *arXiv preprint arXiv:2402.07152*.

. Das, U. (2025). A Graph Neural Network Based Approach for Predicting Wildfire Burned Areas.

Durlević, U., Ilić, V., & Valjarević, A. (2025). Wildfire susceptibility mapping using deep learning and machine learning models based on multi-sensor satellite data fusion: A case study of Serbia. *Fire (Basel, Switzerland), 8*(10), 407. DOI: 10.3390/fire8100407

Ghali, R., & Akhloufi, M. A. (2023). Deep learning approaches for wildland fires using satellite remote sensing data: Detection, mapping, and prediction. *Fire (Basel, Switzerland), 6*(5), 192. DOI: 10.3390/fire6050192

Huot, F., Hu, R. L., Goyal, N., Sankar, T., Ihme, M., & Chen, Y. F. (2022). Next day wildfire spread: A machine learning dataset to predict wildfire spreading from remote-sensing data. *IEEE Transactions on Geoscience and Remote Sensing*, *60*, 1–13. DOI: 10.1109/TGRS.2022.3192974

Illarionova, S., Shadrin, D., Gubanov, F., Shutov, M., Tasuev, U., Evteeva, K., Mironenko, M., & Burnaev, E. (2025). Exploration of geo-spatial data and machine learning algorithms for robust wildfire occurrence prediction. *Scientific Reports*, *15*(1), 10712. DOI: 10.1038/s41598-025-94002-4 PMID: 40155427

Kaggle. (n.d.). https://www.kaggle.com/datasets/abdelghaniaaba/wildfire-prediction-dataset

. Kondylatos, S., Prapas, I., Ronco, M., Papoutsis, I., Camps-Valls, G., Piles, M., ... & Carvalhais, N. (2022). Wildfire danger prediction and understanding with deep learning. *Geophysical Research Letters*, *49*(17), e2022GL099368.

Michail, D., Davalas, C., Chafis, K., Panagiotou, L. I., Prapas, I., Bountos, N. I., Kondylatos, S., & Papoutsis, I. (2025). FireCastNet: Earth-as-a-graph for seasonal fire prediction. *Scientific Reports*, *16*(1), 1006. DOI: 10.1038/s41598-025-30645-7 PMID: 41392185

Pang, B., Cheng, S., Huang, Y., Jin, Y., Guo, Y., Prentice, I. C., Harrison, S. P., & Arcucci, R. (2025). Fire-Image-DenseNet (FIDN) for predicting wildfire burnt area using remote sensing data. *Computers & Geosciences*, *195*, 105783. DOI: 10.1016/j.cageo.2024.105783

Sharma, S., & Khanal, P. (2024). Forest fire prediction: A spatial machine learning and neural network approach. *Fire (Basel, Switzerland)*, *7*(6), 205. DOI: 10.3390/fire7060205

Xu, H., Zlatanova, S., Liang, R., & Canbulat, I. (2025). Generative AI as a Pillar for Predicting 2D and 3D Wildfire Spread: Beyond Physics-Based Models and Traditional Deep Learning. *Fire (Basel, Switzerland)*, *8*(8), 293. DOI: 10.3390/fire8080293

Xu, Z., Li, J., Cheng, S., Rui, X., Zhao, Y., He, H., Guan, H., Sharma, A., Erxleben, M., Chang, R., & Xu, L. L. (2025). Deep learning for wildfire risk prediction: Integrating remote sensing and environmental data. *ISPRS Journal of Photogrammetry and Remote Sensing*, *227*, 632–677. DOI: 10.1016/j.isprsjprs.2025.06.002

Yu, S., & Singh, M. (2025). Deep Learning-Based Remote Sensing Image Analysis for Wildfire Risk Evaluation and Monitoring. *Fire (Basel, Switzerland)*, *8*(1), 19. DOI: 10.3390/fire8010019

. Zhou, Y., Kong, R., Xu, Z., Xu, L., & Cheng, S. (2025). Comparative and interpretative analysis of CNN and transformer models in predicting wildfire spread using remote sensing data. *Journal of Geophysical Research: Machine Learning and Computation, 2*(2), e2024JH000409.

Compilation of References

A, Y., Wang, G., Hu, P., Lai, X., Xue, B., and Fang, Q. (2022). Root-zone soil moisture estimation based on remote sensing data and deep learning. *Environmental Research, 212*(Pt B), 113278. DOI: 10.1016/j.envres.2022.113278

Abdelkader, M., Zhang, Y., & Khan, S. (2023). Temporal transformers for remote sensing change detection. *IEEE Transactions on Geoscience and Remote Sensing*.

Abdollahi, A., & Yebra, M. (2025). Challenges and opportunities in remote sensing-based fuel load estimation for wildfire behavior and management: A comprehensive review. *Remote Sensing (Basel), 17*(3), 415. DOI: 10.3390/rs17030415

Abdulhussain, S. H., Mahmmod, B. M., Alwhelat, A., Shehada, D., Shihab, Z. I., Mohammed, H. J., Abdulameer, T. H., Alsabah, M., Fadel, M. H., Ali, S. K., Abbood, G. H., Asker, Z. A., & Hussain, A. (2025). A comprehensive review of sensor technologies in IoT: Technical aspects, challenges, and future directions. *Computers, 14*(8), 342. DOI: 10.3390/computers14080342

Adab, H., Morbidelli, R., Saltalippi, C., Moradian, M., & Ghalhari, G. A. F. (2020). Machine Learning to Estimate Surface Soil Moisture from Remote Sensing Data. *Water (Basel), 12*(11), 3223. DOI: 10.3390/w12113223

Afridi, W. K., Mukhopadhyay, S. C., & Mainali, B. (2025). Machine Learning Enabled Smart Sensor Node for Evapotranspiration Prediction in Diverse Soil and Environmental Conditions. *IEEE Sensors Journal, 25*(12), 22720–22728. DOI: 10.1109/JSEN.2025.3566935

Agbehadji, I. E., & Obagbuwa, I. C. (2024). Systematic review of machine learning and deep learning techniques for spatiotemporal air quality prediction. *Atmosphere (Basel), 15*(11), 1352. DOI: 10.3390/atmos15111352

Ahmed, A. A. M., Jui, S. J. J., Sharma, E., Ahmed, M. H., Raj, N., & Bose, A. (2024). An advanced deep learning predictive model for air quality index forecasting with remote satellite-derived hydro-climatological variables. *The Science of the Total Environment, 906*, 167234. DOI: 10.1016/j.scitotenv.2023.167234 PMID: 37739083

Ahmed, A. A., Sayed, S., Abdoulhalik, A., Moutari, S., & Oyedele, L. (2024). Applications of machine learning to water resources management: A review of present status and future opportunities. *Journal of Cleaner Production, 441*, 140715. DOI: 10.1016/j.jclepro.2024.140715

Ahmed, Z., Gui, D., Qi, Z., Liu, Y., Liu, Y., & Azmat, M. (2022). Agricultural system modeling: Current achievements, innovations, and future roadmap. *Arabian Journal of Geosciences, 15*(4), 363. DOI: 10.1007/s12517-022-09654-7

Aithal, B. H., Vinay, S., Durgappa, S., & Ramachandra, T. V. (2013, November). Modeling and simulation of urbanisation in greater Bangalore, India. In *Proceedings of the National Spatial Data Infrastructure 2013 Conference* (pp. 34–50). IIT Bombay.

Akanbi, M. B., Adedotun, K. J., Raji, A. K., and Banjoko, I. K. (2025). Optimizing water resource management in agriculture using AI-Powered solar irrigation systems. *Journal of Science Innovation and Technology Research*. DOI: 10.70382/ajsitr.v7i9.011

Akyildiz, I. F., Han, C., Hu, Z., Nie, S., & Jornet, J. M. (2022). Terahertz band communication: An old problem revisited and research directions for the next decade. *IEEE Transactions on Communications, 70*(6), 4250–4285. DOI: 10.1109/TCOMM.2022.3171800

Alam, S. M. R., & Hossain, M. S. (2020). A Rule-Based Classification Method for Mapping Saltmarsh Land-Cover in South-Eastern Bangladesh from Landsat-8 {OLI}. *Canadian Journal of Remote Sensing*, •••, 1–25. DOI: 10.1080/07038992.2020.1789852

Alam, S. M. R., & Hossain, M. S. (2023). Probabilities of acquiring cloud-free and low-tide Landsat observations for mapping saltmarsh over south-eastern Bangladesh from 1980 to 2019. *Marine Geodesy, 46*(6), 562–593. DOI: 10.1080/01490419.2023.2200212

Alam, S. M. R., & Hossain, M. S. (2024). Using a water index approach to mapping periodically inundated saltmarsh land-cover vegetation and eco-zonation using multi-temporal Landsat 8 imagery. *Journal of Coastal Conservation, 28*(1), 19. DOI: 10.1007/s11852-023-01019-w

Aldoseri, A., Al-Khalifa, K. N., & Hamouda, A. M. (2023). Re-Thinking Data Strategy and Integration for Artificial intelligence: Concepts, opportunities, and challenges. *Applied Sciences (Basel, Switzerland)*, *13*(12), 7082. DOI: 10.3390/app13127082

Ali, A., Hussain, T., & Zahid, A. (2025). Smart irrigation technologies and prospects for enhancing water use efficiency for sustainable agriculture. *AgriEngineering*, *7*(4), 106. DOI: 10.3390/agriengineering7040106

Aliqab, K., Dave, K., Sorathiya, V., Alsharari, M., & Armghan, A. (2023). Numerical analysis of phase change material and graphene-based tunable refractive index sensor for infrared frequency spectrum. *Scientific Reports*, *13*(1), 7653. DOI: 10.1038/s41598-023-34859-5 PMID: 37169848

Allen, T. R., Eulie, D. O., Polk, M. A., McLeod, G., Stuart, R., & Garnand, A. (2023). Advancing Estuarine Shoreline Change Analysis Using Small Uncrewed Autonomous Systems. In *IntechOpen eBooks*. IntechOpen., DOI: 10.5772/intechopen.112022

Al-Okby, M. F. R., Neubert, S., Röddelkopf, T., & Thurow, K. (2021). Mobile detection and alarming systems for hazardous gases and volatile chemicals in laboratories and industrial locations. *Sensors (Basel)*, *21*(23), 8128. DOI: 10.3390/s21238128 PMID: 34884132

Alongi, D. M. (2014). Carbon Cycling and Storage in Mangrove Forests. *Annual Review of Marine Science*, *6*(1), 195–219. DOI: 10.1146/annurev-marine-010213-135020

Alvarez-Vanhard, E., Corpetti, T., & Houet, T. (2021). UAV & satellite synergies for optical remote sensing applications: A literature review. *Science of Remote Sensing*, *3*, 100019. DOI: 10.1016/j.srs.2021.100019

Amani, S., Shafizadeh-Moghadam, H., & Morid, S. (2024). Utilizing Machine Learning Models with Limited Meteorological Data as Alternatives for the FAO-56PM Model in Estimating Reference Evapotranspiration. *Water Resources Management*, *38*(6), 1921–1942. DOI: 10.1007/s11269-023-03670-2

Ambika, A. K., Tayal, K., Mishra, V., & Lu, D. (2025). Novel Deep Learning Transformer Model for Short to Sub-Seasonal Streamflow Forecast. *Geophysical Research Letters*, *52*(14), e2025GL116707. Advance online publication. DOI: 10.1029/2025GL116707

Amnuaylojaroen, T. (2025). Advancements and challenges of artificial intelligence in climate modeling for sustainable urban planning. *Frontiers in Artificial Intelligence*, *8*, 1517986. DOI: 10.3389/frai.2025.1517986 PMID: 40463823

Amsalu, A., Garedew, W., Melka, G. A., Çelik, M. A., & Gemeda, D. O. (2026). Trends in climate extreme indices and implications for coffee farming in The Kafa Biosphere Reserve, southwest Ethiopia. *Climate Services*, *42*, 100637. DOI: 10.1016/j.cliser.2026.100637

Andrade, C. de L. B. Costa, M. D. de P., & Freitas, D. M. de. (2024). *Assessing the Carbon Sequestration Potential by Mangroves in a Coastal Metropolitan Region*. DOI: 10.2139/ssrn.5023738

Arfanuzzaman, M. (2024). Bangladesh's pathways to climate-resilient development: A methodical review. *World Development Sustainability*, *4*, 100144. DOI: 10.1016/j.wds.2024.100144

Arystanov, A., Sagin, J., Karabkina, N., Arystanova, R., Yermekov, F., Kabzhanova, G., Bekseitova, R., Aktymbayeva, A., & Kutymova, N. (2025). Automatic classification of agricultural crops using Sentinel-2 data in the rainfed zone of Southern Kazakhstan. *Agronomy (Basel)*, *15*(9), 2040. DOI: 10.3390/agronomy15092040

Asner, G. P., Brodrick, P. G., Philipson, C., Vaughn, N. R., Martin, R. E., Knapp, D. E., Heckler, J., Evans, L. J., Jucker, T., Goossens, B., Stark, D. J., Reynolds, G., Ong, R., Renneboog, N., Kugan, F., & Coomes, D. A. (2018). Mapped aboveground carbon stocks to advance forest conservation and recovery in Malaysian Borneo. *Biological Conservation*, *217*, 289–310. DOI: 10.1016/j.biocon.2017.10.020

Asokan, A., & Anitha, J. (2019). Change detection techniques for remote sensing applications: A survey. *Earth Science Informatics*, *12*(2), 143–160. DOI: 10.1007/s12145-019-00380-5

Atef, I., Ahmed, W., & Abdel-Maguid, R. H. (2023). Modelling of land use land cover changes using machine learning and GIS techniques: A case study in El-Fayoum Governorate, Egypt. *Environmental Monitoring and Assessment*, *195*(6), 637. DOI: 10.1007/s10661-023-11224-7 PMID: 37133528

Babaeian, E., Sadeghi, M., Jones, S. B., Montzka, C., Vereecken, H., & Tuller, M. (2019). Ground, proximal, and satellite remote sensing of soil moisture. *Reviews of Geophysics*, *57*(2), 530–616. DOI: 10.1029/2018RG000618

Bailey, R. T., Abbas, S., Arnold, J. G., & White, M. J. (2025). SWAT+MODFLOW: A new hydrologic model for simulating surface-subsurface flow in managed watersheds. *Geoscientific Model Development*, *18*(17), 5681–5697. DOI: 10.5194/gmd-18-5681-2025

Bakirman, T., & Gumusay, M. U. (2020). Assessment of Machine Learning Methods for Seagrass Classification in the Mediterranean. *Baltic Journal of Modern Computing*, *8*(2). Advance online publication. DOI: 10.22364/bjmc.2020.8.2.07

Balogun, I. A., Adeyewa, D. Z., Balogun, A. A., & Morakinyo, T. E. (2011). Analysis of urban expansion and land use changes in Akure, Nigeria using remote sensing and geographic information system techniques. *Journal of Geography and Regional Planning*, *4*(9), 533–541.

Balsamo, G., Agusti-Parareda, A., Albergel, C., Arduini, G., Beljaars, A., Bidlot, J., & Zeng, X. (2018). Satellite and in situ observations for advancing global Earth surface modelling: A review. *Remote Sensing (Basel)*, *10*(12), 2038. DOI: 10.3390/rs10122038

Bandara, K., Bergmeir, C., & Hewamalage, H. (2021). LSTM-MSNet: Leveraging Forecasts on Sets of Related Time Series With Multiple Seasonal Patterns. *IEEE Transactions on Neural Networks and Learning Systems*, *32*(4), 1586–1599. DOI: 10.1109/TNNLS.2020.2985720

Banerjee, D., Ganguly, S., & Kushwaha, S. (2024). Forecasting Future Groundwater Recharge from Rainfall Under Different Climate Change Scenarios Using Comparative Analysis of Deep Learning and Ensemble Learning Techniques. *Water Resources Management*, *38*(11), 4019–4037. DOI: 10.1007/s11269-024-03850-8

Banerjee, P., & Shankar, R. (2022). Deep learning methods for hydromorphological prediction. *Remote Sensing of Environment*.

Barbier, E. B., Hacker, S. D., Kennedy, C., Koch, E. W., Stier, A. C., & Silliman, B. R. (2011). The value of estuarine and coastal ecosystem services. *Ecological Monographs*, *81*(2), 169–193. DOI: 10.1890/10-1510.1

Bardou, R., Osland, M. J., Scyphers, S. B., Shepard, C. C., Aerni, K. E., Alemu, J. B., Crimian, R., Day, R. H., Enwright, N. M., Feher, L. C., Gibbs, S. L., O'Donnell, K. L., Swinea, S. H., Thorne, K., Truskey, S., Armitage, A. R., Baker, R. J., Breithaupt, J. L., Cavanaugh, K. C., & Hughes, A. R. (2023). Rapidly Changing Range Limits in a Warming World: Critical Data Limitations and Knowledge Gaps for Advancing Understanding of Mangrove Range Dynamics in the Southeastern USA. *Estuaries and Coasts*, *46*(5), 1123–1140. DOI: 10.1007/s12237-023-01209-7

Bargain, A. (2012). Study of Zostera noltii seagrass bed structure and dynamics using multi and hyperspectral remote sensing. *HAL (Le Centre Pour La Communication Scientifique Directe)*. https://theses.hal.science/tel-00799897

Basyuni, M., Aznawi, A. A., Rafli, M., Tinumbunan, J. M. T., Gultom, E. T., Lubis, R. D. A., Sianturi, H. A., Sumarga, E., Mukhtar, E., Slamet, B., Jumilawaty, E., Pribadi, R., Sitinjak, R. R., & Baba, S. (2024). Harnessing Biomass and Blue Carbon Potential: Estimating Carbon Stocks in the Vital Wetlands of Eastern Sumatra, Indonesia. *Land (Basel), 13*(11), 1960. DOI: 10.3390/land13111960

Belgiu, M., & Drăguţ, L. (2016). Random forest in remote sensing: A review of applications and future directions. *ISPRS Journal of Photogrammetry and Remote Sensing, 114*, 24–31. DOI: 10.1016/j.isprsjprs.2016.01.011

Belloli, T. F., de Arruda, D. C., Guasselli, L. A., Cunha, C. S., & Korb, C. C. (2025). Modeling Wetland Biomass and Aboveground Carbon: Influence of Plot Size and Data Treatment Using Remote Sensing and Random Forest. *Land (Basel), 14*(3), 616. DOI: 10.3390/land14030616

Belloli, T. F., Guasselli, L. A., Kuplich, T. M., Ruiz, L. F. C., de Arruda, D. C., Etchelar, C. B., & Simioni, J. D. (2022). Estimation of aboveground biomass and carbon in palustrine wetland using bands and multispectral indices derived from optical satellite imageries PlanetScope and Sentinel-2A. *Journal of Applied Remote Sensing, 16*(3). Advance online publication. DOI: 10.1117/1.JRS.16.034516

Benami, E., Jin, Z., Carter, M. R., Ghosh, A., Hijmans, R. J., Hobbs, A., & Lobell, D. B. (2021). Uniting remote sensing, crop modelling and economics for agricultural risk management. *Nature Reviews. Earth & Environment, 2*(2), 140–159. DOI: 10.1038/s43017-020-00122-y

Bertocci, F., Fort, A., Vignoli, V., Mugnaini, M., & Berni, R. (2017). Optimization of perovskite gas sensor performance: Characterization, measurement and experimental design. *Sensors (Basel), 17*(6), 1352. DOI: 10.3390/s17061352 PMID: 28604587

Bhattacharyya, A., Nambiar, S. M., Ojha, R., Gyaneshwar, A., Chadha, U., & Srinivasan, K. (2023). Machine learning and deep learning powered satellite communications: Enabling technologies, applications, open challenges, and future research directions. *International Journal of Satellite Communications and Networking, 41*(6), 539–588. DOI: 10.1002/sat.1482

Bhowmik, R. T., Jung, Y. S., Aguilera, J. A., Prunicki, M., & Nadeau, K. (2023). A multi-modal wildfire prediction and early-warning system based on a novel machine learning framework. *Journal of Environmental Management, 341*, 117908. DOI: 10.1016/j.jenvman.2023.117908 PMID: 37182403

Biazar, S. M., Golmohammadi, G., Nedhunuri, R. R., Shaghaghi, S., & Mohammadi, K. (2025). Artificial intelligence in Hydrology: Advancements in soil, water resource management, and sustainable development. *Sustainability (Basel)*, *17*(5), 2250. DOI: 10.3390/su17052250

Bindajam, A. A., Mallick, J., Talukdar, S., Shahfahad, , Shohan, A. A. A., & Rahman, A. (2023). Modeling the spatiotemporal heterogeneity of land surface temperature and its relationship with land use land cover using geostatistical techniques and machine learning algorithms. *Environmental Science and Pollution Research International*, *30*(49), 106917–106935. DOI: 10.1007/s11356-022-23211-5 PMID: 36178650

Blanco, A. C., & Pérez, C. (2023). Estimation Of Mangrove Fractional Cover From Multispectral And Hyperspectral Data Using Mixture Tuned Matched Filtering. *The international Archives of the Photogrammetry, Remote Sensing and Spatial Information Sciences. The International Archives of the Photogrammetry, Remote Sensing and Spatial Information Sciences*, *1331*, 1331–1336. Advance online publication. DOI: 10.5194/isprs-archives-XLVIII-1-W2-2023-1331-2023

Blanco-Sacristán, J., Johansen, K., Elías-Lara, M., Tu, Y.-H., Duarte, C. M., & McCabe, M. F. (2024). Quantifying mangrove carbon assimilation rates using UAV imagery. *Scientific Reports*, *14*(1), 4648. Advance online publication. DOI: 10.1038/s41598-024-55090-w PMID: 38409194

Blanka-Végi, V., Tobak, Z., Sipos, G., Barta, K., Szabó, B., & Van Leeuwen, B. (2025). Estimation of the Spatiotemporal Variability of Surface soil Moisture Using Machine Learning Methods Integrating Satellite and Ground-based Soil Moisture and Environmental Data. *Water Resources Management*, *39*(5), 2317–2334. DOI: 10.1007/s11269-024-04069-3

Blume, A., Pertiwi, A. P., Lee, C. B., & Traganos, D. (2023). Bahamian seagrass extent and blue carbon accounting using Earth Observation. *Frontiers in Marine Science*, *10*, 1058460. Advance online publication. DOI: 10.3389/fmars.2023.1058460

Bounoua, I., Saidi, Y., Yaagoubi, R., & Bouziani, M. (2024). Deep learning approaches for water stress forecasting in arboriculture using time series of remote sensing images: Comparative study between ConvLSTM and CNN-LSTM models. *Technologies*, *12*(6), 77. DOI: 10.3390/technologies12060077

Buckley, D. J., Black, N. C., Castanon, E. G., Melios, C., Hardman, M., & Kazakova, O. (2020). Frontiers of graphene and 2D material-based gas sensors for environmental monitoring. *2D Materials*, *7*(3), 032002.

Caron, N., Noura, H. N., Nakache, L., Guyeux, C., & Aynes, B. (2025). AI for Wildfire Management: From Prediction to Detection, Simulation, and Impact Analysis—Bridging Lab Metrics and Real-World Validation. *AI*, *6*(10), 253. DOI: 10.3390/ai6100253

Ceccherelli, G., Chiabrando, F., Gallitto, F., Lingua, A. M., Longhi, V., Maschio, P. F., Matrone, F., Menna, F., Nocerino, E., Scalici, M., Secco, S., & Tanduo, B. (2024). Multitemporal Seagrass Mapping and Monitoring of Posidonia Meadows and Banquettes for Blue Carbon Conservation: The Poseidon Project. *The œinternational Archives of the Photogrammetry, Remote Sensing and Spatial Information Sciences. The International Archives of the Photogrammetry, Remote Sensing and Spatial Information Sciences*, *65*, 65–71. Advance online publication. DOI: 10.5194/isprs-archives-XLVIII-1-2024-65-2024

Çelik, M. A. (2026). Agro-climatic classification of Türkiye's Agricultural Basins using long-term ERA5-Land temperature data. *Physics and Chemistry of the Earth Parts A/B/C*, *142*, 104307. DOI: 10.1016/j.pce.2026.104307

Çelik, M. A., Kıraç, A., & Gouveia, C. M. (2026). North Atlantic Oscillation modulation of drought and vegetation activity in Türkiye. *Theoretical and Applied Climatology*, *157*(2), 74. Advance online publication. DOI: 10.1007/s00704-025-05971-5

Chafe, O., Broz, A., Levenson, E. S., Farinacci, M. D., Anderson, R. O., & Silva, L. C. R. (2024). The spatiotemporal domains of natural climate solutions research and strategies for implementation in the Pacific Northwest, USA. *Frontiers in Climate*, *6*, 1273632. Advance online publication. DOI: 10.3389/fclim.2024.1273632

Chang, Y., Zhang, C., Huang, J., Chang, H., Wang, C., & Huo, Z. (2025). Machine Learning for Reference Crop Evapotranspiration Modeling: A State-of-the-Art Review and Future Directions. *Agronomy (Basel)*, *15*(9), 2038. DOI: 10.3390/agronomy15092038

Chaube, N. R., Lele, N., Misra, A., Murthy, T. V. R., Manna, S., Hazra, S., Panda, M. R., & Samal, R. N. (2019). Mangrove Species Discrimination and Health Assessment using AVIRIS-NG Hyperspectral Data. *Current Science*, *116*(7), 1136. DOI: 10.18520/cs/v116/i7/1136-1142

Chau, K. (2006). A review on the integration of artificial intelligence into coastal modeling. *Journal of Environmental Management*, *80*(1), 47–57. DOI: 10.1016/j.jenvman.2005.08.012

Chávez, D. E., López-Portillo, J., Gallardo-Cruz, J. A., & Meave, J. A. (2023). Approaches, potential, and challenges in the use of remote sensing to study mangrove and other tropical wetland forests. *Botanical Sciences*, *102*(1), 1–25. DOI: 10.17129/botsci.3358

Chen, D., Cheng, S., Hu, J., Kasoar, M., & Arcucci, R. (2024). Explainable global wildfire prediction models using graph neural networks. *arXiv preprint arXiv:2402.07152*.

Chen, Y., Shen, X., Jing, F., & Xiong, P. (2010). Application of outgoing longwave radiation data for earthquake research. In Proceedings - 2010 IEEE International Conference on Intelligent Computing and Intelligent Systems, ICIS 2010, 2, 46-48.

Cheng, K., Zhang, C., Fan, Y., Diao, H., & Xia, S. (2025). AI-based approaches for wetland mapping and classification: A review of current practices and future perspectives. *Artificial Intelligence in Geosciences*, *6*(2), 100165. DOI: 10.1016/j.aiig.2025.100165

Chen, L., Šimůnek, J., Bradford, S. A., Ajami, H., & Meles, M. B. (2022). A computationally efficient hydrologic modeling framework to simulate surface-subsurface hydrological processes at the hillslope scale. *Journal of Hydrology (Amsterdam)*, *614*, 128539. DOI: 10.1016/j.jhydrol.2022.128539

Chen, W., Chen, P., Zhang, H., He, Y., Tang, J., & Wu, S. (2023). Review of airborne oceanic lidar remote sensing. *Intelligent Marine Technology and Systems*, *1*(1), 10. Advance online publication. DOI: 10.1007/s44295-023-00007-y

Chen, X., Zhao, L., Zheng, F., Li, J., Li, L., Ding, H., Zhang, K., Liu, S., Li, D., & de Leeuw, G. (2022). Neural network aerosol retrieval for geostationary satellite (NNAeroG) based on temporal, spatial and spectral measurements. *Remote Sensing (Basel)*, *14*(4), 980. DOI: 10.3390/rs14040980

Chen, Z., Zhu, Z., Jiang, H., & Sun, S. (2020). Estimating daily reference evapotranspiration based on limited meteorological data using deep learning and classical machine learning methods. *Journal of Hydrology (Amsterdam)*, *591*, 125286. DOI: 10.1016/j.jhydrol.2020.125286

Cherian, S. M., & K, R. (2024). Random forest and support vector machine classifiers for coastal wetland characterization using the combination of features derived from optical data and synthetic aperture radar dataset. *Journal of Water and Climate Change*, *15*(1), 29–49. DOI: 10.2166/wcc.2023.238

Chin, R. J., & Lai, S. H. (2025). EDITORIAL: INTEGRATION OF HYDROLOGICAL MODELS AND MACHINE LEARNING TECHNIQUES FOR WATER RESOURCES MANAGEMENT. *Journal of Civil Engineering Science and Technology*, *16*(1), 1–5. DOI: 10.33736/jcest.9191.2025

Chowdhury, K. J., Ali, M. R., Chowdhury, M. A., & Islam, S. L. U. (2025). Climate change induced risks assessment of a coastal area: A "socioeconomic and livelihood vulnerability index" based study in coastal Bangladesh. *Natural Hazards Review*, *5*(1), 75–87. DOI: 10.1016/j.nhres.2024.06.005

Chowdhury, M., Martínez-Sansigre, A., Mole, M., Alonso-Peleato, E., Basos, N., Blanco, J. M., Ramírez-Nicolás, M., Caballero, I., & Pérez, I. de la C. (2024). AI-driven remote sensing enhances Mediterranean seagrass monitoring and conservation to combat climate change and anthropogenic impacts. *Scientific Reports*, *14*(1), 8360. Advance online publication. DOI: 10.1038/s41598-024-59091-7 PMID: 38600271

Christofi, D., Mettas, C., Evagorou, E., Stylianou, N., Eliades, M., Theocharidis, C., Chatzipavlis, A., Hasiotis, T., & Hadjimitsis, D. (2025). A Review of Open Remote Sensing Data with GIS, AI, and UAV Support for Shoreline Detection and Coastal Erosion Monitoring. *Applied Sciences (Basel, Switzerland)*, *15*(9), 4771. DOI: 10.3390/app15094771

Cisternas, M., Atwater, B. F., Torrejón, F., Sawai, Y., Machuca, G., Lagos, M., Eipert, A., Youlton, C., Salgado, I., Kamataki, T., Shishikura, M., Rajendran, C. P., Malik, J. K., Rizal, Y., & Husni, M. (2005). Predecessors of the giant 1960 Chile earthquake. *Nature*, *437*(7057), 404–407. DOI: 10.1038/nature03943 PMID: 16163355

Cloete, D. N., Shoko, C., Dube, T., & Clarke, S. (2024). Remote sensing-based land use land cover classification for the Heuningnes Catchment, Cape Agulhas, South Africa. *Physics and Chemistry of the Earth Parts A/B/C*, *134*, 103559. DOI: 10.1016/j.pce.2024.103559

Colaço, R., & de Abreu e Silva, J. (2021). Commercial classification and location modelling: Integrating different perspectives on commercial location and structure. *Land (Basel)*, *10*(6), 567. DOI: 10.3390/land10060567

Costanza, R., de Groot, R., Sutton, P., van der Ploeg, S., Anderson, S. J., Kubiszewski, I., Farber, S., & Turner, R. K. (2014). Changes in the global value of ecosystem services. *Global Environmental Change*, *26*, 152–158. https://doi.org/https://doi.org/10.1016/j.gloenvcha.2014.04.002. DOI: 10.1016/j.gloenvcha.2014.04.002

d'Andrimont, R., Verhegghen, A., Lemoine, G., Kempeneers, P., Meroni, M., & van der Velde, M. (2021). From parcel to continental scale—A first European crop type map based on Sentinel-1 and LUCAS Copernicus in situ observations. *Remote Sensing of Environment, 266*, 112708. DOI: 10.1016/j.rse.2021.112708

Dadashpoor, H., Azizi, P., & Moghadasi, M. (2019). Land use change, urbanization, and change in landscape pattern in a metropolitan area. *The Science of the Total Environment, 655*, 707–719. DOI: 10.1016/j.scitotenv.2018.11.267 PMID: 30476851

Dai, X., Guo, Z., & Chen, C. (2016, July). Improvement of mono-window algorithm for land surface temperature retrieval integrated with subpixel mapping for Landsat imagery. In *2016 4th International Workshop on Earth Observation and Remote Sensing Applications (EORSA)* (pp. 24–27). IEEE. DOI: 10.1109/EORSA.2016.7552759

Deeks, E., Magalhães, K. M., Traganos, D., Ward, R. D., Normande, I. C., Dawson, T. P., & Kratina, P. (2024). Seagrass mapping of north-eastern Brazil using Google Earth Engine and Sentinel-2 imagery. *Environmental and Sustainability Indicators, 100489*, 100489. Advance online publication. DOI: 10.1016/j.indic.2024.100489

Dembélé, M., Schaefli, B., Van De Giesen, N., & Mariéthoz, G. (2020). Suitability of 17 gridded rainfall and temperature datasets for large-scale hydrological modelling in West Africa. *Hydrology and Earth System Sciences, 24*(11), 5379–5406. DOI: 10.5194/hess-24-5379-2020

Deng, S., Yang, F., Wang, Y., We, Z., & Wang, W. (2024). Progress in Remote Sensing Monitoring of Mangrove Carbon Storage. *Yaogan Xuebao, 1*(10), 2448–2468. Advance online publication. DOI: 10.11834/jrs.20243293

Deng, X., Gao, F., Liao, S., Liu, Y., & Chen, W. (2023). Spatiotemporal evolution patterns of urban heat island and its relationship with urbanization in Guangdong–Hong Kong–Macao Greater Bay Area of China from 2000 to 2020. *Ecological Indicators, 146*, 109817. DOI: 10.1016/j.ecolind.2022.109817

Dhillon, M. S., Kübert-Flock, C., Dahms, T., Rummler, T., Arnault, J., Steffan-Dewenter, I., & Ullmann, T. (2023). Evaluation of MODIS, Landsat 8 and Sentinel-2 data for accurate crop yield predictions: A case study using STARFM NDVI in Bavaria, Germany. *Remote Sensing (Basel), 15*(7), 1830. DOI: 10.3390/rs15071830

Dokht, R. M. H., Gu, Y. J., & Sacchi, M. D. (2017). Singular spectrum analysis and its applications in mapping mantle seismic structure. *Geophysical Journal International, 208*(3), 1430–1442. DOI: 10.1093/gji/ggw473

Donnelly, J., Daneshkhah, A., & Abolfathi, S. (2023). Physics-informed neural networks as surrogate models of hydrodynamic simulators. *The Science of the Total Environment, 912*, 168814. DOI: 10.1016/j.scitotenv.2023.168814 PMID: 38016570

Doorn, N. (2020). Artificial intelligence in the water domain: Opportunities for responsible use. *The Science of the Total Environment, 755*(Pt 1), 142561. DOI: 10.1016/j.scitotenv.2020.142561 PMID: 33039891

Doswald, N., Munroe, R., Roe, D., Giuliani, A., Castelli, I., Stephens, J., Möller, I., Spencer, T., Vira, B., & Reid, H. (2014). Effectiveness of ecosystem-based approaches for adaptation: Review of the evidence-base. *Climate and Development, 6*(2), 185–201. DOI: 10.1080/17565529.2013.867247

Duan, X., Aslam, R. W., Naqvi, S. A. A., Kucher, D. E., Afzal, Z., Raza, D., Zulqarnain, R. M., & Said, Y. (2025). Multi-Index assessment and machine learning integration for drought monitoring using Google Earth Engine. *IEEE Journal of Selected Topics in Applied Earth Observations and Remote Sensing, 18*, 16497–16516. DOI: 10.1109/JSTARS.2025.3580652

Duarte, C. M., Losada, I. J., Hendriks, I. E., Mazarrasa, I., & Marbà, N. (2013). The role of coastal plant communities for climate change mitigation and adaptation. *Nature Climate Change, 3*(11), 961–968. DOI: 10.1038/nclimate1970

Du, C., Jiang, S., Chen, C., Guo, Q., He, Q., & Zhan, C. (2024). Machine Learning-Based estimation of daily cropland evapotranspiration in diverse climate zones. *Remote Sensing (Basel), 16*(5), 730. DOI: 10.3390/rs16050730

Durlević, U., Ilić, V., & Valjarević, A. (2025). Wildfire susceptibility mapping using deep learning and machine learning models based on multi-sensor satellite data fusion: A case study of Serbia. *Fire (Basel, Switzerland), 8*(10), 407. DOI: 10.3390/fire8100407

El-Masry, E. A. (2024). A Non-destructive Rapid Assessment of Blue Carbon Sequestration Potential in Mangrove Forests of the Red Sea Region. *Egyptian Journal of Aquatic Biology and Fisheries, 28*(1), 1119–1152. DOI: 10.21608/ejabf.2024.339764

Eskibağlar, B. K., Yavuz, Y. P., Doğan, G. K., Algarni, A., Cakmak, Y., & Pacal, I. (2026). Denta-HybridoNet: A hybrid CNN-transformer architecture for automated detection of developmental dental anomalies in pediatric panoramic radiographs. *Biomedical Signal Processing and Control, 118*, 109784. DOI: 10.1016/j.bspc.2026.109784

Fakhrurrozi, F., Laurance, L., Fauzi, M. N. R., Ramadhan, F., Idris, I., Yusri, S., Paryono, P., Lestariningsih, W. A., & Damayanti, A. A. (2023). Estimating Blue Carbon Stock: Mangrove and Seagrass Biomass Assessment in Tunda Island, Serang, Banten, Indonesia. *BIO Web of Conferences, 70*, 1009. DOI: 10.1051/bioconf/20237001009

Famine Early Warning Systems Network (FEWS NET). (2022). *FEWS NET methodology and analysis framework*. USAID. https://fews.net

Fan, J., Zhang, X., Zhao, C., Qin, Z., De Vroey, M., & Defourny, P. (2021). Evaluation of crop type classification with different high-resolution satellite data sources. *Remote Sensing (Basel)*, *13*(5), 911. DOI: 10.3390/rs13050911

FAO. (2021). *The state of food and agriculture 2021: Making agrifood systems more resilient to shocks and stresses*. Rome: Food and Agriculture Organization of the United Nations. https://www.fao.org/3/cb4476en/cb4476en.pdf

Febrianto, S., Rahman, A., Jati, O. E., Wirasatriya, A., Muskananfola, M. R., & Latifah, N. (2025). Machine learning for mangrove species distribution using Sentinel 2 satellite image in Segara Anakan, Cilacap Region, Indonesia. *Regional Studies in Marine Science*, *81*, 103984. DOI: 10.1016/j.rsma.2024.103984

Feng, D., Beck, H., Lawson, K., & Shen, C. (2023). The suitability of differentiable, physics-informed machine learning hydrologic models for ungauged regions and climate change impact assessment. *Hydrology and Earth System Sciences*, *27*(12), 2357–2373. DOI: 10.5194/hess-27-2357-2023

Fereshtehpour, M., Bashir, R., & Tandon, N. F. (2025). Risk-based framework to determine climate-informed design storms for road drainage infrastructure. *The Science of the Total Environment*, *1001*, 180427. DOI: 10.1016/j.scitotenv.2025.180427

Fernández, N., García, M., Gil, E., Serrano, E., & Delibes, M. (2014). Bottom-up resource limitation: the ecosystem energy balance predicts the quality of nutrition in a herbivore prey population. *Research Portal Denmark*, 21. https://local.forskningsportal.dk/local/dki-cgi/ws/cris-link?src=dtu&id=dtu-62123df4-c492-4e0f-967d-94cd12f34e31&ti=Bottom-up%20resource%20limitation%3A%20the%20ecosystem%20energy%20balance%20predicts%20the%20quality%20of%20nutrition%20in%20a%20herbivore%20prey%20population

Ferrario, F., Beck, M. W., Storlazzi, C. D., Micheli, F., Shepard, C. C., & Airoldi, L. (2014). The effectiveness of coral reefs for coastal hazard risk reduction and adaptation. *Nature Communications*, *5*(1), 3794. DOI: 10.1038/ncomms4794

Fersch, B., Francke, T., Heistermann, M., Schrön, M., Döpper, V., Jakobi, J., Baroni, G., Blume, T., Bogena, H., Budach, C., Gränzig, T., Förster, M., Güntner, A., Franssen, H. J. H., Kasner, M., Köhli, M., Kleinschmit, B., Kunstmann, H., Patil, A., & Oswald, S. (2020). A dense network of cosmic-ray neutron sensors for soil moisture observation in a highly instrumented pre-Alpine headwater catchment in Germany. *Earth System Science Data*, *12*(3), 2289–2309. DOI: 10.5194/essd-12-2289-2020

Filipović, N., Brdar, S., Mimić, G., Marko, O., & Crnojević, V. (2021). Regional soil moisture prediction system based on Long Short-Term Memory network. *Biosystems Engineering*, *213*, 30–38. DOI: 10.1016/j.biosystemseng.2021.11.019

Gao, N., Xue, H., Shao, W., Zhao, S., Qin, K. K., Prabowo, A., Rahaman, M. S., & Salim, F. D. (2022). Generative adversarial networks for spatio-temporal data: A survey. [TIST]. *ACM Transactions on Intelligent Systems and Technology*, *13*(2), 1–25. DOI: 10.1145/3474838

Gelete, G. (2023). Application of hybrid machine learning-based ensemble techniques for rainfall-runoff modeling. *Earth Science Informatics*, *16*(3), 2475–2495. DOI: 10.1007/s12145-023-01041-4

Ghali, R., & Akhloufi, M. A. (2023). Deep learning approaches for wildland fires using satellite remote sensing data: Detection, mapping, and prediction. *Fire (Basel, Switzerland)*, *6*(5), 192. DOI: 10.3390/fire6050192

Ghosh, S., Dasgupta, A., & Swetapadma, A. (2019, February). A study on support vector machine based linear and non-linear pattern classification. In *2019 International Conference on Intelligent Sustainable Systems (ICISS)* (pp. 24–28). IEEE. DOI: 10.1109/ISS1.2019.8908018

Giri, C. (2023). Frontiers in Global Mangrove Forest Monitoring. *Remote Sensing*, *15*(15), 3852. DOI: 10.3390/rs15153852 PMID: 39749006

Gobinath, C., Thanjaivadivel, M., Rangarajan, S., & Kalaivanan, K. (2025). Impact and Assessment of Artificial Intelligence-Enabled UAV for Real-Time Data Streaming Application. In *Drones for Transportation Logistics and Disaster Management* (pp. 115–144). Wiley., DOI: 10.1002/9781394386451.ch6

Golyandina, A. K., Nekrutkin, N. N., & Zhigljavsky, V. (2014). Basic singular spectrum analysis and forecasting with R. *Computational Statistics & Data Analysis*, *71*, 934–954. DOI: 10.1016/j.csda.2013.04.009

Gómez, J. A., Patiño, J. E., Duque, J. C., & Passos, S. (2019). Spatiotemporal modeling of urban growth using machine learning. *Remote Sensing (Basel)*, *12*(1), 109. DOI: 10.3390/rs12010109

Gorelick, N., Hancher, M., Dixon, M., Ilyushchenko, S., Thau, D., & Moore, R. (2017). Google Earth Engine: Planetary-scale geospatial analysis for environmental monitoring. *Remote Sensing of Environment*. Advance online publication. DOI: 10.1016/j.rse.2017.06.031

Gorman, D., Jayakody, S., Kodikara, S., Udagedara, S., Dahanayaka, D. D. G. L., Ranawana, K. B., Kumara, M. P., Pahalawattaarachchi, V., Mathiventhan, T., Steven, A., Gunawardana, J. A. A. L., & Vanderklift, M. A. (2022). Past, Present and Future of Sri Lankan Coastal Macrophyte-Dominated Ecosystems: Blue Carbon, Conservation, Restoration and Policy. In *CRC Press eBooks* (p. 301). Informa. DOI: 10.1201/9781003288602-6

Gounaridis, L., Groumas, P., Schreuder, E., Tsekenis, G., Marousis, A., Heideman, R., Avramopoulos, H., & Kouloumentas, C. (2017). High performance refractive index sensor based on low Q-factor ring resonators and FFT processing of wavelength scanning data. *Optics Express*, *25*(7), 7483–7495. DOI: 10.1364/OE.25.007483 PMID: 28380870

Guo, X., Zhang, Y. S., Zhong, M. J., Shen, W. R., & Wei, C. X. (2010). Variation characteristics of OLR for the Wenchuan earthquake. *Diqiu Wulixue Bao*, *53*(11), 2688–2695.

Handschuh, J., Erbertseder, T., & Baier, F. (2023). Systematic evaluation of four satellite AOD datasets for estimating PM2.5 using a random forest approach. *Remote Sensing (Basel)*, *15*(8), 2064. DOI: 10.3390/rs15082064

Han, S., Gao, R., Waseem, L. A., Shitu, K., Rebouh, N. Y., Zulqarnain, R. M., & Said, Y. (2025). High-Accuracy Mapping of Coastal and Wetland Areas Using Multisensor Data Fusion and Deep Feature Learning. *IEEE Journal of Selected Topics in Applied Earth Observations and Remote Sensing*, *18*, 24546–24559. DOI: 10.1109/JSTARS.2025.3606476

Hackip, S. C., Rajwade, Y. A., Rao, K. V. R., Kumar, S. P., Marak, A. B., & Srivastava, A. (2025). Approaches for Assessment of Soil Moisture with Conventional Methods, Remote Sensing, UAV, and Machine Learning Methods—A Review. *Water (Basel)*, *17*(16), 2388. DOI: 10.3390/w17162388

Hao, Y., Mao, J., Bachmann, C. M., Hoffman, F. M., Koren, G., Chen, H., Tian, H., Liu, J., Tao, J., Tang, J., Li, L., Liu, L., Apple, M., Shi, M., Jin, M., Zhu, Q., Kannenberg, S., Shi, X., Zhang, X., & Dai, Y. (2025). Soil moisture controls over carbon sequestration and greenhouse gas emissions: A review. *NPJ Climate and Atmospheric Science*, *8*(1), 16. Advance online publication. DOI: 10.1038/s41612-024-00888-8

Haque, S. A., & Al Jufaili, S. M. (2026). Applications of Artificial Intelligence in Fisheries: From Data to Decisions. *Big Data and Cognitive Computing*, *10*(1), 19. DOI: 10.3390/bdcc10010019

Harsa, H., Winarso, G., Setiawan, K. T., & Asriningrum, W. (2023). Satellite-based Mangrove species abundance estimate using Machine Learning ensemble. *IOP Conference Series. Earth and Environmental Science*, *1276*(1), 12055. DOI: 10.1088/1755-1315/1276/1/012055

Hart, A., Rodríguez, Á. M., Carvajal, J. R., & Ciottone, G. R. (2022). Earthquake response in Chile: A case study in health emergency and disaster risk management. *American Journal of Disaster Medicine*, *14*(4), 313–318. PMID: 35325465

Heaviside, C., Macintyre, H., & Vardoulakis, S. (2017). The urban heat island: Implications for health in a changing environment. *Current Environmental Health Reports*, *4*(3), 296–305. DOI: 10.1007/s40572-017-0150-3 PMID: 28695487

Hemati, M., Mahdianpari, M., Shiri, H., & Mohammadimanesh, F. (2025). Comparative Analysis of UAVSAR Polarimetric Decompositions for Wetland Aboveground Biomass Mapping Using Machine Learning Models. *The œinternational Archives of the Photogrammetry, Remote Sensing and Spatial Information Sciences. The International Archives of the Photogrammetry, Remote Sensing and Spatial Information Sciences*, *19*, 19–25. Advance online publication. DOI: 10.5194/isprs-archives-XLVIII-M-8-2025-19-2025

Hendrickx, M. G. A., Vanderborght, J., Janssens, P., Bombeke, S., Matthyssen, E., Waverijn, A., & Diels, J. (2025). Pooled error variance and covariance estimation of sparse in situ soil moisture sensor measurements in agricultural fields in Flanders. *Soil (Göttingen)*, *11*(1), 435–456. DOI: 10.5194/soil-11-435-2025

Herath, H. M. V. V., Chadalawada, J., & Babovic, V. (2020). Hydrologically Informed Machine Learning for Rainfall-Runoff Modelling: Towards Distributed Modelling. *Hydrology and Earth System Sciences*. Advance online publication. DOI: 10.5194/hess-2020-487

Hiles, W., Miller, L. C., Smeaton, C., & Austin, W. E. N. (2024). UAV approaches for improved mapping of vegetation cover and estimation of carbon storage of small saltmarshes: Examples from Loch Fleet, northeast Scotland. *Biogeosciences*, *21*(4), 929–948. DOI: 10.5194/bg-21-929-2024

Hill, V., Zimmerman, R. C., Byron, D., & Heck, K. L.Jr. (2024). Mapping Seagrass Distribution and Abundance: Comparing Areal Cover and Biomass Estimates Between Space-Based and Airborne Imagery. *Remote Sensing (Basel)*, *16*(23), 4351. DOI: 10.3390/rs16234351

Holloway, J., & Mengersen, K. (2018). Statistical Machine Learning Methods and Remote Sensing for Sustainable Development Goals: A Review. *Remote Sensing (Basel), 10*(9), 1365. DOI: 10.3390/rs10091365

Hossain, M. S., Bujang, J. S., Zakaria, M. H., & Hashim, M. (2015). The application of remote sensing to seagrass ecosystems: An overview and future research prospects. *International Journal of Remote Sensing*, *36*(1), 61–113. DOI: 10.1080/01431161.2014.990649

Hossain, M. S., Bujang, J. S., Zakaria, M. H., & Hashim, M. (2016). Marine and human habitat mapping for the Coral Triangle Initiative region of Sabah using Landsat and Google Earth imagery. *Marine Policy*, *72*, 176–191. DOI: 10.1016/j.marpol.2016.07.003

Hossain, M. S., & Hashim, M. (2019). Potential of Earth Observation (EO) technologies for seagrass ecosystem service assessments. *International Journal of Applied Earth Observation and Geoinformation*, *77*, 15–29. DOI: 10.1016/j.jag.2018.12.009

Hossain, M. S., Hashim, M., Bujang, J. S., Zakaria, M. H., & Muslim, A. M. (2019). Assessment of the impact of coastal reclamation activities on seagrass meadows in Sungai Pulai estuary, Malaysia, using Landsat data (1994–2017). *International Journal of Remote Sensing*, *40*(9), 3571–3605. DOI: 10.1080/01431161.2018.1547931

Hossain, M. S., Muslim, A. M., Nadzri, M. I., Sabri, A. W., Khalil, I., Mohamad, Z., & Pour, A. B. (2021). Coral habitat mapping: A comparison between maximum likelihood, Bayesian and Dempster–Shafer classifiers. *Geocarto International*, *36*(11), 1217–1235. DOI: 10.1080/10106049.2019.1637466

Hossain, M. S., Muslim, A. M., Nadzri, M. I., Teruhisa, K., David, D., Khalil, I., & Mohamad, Z. (2020). Can ensemble techniques improve coral reef habitat classification accuracy using multispectral data? *Geocarto International*, *35*(11), 1214–1232. DOI: 10.1080/10106049.2018.1557263

Hosseini, S. M., & Mahjouri, N. (2015). Integrating Support Vector Regression and a geomorphologic Artificial Neural Network for daily rainfall-runoff modeling. *Applied Soft Computing*, *38*, 329–345. DOI: 10.1016/j.asoc.2015.09.049

Huang, J., Sehgal, V., Alvarez, L. V., Brocca, L., Cai, S., Cheng, R., Cheng, X., Du, J., Masri, B. E., Endsley, K. A., Fang, Y., Hu, J., Jampani, M., Kibria, M. G., Koren, G., Li, L., Liu, L., Mao, J., Moreno, H. A., & Fisher, J. B. (2025). Remotely Sensed High-Resolution Soil Moisture and evapotranspiration: Bridging the gap between science and society. *Water Resources Research*, *61*(5), e2024WR037929. Advance online publication. DOI: 10.1029/2024WR037929

Huang, X., Shi, Z. H., Zhu, H. D., Zhang, H. Y., Ai, L., & Yin, W. (2015). Soil moisture dynamics within soil profiles and associated environmental controls. *Catena, 136*, 189–196. DOI: 10.1016/j.catena.2015.01.014

Hu, C., Wu, Q., Li, H., Jian, S., Li, N., & Lou, Z. (2018). Deep Learning with a Long Short-Term Memory Networks Approach for Rainfall-Runoff Simulation. *Water (Basel), 10*(11), 1543. DOI: 10.3390/w10111543

Hu, K., Xu, Q., Zhu, K., Wang, Z., Chu, Y., Qian, Y., Zhang, X., Wang, B., & Zhang, H. (2026). Recent Advances of Artificial Intelligence in Aquatic Bioindicators and Ecological Assessment. *Water, Air, and Soil Pollution, 237*(5), 306. DOI: 10.1007/s11270-025-08907-x

Hu, L., Xie, Y., Liu, Y., Chen, S., Yu, H., Bie, Q., Zhao, F., & Zhao, Y. (2025). Mapping the primary factors driving spatiotemporal variations of surface soil moisture in the Yellow River Basin of China. *Scientific Reports, 15*(1), 12359. DOI: 10.1038/s41598-025-96129-w PMID: 40211021

Huot, F., Hu, R. L., Goyal, N., Sankar, T., Ihme, M., & Chen, Y. F. (2022). Next day wildfire spread: A machine learning dataset to predict wildfire spreading from remote-sensing data. *IEEE Transactions on Geoscience and Remote Sensing, 60*, 1–13. DOI: 10.1109/TGRS.2022.3192974

Hussain, M., & Imitiyaz, I. (2018). Urbanization concepts, dimensions and factors. *International Journal of Recent Scientific Research, 9*(1), 23513–23523.

Hu, Y., Tian, B., Yuan, L., Li, X., Huang, Y., Shi, R., Jiang, X., Wang, , & Sun, C. (2021). Mapping coastal salt marshes in China using time series of Sentinel-1 SAR. *ISPRS Journal of Photogrammetry and Remote Sensing, 173*, 122–134. DOI: 10.1016/j.isprsjprs.2021.01.003

Huynh, N. N. T., Garambois, P.-A., Renard, B., Colleoni, F., Monnier, J., & Roux, H. (2025). A distributed hybrid physics–AI framework for learning corrections of internal hydrological fluxes and enhancing high-resolution regionalized flood modeling. *Hydrology and Earth System Sciences, 29*(15), 3589–3613. DOI: 10.5194/hess-29-3589-2025

Hyyppä, J., Yu, X., Hyyppä, H., Vastaranta, M., Holopainen, M., Kukko, A., Kaartinen, H., Jaakkola, A., Vaaja, M., Koskinen, J., & Alho, P. (2012). Advances in Forest Inventory Using Airborne Laser Scanning. *Remote Sensing (Basel), 4*(5), 1190–1207. DOI: 10.3390/rs4051190

Illarionova, S., Shadrin, D., Gubanov, F., Shutov, M., Tasuev, U., Evteeva, K., Mironenko, M., & Burnaev, E. (2025). Exploration of geo-spatial data and machine learning algorithms for robust wildfire occurrence prediction. *Scientific Reports*, *15*(1), 10712. DOI: 10.1038/s41598-025-94002-4 PMID: 40155427

Ioras, F., & Bandara, I. (2025). Artificial Intelligence and Sustainable Practices in Coastal Marinas: A Comparative Study of Monaco and Ibiza. *Sustainability (Basel)*, *17*(16), 7404. DOI: 10.3390/su17167404

Islam, F. S. (2025). A multi-dimensional AI framework for sustainable drinking water management: Integrating federated learning, digital twins, and blockchain. *Journal of Engineering Research and Reports*, *27*(6), 466–492. DOI: 10.9734/jerr/2025/v27i61558

Islam, M. T., Roy, S. C., Jahan, N., Al-Mahmud, , Islam, M. M., Ferdaus, A. A., Fujisawa, K., & Adham, A. K. M. (2025). Comparative evaluation of machine learning models for extreme river water level forecasting in Bangladesh: Implications for flood and drought resilience. *Progress in Disaster Science*, *27*, 100449. DOI: 10.1016/j.pdisas.2025.100449

Islam, T., Mazumder, T., Shahriair Roni, M. N., & Nur, M. S. (2024). A comparative study of machine learning models for predicting Aman rice yields in Bangladesh. *Heliyon*, *10*(23), e40764. DOI: 10.1016/j.heliyon.2024.e40764 PMID: 39691189

Jabed, M. A., & Murad, M. A. A. (2024). Crop yield prediction in agriculture: A comprehensive review of machine learning and deep learning approaches, with insights for future research and sustainability. *Heliyon*, *10*(24), e40836. DOI: 10.1016/j.heliyon.2024.e40836 PMID: 39720079

James, D., Collin, A., Mury, A., & Letard, M. (2021). ENHANCING UAV COASTAL MAPPING USING INFRARED PANSHARPENING. *The œinternational Archives of the Photogrammetry, Remote Sensing and Spatial Information Sciences. The International Archives of the Photogrammetry, Remote Sensing and Spatial Information Sciences*, *257*. Advance online publication. DOI: 10.5194/isprs-archives-XLIII-B3-2021-257-2021

Jayakumar, D., Bouhoula, A., & Al-Zubari, W. K. (2024). Unlocking the potential of artificial intelligence for sustainable water management focusing operational applications. *Water (Basel)*, *16*(22), 3328. DOI: 10.3390/w16223328

Jeon, E., Kim, S., Park, S., Kwak, J., & Choi, I. (2021). Semantic segmentation of seagrass habitat from drone imagery based on deep learning: A comparative study. *Ecological Informatics*, *66*, 101430. DOI: 10.1016/j.ecoinf.2021.101430

Jessen, B., & Hill, K. (2024). Sea Grant at the Blue Carbon Frontier: Integrating Law, Science, Community Values, and Economics. *Oceanography (Washington, D.C.)*, *37*(1), 84–90. DOI: 10.5670/oceanog.2024.241

Jia, L., Yen, N., & Pei, Y. (2025). A Physics-Coupled deep learning framework for hydrodynamic diffusion modeling in watershed systems: Integrating spatiotemporal networks and environmental constraints. *IEEE Access : Practical Innovations, Open Solutions*, *13*, 34985–35003. DOI: 10.1109/ACCESS.2025.3542173

Jiang, Z.-Y., Yang, Z.-G., Zhang, S.-Y., Liao, C.-M., Hu, Z.-M., Cao, R.-C., & Wu, H.-W. (2020). Revealing the spatio-temporal variability of evapotranspiration and its components based on an improved Shuttleworth-Wallace model in the Yellow River Basin. *Journal of Environmental Management*, *262*, 110310. DOI: 10.1016/j.jenvman.2020.110310 PMID: 32250793

Jing, F., Gu, X., & Shen, X. (2009). Study on outgoing longwave radiation variations associated with strong earthquake. *Proceedings of SPIE - the International Society for Optical Engineering*, *7651*, 765113. DOI: 10.1117/12.855573

Jones, A. R., Segaran, R. R., Clarke, K., Waycott, M., Goh, W. S. H., & Gillanders, B. M. (2020). Estimating Mangrove Tree Biomass and Carbon Content: A Comparison of Forest Inventory Techniques and Drone Imagery. *Frontiers in Marine Science*, *6*, 784. Advance online publication. DOI: 10.3389/fmars.2019.00784

Jones, A., Kuehnert, J., Fraccaro, P., Meuriot, O., Ishikawa, T., Edwards, B., Stoyanov, N., Remy, S. L., Weldemariam, K., & Assefa, S. (2023). AI for climate impacts: Applications in flood risk. *NPJ Climate and Atmospheric Science*, *6*(1), 63. Advance online publication. DOI: 10.1038/s41612-023-00388-1

Joshi, A., Pradhan, B., Gite, S., & Chakraborty, S. (2023). Remote-sensing data and deep-learning techniques in crop mapping and yield prediction: A systematic review. *Remote Sensing (Basel)*, *15*(8), 2014. DOI: 10.3390/rs15082014

Joyce, K. E., Fickas, K. C., & Kalamandeen, M. (2022). The unique value proposition for using drones to map coastal ecosystems. *Cambridge Prisms: Coastal Futures*, *1*, e6. Advance online publication. DOI: 10.1017/cft.2022.7

Jung, C.-R., Chen, W.-T., & Nakayama, S. F. (2021). A national-scale 1-km resolution PM2.5 estimation model over Japan using MAIAC AOD and a two-stage random forest model. *Remote Sensing (Basel)*, *13*(18), 3657. DOI: 10.3390/rs13183657

Kaggle. (n.d.). https://www.kaggle.com/datasets/abdelghaniaaba/wildfire-prediction-dataset

Kalasampath, K., Spoorthi, K. N., Sajeev, S., Kuppa, S. S., Ajay, K., & Maruthamuthu, A. (2025). A Literature Review on Applications of Explainable Artificial Intelligence (XAI). *IEEE Access : Practical Innovations, Open Solutions, 13*, 41111–41140. DOI: 10.1109/ACCESS.2025.3546681

Kamilaris, A., & Prenafeta-Boldú, F. X. (2018). Deep learning in agriculture: A survey. *Computers and Electronics in Agriculture, 147*, 70–90. DOI: 10.1016/j.compag.2018.02.016

Kamyab, H., Khademi, T., Chelliapan, S., SaberiKamarposhti, M., Rezania, S., Yusuf, M., Farajnezhad, M., Abbas, M., Hun Jeon, B., & Ahn, Y. (2023). The latest innovative avenues for the utilization of artificial Intelligence and big data analytics in water resource management. *Results in Engineering, 20*, 101566. DOI: 10.1016/j.rineng.2023.101566

Kandrot, S., Hayes, S., & Holloway, P. (2021). Applications of Uncrewed Aerial Vehicles (UAV) Technology to Support Integrated Coastal Zone Management and the UN Sustainable Development Goals at the Coast [Review of *Applications of Uncrewed Aerial Vehicles (UAV) Technology to Support Integrated Coastal Zone Management and the UN Sustainable Development Goals at the Coast*]. *Estuaries and Coasts, 45*(5), 1230. Springer Science+Business Media. DOI: 10.1007/s12237-021-01001-5

Kang, M., Cho, I., Park, J., Jeong, J., Lee, K., Lee, B., Del Orbe Henriquez, D., Yoon, K., & Park, I. (2022). High accuracy real-time multi-gas identification by a batch-uniform gas sensor array and deep learning algorithm. *ACS Sensors, 7*(2), 430–440. DOI: 10.1021/acssensors.1c01204 PMID: 35041384

Kang, Y., Choi, H., Im, J., Park, S., Shin, M., Song, C.-K., & Kim, S. (2021). Estimation of surface-level NO_2 and O_3 concentrations using TROPOMI data and machine learning over East Asia. *Environmental Pollution, 288*, 117711. DOI: 10.1016/j.envpol.2021.117711 PMID: 34329053

Kar, K. K., Haggerty, R., Sharma, H., Dwivedi, D., & Roy, T. (2025). Evapotranspiration partitioning using flux tower data in a Semi-Arid ecosystem. *Hydrological Processes, 39*(3), e70083. Advance online publication. DOI: 10.1002/hyp.70083

Karthikeyan, L., & Mishra, A. K. (2021). Multi-layer high-resolution soil moisture estimation using machine learning over the United States. *Remote Sensing of Environment, 266*, 112706. DOI: 10.1016/j.rse.2021.112706

Kasani, S., Curtin, K., & Wu, N. (2019). A review of 2D and 3D plasmonic nanostructure array patterns: Fabrication, light management and sensing applications. *Nanophotonics, 8*(12), 2065–2089. DOI: 10.1515/nanoph-2019-0158

Kashem, M. A. (2025). Exploring climate change and sustainability issues in Bangladesh through system dynamics. *Discover Sustainability*, *6*(1), 1073. DOI: 10.1007/s43621-025-01652-5

Kayizzi-Mugerwa, S., Shimeles, A., & Yameogo, N. (2016). *Urbanization and socio-economic development in Africa*. Routledge.

Khan, A., & Sudheer, M. (2022). Machine learning-based monitoring and modeling for spatio-temporal urban growth of Islamabad. *The Egyptian Journal of Remote Sensing and Space Sciences*, *25*(2), 541–550. DOI: 10.1016/j.ejrs.2022.03.012

Khojasteh, D., Glamore, W., Heimhuber, V., & Felder, S. (2021). Sea level rise impacts on estuarine dynamics: A review. *The Science of the Total Environment*, *780*, 146470. DOI: 10.1016/j.scitotenv.2021.146470

Khurana, Y., Soni, P. K., & Bhatt, D. P. (2023). SVM-based classification of multi-temporal Sentinel-2 imagery of dense urban land cover of Delhi-NCR region. *Earth Science Informatics*, *16*(2), 1765–1777. DOI: 10.1007/s12145-023-01008-5

Kibirige, G. W., Yang, M.-C., Liu, C.-L., & Chen, M.-C. (2023). Using satellite data on remote transportation of air pollutants for PM2.5 prediction in northern Taiwan. *PLoS One*, *18*(3), e0282471. DOI: 10.1371/journal.pone.0282471 PMID: 36897845

Kikon, A., & Deka, P. C. (2021). Artificial intelligence application in drought assessment, monitoring and forecasting: A review. *Stochastic Environmental Research and Risk Assessment*, *36*(5), 1197–1214. DOI: 10.1007/s00477-021-02129-3

Kim, M., Kim, J., Lim, H., Lee, S., Cho, Y., Lee, Y.-G., Go, S., & Lee, K. (2024). Aerosol optical depth data fusion with geostationary Korea multi-purpose satellite (GEO-KOMPSAT-2) instruments GEMS, AMI, and GOCI-II: Statistical and deep neural network methods. *Atmospheric Measurement Techniques*, *17*(14), 4317–4335. DOI: 10.5194/amt-17-4317-2024

Kim, T., Yang, T., Gao, S., Zhang, L., Ding, Z., Wen, X., Gourley, J. J., & Hong, Y. (2021). Can artificial intelligence and data-driven machine learning models match or even replace process-driven hydrologic models for streamflow simulation?: A case study of four watersheds with different hydro-climatic regions across the CONUS. *Journal of Hydrology (Amsterdam)*, *598*, 126423. DOI: 10.1016/j.jhydrol.2021.126423

Kokkalis, P., Jassar, H. K. A., Sarraf, H. A., Nair, R., & Hendi, H. A. (2024). Evaluation of ERA5 and NCEP reanalysis climate models for precipitation and soil moisture over a semi-arid area in Kuwait. *Climate Dynamics*, *62*(6), 4893–4904. DOI: 10.1007/s00382-024-07141-1

Koldasbayeva, D., Tregubova, P., Gasanov, M., Zaytsev, A., Petrovskaia, A., & Burnaev, E. (2024). Challenges in data-driven geospatial modeling for environmental research and practice. *Nature Communications*, *15*(1), 10700. DOI: 10.1038/s41467-024-55240-8 PMID: 39702456

Kong, X., Li, N., Lin, L., Xiong, P., & Qi, J. (2018). Relationship of Stress Changes and Anomalies in OLR Data of the Wenchuan and Lushan Earthquakes. *IEEE Journal of Selected Topics in Applied Earth Observations and Remote Sensing*, *11*(8), 2966–2976. DOI: 10.1109/JSTARS.2018.2839089

Kpienbaareh, D., Sun, X., Wang, J., Luginaah, I., Bezner Kerr, R., Lupafya, E., & Dakishoni, L. (2021). Crop type and land cover mapping in Northern Malawi using the integration of Sentinel-1, Sentinel-2, and PlanetScope satellite data. *Remote Sensing (Basel)*, *13*(4), 700. DOI: 10.3390/rs13040700

Krid, H. B., Wertani, H., & Hlali, A.. (2025). Optimization of a hybrid graphene–copper terahertz gas sensor using machine learning. *IEEE Transactions on Plasma Science*. Advance online publication. DOI: 10.1109/TPS.2025.3644784

Krid, H. B., Wertani, H., Hlali, A., & Zairi, H. (2025). High-sensitivity graphene-based terahertz biosensor for cervical cancer diagnosis: SVM-assisted optimization. *IEEE Transactions on Plasma Science*, *53*(12), 4000–4008. DOI: 10.1109/TPS.2025.3632754

Krid, H. B., Wertani, H., Hlali, A., & Zairi, H. (2026). Graphene–copper hybrid terahertz biosensor with KNN sensitivity optimization for early skin cancer (melanoma) detection. *Sensing and Imaging*, *27*(1), 17. DOI: 10.1007/s11220-026-00722-w

Krishnan, A., Ramabhadran, R., Divakaran, G., & Muraleedharan, K. (2023). Analysis and synthesis of East Himalayan earthquakes using a hybrid wavelet synchrosqueezed transform. *Arabian Journal of Geosciences*, *16*(12), 637. DOI: 10.1007/s12517-023-11673-x

Kuenzer, C., Bluemel, A., Gebhardt, S., Quoc, T. V., & Dech, S. (2011). Remote Sensing of Mangrove Ecosystems: A Review. *Remote Sensing (Basel)*, *3*(5), 878–928. DOI: 10.3390/rs3050878

Kuenzer, C., Ottinger, M., Liu, G., Sun, B., Baumhauer, R., & Dech, S. (2014). Earth observation-based coastal zone monitoring of the Yellow River Delta: Dynamics in China's second largest oil producing region over four decades. *Applied Geography (Sevenoaks, England)*, *55*, 92–107. DOI: 10.1016/j.apgeog.2014.08.015

Kuffour, B. N. O., Engdahl, N. B., Woodward, C. S., Condon, L. E., Kollet, S., & Maxwell, R. M. (2020). Simulating coupled surface–subsurface flows with ParFlow v3.5.0: Capabilities, applications, and ongoing development of an open-source, massively parallel, integrated hydrologic model. *Geoscientific Model Development, 13*(3), 1373–1397. DOI: 10.5194/gmd-13-1373-2020

Kuhn, L., Beirle, S., Osipov, S., Pozzer, A., & Wagner, T. (2024). NitroNet – A machine learning model for the prediction of tropospheric NO_2 profiles from TROPOMI observations. *Atmospheric Measurement Techniques, 17*(21), 6485–6516. DOI: 10.5194/amt-17-6485-2024

Kurtz, W., Lapin, A., Schilling, O. S., Tang, Q., Schiller, E., Braun, T., Hunkeler, D., Vereecken, H., Sudicky, E., Kropf, P., Hendricks Franssen, H.-J., & Brunner, P. (2017). Integrating hydrological modelling, data assimilation and cloud computing for real-time management of water resources. *Environmental Modelling & Software, 93*, 418–435. DOI: 10.1016/j.envsoft.2017.03.011

LaGrange, T. C. (1999). The impact of neighborhoods, schools, and malls on the spatial distribution of property damage. *Journal of Research in Crime and Delinquency, 36*(4), 393–422. DOI: 10.1177/0022427899036004003

Laignel, B., Vignudelli, S., Almar, R., Becker, M., Bentamy, A., Benveniste, J., Birol, F., Frappart, F., Idier, D., Salameh, E., Passaro, M., Menende, M., Simard, M., Turki, I., & Verpoorter, C. (2023). Observation of the Coastal Areas, Estuaries and Deltas from Space. *Surveys in Geophysics, 44*(5), 1309–1356. DOI: 10.1007/s10712-022-09757-6

Lanceman, D., Sadat-Noori, M., Gaston, T. F., Drummond, C., & Glamore, W. (2022). Blue carbon ecosystem monitoring using remote sensing reveals wetland restoration pathways. *Frontiers in Environmental Science, 10*, 924221. Advance online publication. DOI: 10.3389/fenvs.2022.924221

Lecours, V., Disney, M., He, K. S., Pettorelli, N., Rowcliffe, J. M., Sankey, T. T., & Scales, K. L. (2021). Remote sensing and the UN Ocean Decade: High expectations, big opportunities. *Remote Sensing in Ecology and Conservation, 8*(3), 267–271. DOI: 10.1002/rse2.241

LeCun, Y., Bengio, Y., & Hinton, G. (2015). Deep learning. *Nature, 521*(7553), 436–444. DOI: 10.1038/nature14539

Le, D. N., Le, V. H., & Roy, P. (2020). Graphene under uniaxial inhomogeneous strain and an external electric field: Landau levels, electronic, magnetic and optical properties. *The European Physical Journal B, 93*(8), 158. DOI: 10.1140/epjb/e2020-10222-3

Lemenkova, P. (2019). *Seagrass Mapping and Monitoring Along the Coasts of Crete, Greece*. DOI: 10.31237/osf.io/p4h9v

Leonelli, S. (2020). Scientific research and big data. In Zalta, E. N. (Ed.), *The Stanford Encyclopedia of Philosophy* (Fall 2020 Edition)., https://plato.stanford.edu/entries/science-big-data/

Lesiv, M., Laso Bayas, J. C., See, L., Duerauer, M., Dahlia, D., Durando, N., Hazarika, R., Kumar Sahariah, P., Vakolyuk, M., Blyshchyk, V., Bilous, A., Perez-Hoyos, A., Gengler, S., Prestele, R., Bilous, S., Akhtar, I. H., Singha, K., Choudhury, S. B., Chetri, T., & Fritz, S. (2019). Estimating the global distribution of field size using crowdsourcing. *Global Change Biology*, *25*(1), 174–186. DOI: 10.1111/gcb.14492 PMID: 30549201

Li, L., Wang, J., Franklin, M., Yin, Q., Wu, J., Camps-Valls, G., Zhu, Z., Wang, C., Ge, Y., & Reichstein, M. (2023). Improving air quality assessment using physics-inspired deep graph learning. *npj Climate and Atmospheric Science, 6*, 152. DOI: 10.1038/s41612-023-00475-3

Li, L., Xu, H., Zhang, Q., Zhan, Z., Liang, X., & Xing, J. (2024). Estimation methods of wetland carbon sink and factors influencing wetland carbon cycle: a review [Review of *Estimation methods of wetland carbon sink and factors influencing wetland carbon cycle: a review*]. *Carbon Research, 3*(1). Springer Science+Business Media. DOI: 10.1007/s44246-024-00135-y

Liao, C., Wang, J., Shan, B., Shang, J., Dong, T., & He, Y. (2023). Near real time detection and forecasting of within-field phenology of winter wheat and corn using Sentinel-2 time series data. *ISPRS Journal of Photogrammetry and Remote Sensing*, *196*, 105–119. DOI: 10.1016/j.isprsjprs.2022.12.025

Li, B., Sun, T., Tian, F., Tudaji, M., Qin, L., & Ni, G. (2024). Hybrid hydrological modeling for large alpine basins: A semi-distributed approach. *Hydrology and Earth System Sciences*, *28*(20), 4521–4538. DOI: 10.5194/hess-28-4521-2024

Li, C., Joiner, J., Liu, F., Krotkov, N. A., Fioletov, V., & McLinden, C. (2022). A new machine-learning-based analysis for improving satellite-retrieved atmospheric composition data: OMI SO_2 as an example. *Atmospheric Measurement Techniques*, *15*(18), 5497–5514. DOI: 10.5194/amt-15-5497-2022

Li, D., Sun, L., Yu, Y., & Tian, P. (2025). Geospatial micro-estimates of slum populations in 129 Global South countries using machine learning and public data. *Earth System Science Data Discussions*, *2025*, 1–33. DOI: 10.5194/essd-2025-260

Li, L., Mu, X., Chianucci, F., Qi, J., Jiang, J., Zhou, J., Chen, L., Huang, H., Yan, G., & Liu, S. (2022). Ultrahigh-resolution boreal forest canopy mapping: Combining UAV imagery and photogrammetric point clouds in a deep-learning-based approach. *International Journal of Applied Earth Observation and Geoinformation*, *107*, 102686. DOI: 10.1016/j.jag.2022.102686

Lin, L., Liang, Y., Liu, L., Zhang, Y., Xie, D., Yin, F., & Ashraf, T. (2022). Estimating PM2.5 concentrations using the machine learning RF-XGBoost model in Guanzhong urban agglomeration, China. *Remote Sensing (Basel)*, *14*(20), 5239. DOI: 10.3390/rs14205239

Lin, Y., Wang, D., Wang, G., Qiu, J., Long, K., Du, Y., Xie, H., Wei, Z., Shangguan, W., & Dai, Y. (2021). A hybrid deep learning algorithm and its application to streamflow prediction. *Journal of Hydrology (Amsterdam)*, *601*, 126636. DOI: 10.1016/j.jhydrol.2021.126636

Liu, D., Jiang, W., Mu, L., & Wang, S. (2020). Streamflow prediction using Deep Learning Neural Network: Case study of Yangtze River. *IEEE Access : Practical Innovations, Open Solutions*, *8*, 90069–90086. DOI: 10.1109/ACCESS.2020.2993874

Liu, D., Wang, Z., Wang, L., Chen, J., Li, C., & Shi, Y. (2024). Improved remote sensing reference evapotranspiration estimation using simple satellite data and machine learning. *The Science of the Total Environment*, *947*, 174480. DOI: 10.1016/j.scitotenv.2024.174480 PMID: 38972400

Liu, Q., Yuan, F., Adams, C., Rebelo, L.-M., & Wellington, M. (2024). Dynamic Coastal Mapping Using Sentinel-1 and Sentinel-2 Data Through Digital Earth Africa. *The International Archives of the Photogrammetry, Remote Sensing and Spatial Information Sciences*, *XLVIII-4–2024*, 325–330. DOI: 10.5194/isprs-archives-XLVIII-4-2024-325-2024

Liu, Z., Zhou, J., Yang, X., Zhao, Z., & Lv, Y. (2024). Research on water resource modeling based on machine learning technologies. *Water (Basel)*, *16*(3), 472. DOI: 10.3390/w16030472

Li, X., Chen, Y., & Ma, J. (2024). Multimodal fusion networks for climate-driven landform dynamics. [case study of multimodal fusion]. *International Journal of Applied Earth Observation and Geoinformation*.

Li, Y., Yuan, L., Song, Z., Yu, S., Zhang, X., Tian, B., & Liu, M. (2024). Salt marsh carbon stock estimation using deep learning with Sentinel-1 SAR of the Yangtze River estuary, China. *International Journal of Applied Earth Observation and Geoinformation*, *133*, 104138. DOI: 10.1016/j.jag.2024.104138

Long, S., Wei, X., Zhang, F., Zhang, R., Xu, J., Wu, K., Li, Q., & Li, W. (2022). Estimating daily ground-level NO_2 concentrations over China based on TROPOMI observations and machine learning approach. *Atmospheric Environment*, *289*, 119310. DOI: 10.1016/j.atmosenv.2022.119310

Lu, D., Konapala, G., Painter, S. L., Kao, S.-C., & Gangrade, S. (2021). Streamflow simulation in data-scarce basins using Bayesian and physics-informed machine learning models. *Journal of Hydrometeorology*. Advance online publication. DOI: 10.1175/JHM-D-20-0082.1

Lungu, O. N., Chabala, L. M., & Shepande, C. (2020). Satellite-based crop monitoring and yield estimation—A review. *Journal of Agricultural Science (Toronto)*, *13*(1), 180. DOI: 10.5539/jas.v13n1p180

Luo, W., Xu, X., Liu, W., Liu, M., Li, Z., Peng, T., Xu, C., Zhang, Y., & Zhang, R. (2018). UAV based soil moisture remote sensing in a karst mountainous catchment. *Catena*, *174*, 478–489. DOI: 10.1016/j.catena.2018.11.017

Lu, T., Jiang, T., & Liu, Y. (2025). Research on an optimization model for water resource management based on deep learning. *IEEE Access : Practical Innovations, Open Solutions*, *13*, 147906–147924. DOI: 10.1109/ACCESS.2025.3600664

Macreadie, P. I., Antón, A., Raven, J. A., Beaumont, N., Connolly, R. M., Friess, D. A., Kelleway, J. J., Kennedy, H., Kuwae, T., Lavery, P. S., Lovelock, C. E., Smale, D. A., Apostolaki, E. T., Atwood, T. B., Baldock, J., Bianchi, T. S., Chmura, G. L., Eyre, B. D., Fourqurean, J. W., … Duarte, C. M. (2019). The future of Blue Carbon science [Review of *The future of Blue Carbon science*]. *Nature Communications*, *10*(1). Nature Portfolio. DOI: 10.1038/s41467-019-11693-w

Macreadie, P. I., Costa, M. D. P., Atwood, T. B., Friess, D. A., Kelleway, J. J., Kennedy, H., Lovelock, C. E., Serrano, O., & Duarte, C. M. (2021). Blue carbon as a natural climate solution. *Nature Reviews. Earth & Environment*, 2(12), 826–839. DOI: 10.1038/s43017-021-00224-1

Mahmoud, M. B. R., Hemdan, D. I. I., Alajmani, S. H., Alyami, R. Y., Elmarhomy, G., Hashim, H., & Atlam, E.-S. (2025). Optimized AI and IoT-Driven framework for intelligent water resource Management. *IEEE Access : Practical Innovations, Open Solutions*, *13*, 97628–97646. DOI: 10.1109/ACCESS.2025.3572067

Mahmoudvand, P. C. R., & Rahim. (2021). Prediction intervals for the vector singular spectrum analysis forecasting algorithm in a median-based singular spectrum analysis. Computational Mathematics and Methods, 22(2), e2404. DOI: <ALIGNMENT.qj></ALIGNMENT>10.1002/cmm4.2404

Mainali, K., Evans, M., Saavedra, D., Mills, E., Madsen, B., & Minnemeyer, S. (2023). Convolutional neural network for high-resolution wetland mapping with open data: Variable selection and the challenges of a generalizable model. *The Science of the Total Environment*, *861*, 160622. DOI: 10.1016/j.scitotenv.2022.160622

Malerba, M. E., Duarte de Paula Costa, M., Friess, D. A., Schuster, L., Young, M. A., Lagomasino, D., Serrano, O., Hickey, S. M., York, P. H., Rasheed, M., Lefcheck, J. S., Radford, B., Atwood, T. B., Ierodiaconou, D., & Macreadie, P. (2023). Remote sensing for cost-effective blue carbon accounting. *Earth-Science Reviews*, *238*, 104337. DOI: 10.1016/j.earscirev.2023.104337

Mallick, S. K., Das, P., Maity, B., Rudra, S., Pramanik, M., Pradhan, B., & Sahana, M. (2021). Understanding future urban growth, urban resilience and sustainable development of small cities using prediction–adaptation–resilience approach. *Sustainable Cities and Society*, *74*, 103196. DOI: 10.1016/j.scs.2021.103196

Mancini, A., Solfanelli, F., Coviello, L., Martini, F. M., Mandolesi, S., & Zanoli, R. (2024). Time series from Sentinel-2 for organic durum wheat yield prediction using functional data analysis and deep learning. *Agronomy (Basel)*, *14*(1), 109. DOI: 10.3390/agronomy14010109

Manenti, R. R., Souza, W. E., & Porsani, M. J. (2019). Deep structures seismic enhancement using singular spectral analysis in time and frequency domain: Application in the regional transect of Parnaíba basin-Brazil. *Journal of Applied Geophysics*, *164*, 19–25. DOI: 10.1016/j.jappgeo.2019.02.020

Marino, M., Baptist, M. J., Alkharoubi, A. I. K., Nasca, S., Cavallaro, L., Foti, E., & Musumeci, R. E. (2025). Nature-based Solutions as Building Blocks for coastal flood risk reduction: A model-based ecosystem service assessment. *Scientific Reports*, *15*(1), 12070. DOI: 10.1038/s41598-025-95230-4

Marshak, A. H., & Van Vliet, C. M. (1984). Electrical current and carrier density in degenerate materials with nonuniform band structure. *Proceedings of the IEEE*, *72*(2), 148–164. DOI: 10.1109/PROC.1984.12836

Masupha, T. E., Moeletsi, M. E., & Tsubo, M. (2024). A transformative framework reshaping sustainable drought risk management through advanced early warning systems. *iScience*, *27*(7), 110066. DOI: 10.1016/j.isci.2024.110066 PMID: 38989469

Matabber, A., Rhuhanga, L. L. N., Agehara, S., & Mozafarian, M. (2026). Artificial intelligence (AI) in detection of abiotic stress in plants: A review. *Sensors (Basel)*, *26*(4), 1122. DOI: 10.3390/s26041122 PMID: 41755064

Materia, S., García, L. P., Van Straaten, C. O. S., Mamalakis, A., Cavicchia, L., Coumou, D., De Luca, P., Kretschmer, M., & Donat, M. (2024). Artificial intelligence for climate prediction of extremes: State of the art, challenges, and future perspectives. *Wiley Interdisciplinary Reviews: Climate Change*, *15*(6), e914. Advance online publication. DOI: 10.1002/wcc.914

Maurya, K., Mahajan, S., & Chaube, N. R. (2021). Remote sensing techniques: Mapping and monitoring of mangrove ecosystem—a review. *Complex & Intelligent Systems*, *7*(6), 2797–2818. DOI: 10.1007/s40747-021-00457-z

Maxwell, A. E., Warner, T. A., & Fang, F. (2018). Implementation of machine-learning classification in remote sensing: An applied review. *International Journal of Remote Sensing*, *39*(9), 2784–2817. DOI: 10.1080/01431161.2018.1433343

Mazzia, V., Khaliq, A., & Chiaberge, M. (2020). Improvement in land cover and crop classification based on temporal features learning from Sentinel-2 data using recurrent-convolutional neural network (R-CNN). *Applied Sciences (Basel, Switzerland)*, *10*(1), 238. DOI: 10.3390/app10010238

McFeeters, S. K. (1996). The use of the normalized difference water index (NDWI) in the delineation of open water features. *International Journal of Remote Sensing*, *17*(7), 1425–1432. DOI: 10.1080/01431169608948714

Mcleod, E., Chmura, G. L., Bouillon, S., Salm, R., Björk, M., Duarte, C. M., Lovelock, C. E., Schlesinger, W. H., & Silliman, B. R. (2011). A blueprint for blue carbon: Toward an improved understanding of the role of vegetated coastal habitats in sequestering CO_2. *Frontiers in Ecology and the Environment*, *9*(10), 552–560. DOI: 10.1890/110004

Mehdary, A., Chehri, A., Jakimi, A., & Saadane, R. (2024). Hyperparameter optimization with genetic algorithms and XGBoost: A step forward in smart grid fraud detection. *Sensors (Basel)*, *24*(4), 1230. DOI: 10.3390/s24041230 PMID: 38400385

Mengis, N., Paul, A. J., & Fernández-Méndez, M. (2023). Counting (on) blue carbon—Challenges and ways forward for carbon accounting of ecosystem-based carbon removal in marine environments. *PLOS Climate*, *2*(8), e0000148. Advance online publication. DOI: 10.1371/journal.pclm.0000148

Meng, Z., Hu, Y., Ren, G., Zhu, W., Wang, J., Liu, S., & Ma, Y. (2024). Remote sensing monitoring of seagrass bed dynamics using cross-temporal-spatial domain transfer learning in Yellow river Delta. *International Journal of Remote Sensing*, *45*(6), 1972–1996. DOI: 10.1080/01431161.2024.2321467

Mhanna, S., Halloran, L. J., Zwahlen, F., Asaad, A. H., & Brunner, P. (2023). Using machine learning and remote sensing to track land use/land cover changes due to armed conflict. *The Science of the Total Environment, 898*, 165600. DOI: 10.1016/j.scitotenv.2023.165600 PMID: 37467974

Michail, D., Davalas, C., Chafis, K., Panagiotou, L. I., Prapas, I., Bountos, N. I., Kondylatos, S., & Papoutsis, I. (2025). FireCastNet: Earth-as-a-graph for seasonal fire prediction. *Scientific Reports, 16*(1), 1006. DOI: 10.1038/s41598-025-30645-7 PMID: 41392185

Miller, L., Pelletier, C., & Webb, G. I. (2024). Deep Learning for Satellite Image Time-Series Analysis: A review. *IEEE Geoscience and Remote Sensing Magazine, 12*(3), 81–124. DOI: 10.1109/MGRS.2024.3393010

Min, Y., Cui, L., Li, J., Han, Y., Zhuo, Z., Yin, X., Zhou, D., & Ke, Y. (2023). Detection of large-scale Spartina alterniflora removal in coastal wetlands based on Sentinel-2 and Landsat 8 imagery on Google Earth Engine. *International Journal of Applied Earth Observation and Geoinformation, 125*, 103567. DOI: 10.1016/j.jag.2023.103567

Mir, Y. H., Mir, S., Ganie, M. A., Bhat, J. A., Shah, A. M., Mushtaq, M., & Irshad, I. (2025). Overview of land use and land cover change and its impacts on natural resources. In *Ecologically mediated development: Promoting biodiversity conservation and food security* (pp. 101–130). Springer Nature. DOI: 10.1007/978-981-96-2413-3_5

Mohamed, E. K., Gahalaut, V. K., Sekertekin, A., & Inyurt, S. (2021). Atmospheric, ionospheric and earth-related variations associated with the 11th August 2012 earthquakes, Ahar, Iran. *Journal of Atmospheric and Solar-Terrestrial Physics, 216*, 105595. DOI: 10.1016/j.jastp.2021.105595

Mohan, M. (2025). Reimagining mangrove conservation through AI and remote sensing in an era of conflict. *Discover Conservation, 2*(1), 32. DOI: 10.1007/s44353-025-00054-y

Monaca, A. (2024). Greetings from HYDRO 2023 Organizing Committee. *The International Hydrographic Review, 30*(1), 125–126. DOI: 10.58440/ihr-30-1-e02

Montaggioni, L. F., & Braithwaite, C. J. R. (2009). Structure, Zonation and Dynamic Patterns of Coral Reef Communities. In *Developments in Marine Geology* (Vol. 5, pp. 67–122). Elsevier., DOI: 10.1016/S1572-5480(09)05003-9

Moreaux, V., Papale, D., Gielen, B., & Loustau, D. (2019). Optimising observation networks for the early detection and unequivocal attribution of environmental effects on European forests. *HAL (Le Centre Pour La Communication Scientifique Directe)* https://hal.inrae.fr/hal-02735529

Moreno, S. R., & dos Santos Coelho, L. (2018). Wind speed forecasting approach based on singular spectrum analysis and adaptive neuro fuzzy inference system. *Renewable Energy, 126*, 736–754.

Moses, W. J., Woude, A. V., & Palacios, S. L. (2022). Editorial: Emerging technologies and techniques for remote sensing of coastal and inland waters. *Frontiers in Environmental Science, 10*, 1028307. Advance online publication. DOI: 10.3389/fenvs.2022.1028307

Mukaromah, L., As-syakur, A. R., & Prastowo, E. (2019). Opportunities for the application of Red-edge reflectance-based indices in coastal vegetation and agro-ecological studies. *Journal of Marine and Aquatic Sciences, 5*(2), 251. DOI: 10.24843/jmas.2019.v05.i02.p13

Muñoz-Sabater, J., Dutra, E., Agustí-Panareda, A., Albergel, C., Arduini, G., Balsamo, G., Boussetta, S., Choulga, M., Harrigan, S., Hersbach, H., Martens, B., Miralles, D. G., Piles, M., Rodríguez-Fernández, N. J., Zsoter, E., Buontempo, C., & Thépaut, J.-N. (2021). ERA5-Land: A state-of-the-art global reanalysis dataset for land applications. *Earth System Science Data, 13*(9), 4349–4383. DOI: 10.5194/essd-13-4349-2021

Murguia-Cozar, A., Macedo-Cruz, A., Fernandez-Reynoso, D. S., & Salgado-Transito, J. A. (2021), . Recognition of maize phenology in Sentinel images with machine learning. *Sensors (Basel), 22*(1), 94. DOI: 10.3390/s22010094 PMID: 35009637

Muruganantham, P., Wibowo, S., Grandhi, S., Samrat, N. H., & Islam, N. (2022). A systematic literature review on crop yield prediction with deep learning and remote sensing. *Remote Sensing (Basel), 14*(9), 1990. DOI: 10.3390/rs14091990

Muslim, A. M., Chong, W. S., Safuan, C. D. M., Khalil, I., & Hossain, M. S. (2019). Coral Reef Mapping of UAV: A Comparison of Sun Glint Correction Methods. *Remote Sensing (Basel), 11*(20), 2422. DOI: 10.3390/rs11202422

Muthukumar, P., Cocom, E., Nagrecha, K., Comer, D., Burga, I., Taub, J., Calvert, C. F., Holm, J., & Pourhomayoun, M. (2022). Predicting PM2.5 atmospheric air pollution using deep learning with meteorological data and ground-based observations and remote-sensing satellite big data. *Air Quality, Atmosphere & Health, 15*(7), 1221–1234. DOI: 10.1007/s11869-021-01126-3 PMID: 34840624

Naess, P. (2001). Urban planning and sustainable development. *European Planning Studies*, *9*(4), 503–524. DOI: 10.1080/713666490

Nalakurthi, N. V. S. R., Abimbola, I., Ahmed, T., Anton, I., Riaz, K., Ibrahim, Q., Banerjee, A., Tiwari, A., & Gharbia, S. (2024). Challenges and Opportunities in Calibrating Low-Cost Environmental Sensors. *Sensors (Basel)*, *24*(11), 3650. DOI: 10.3390/s24113650

Nduku, L., Munghemezulu, C., Mashaba-Munghemezulu, Z., Ratshiedana, P. E., Sibanda, S., & Chirima, J. G. (2024). Synergetic Use of Sentinel-1 and Sentinel-2 Data for Wheat-Crop Height Monitoring Using Machine Learning. *AgriEngineering*, *6*(2), 1093–1116. DOI: 10.3390/agriengineering6020063

Nearing, G. S., Kratzert, F., Sampson, A. K., Pelissier, C. S., Klotz, D., Frame, J. M., Prieto, C., & Gupta, H. V. (2020). What role does hydrological science play in the age of machine learning? *Water Resources Research*, *57*(3), e2020WR028091. Advance online publication. DOI: 10.1029/2020WR028091

Ngo, T. D., Quoc, K. N., Dang, N. T., Dang, C. T. M., Tran, L. L., & Nguyen, H. D. (2024). Analysis Of The Mangrove Structure In The Dong Rui Commune Based On Multispectral Unmanned Aerial Vehicle Image Data. *Geography, Environment, Sustainability*, *16*(4), 14–25. DOI: 10.24057/2071-9388-2023-2641

Nguyen, T., Liquet, B., Mengersen, K., & Sous, D. (2021). Mapping of Coral Reefs with Multispectral Satellites: A Review of Recent Papers. *Remote Sensing (Basel)*, *13*(21), 4470. DOI: 10.3390/rs13214470

Nicholls, R. J., Hutton, C. W., Adger, W. N., Hanson, S. E., & Rahman, Md. M., & Salehin, M. (2018). Integrative Analysis for the Ganges-Brahmaputra-Meghna Delta, Bangladesh. In *Ecosystem Services for Well-Being in Deltas* (pp. 71–90). Springer International Publishing. DOI: 10.1007/978-3-319-71093-8_4

Nuissl, H., & Siedentop, S. (2021). Urbanisation and land use change. In *Sustainable land management in a European context* (pp. 75–99). DOI: 10.1007/978-3-030-50841-8_5

O'Donncha, F., Hu, Y., Palmes, P., Burke, M., Filgueira, R., & Grant, J. (2022). A spatio-temporal LSTM model to forecast across multiple temporal and spatial scales. *Ecological Informatics*, *69*, 101687. DOI: 10.1016/j.ecoinf.2022.101687

Olawade, D. B., Wada, O. Z., Ige, A. O., Egbewole, B. I., Olojo, A., & Oladapo, B. I. (2024). Artificial intelligence in environmental monitoring: Advancements, challenges, and future directions. *Hygiene and Environmental Health Advances*, *12*, 100114. DOI: 10.1016/j.heha.2024.100114

Oliver, T. H., & Morecroft, M. D. (2014). Interactions between climate change and land use change on biodiversity: Attribution problems, risks, and opportunities. *Wiley Interdisciplinary Reviews: Climate Change*, *5*(3), 317–335. DOI: 10.1002/wcc.271

Olson, D., & Anderson, J. (2021). Review on unmanned aerial vehicles, remote sensors, imagery processing, and their applications in agriculture. *Agronomy Journal*, *113*(2), 971–992. DOI: 10.1002/agj2.20595

Otamendi, U., Maiza, M., Olaizola, I. G., Sierra, B., Florez, M., & Quartulli, M. (2024). Integrated water resource management in the Segura Hydrographic Basin: An artificial intelligence approach. *Journal of Environmental Management*, *370*, 122526. DOI: 10.1016/j.jenvman.2024.122526 PMID: 39357444

Ougahi, J. H., & Rowan, J. S. (2025). Enhanced streamflow forecasting using hybrid modelling integrating glacio-hydrological outputs, deep learning and wavelet transformation. *Scientific Reports*, *15*(1), 2762. DOI: 10.1038/s41598-025-87187-1 PMID: 39843529

Ouzounov, D., Pulinets, S., Liu, J. Y., Hattori, K., & Han, P. (2018). Multiparameter Assessment of Pre-Earthquake Atmospheric Signals. In Liu, D., & Liu, J. Y. (Eds.), *Geophysical Monograph Series, 234, 339-359.*, DOI: 10.1002/9781119156949.ch20

Ozdemir, B., Sermet, F., & Pacal, I. (2025). Attention-enhanced ConvNeXt for accurate, efficient, and interpretable crack detection. *Expert Systems with Applications*, •••, 129165.

Pacal, I., Algarni, A., & Kunduracioglu, I. (2026). Adaptive and efficient deep learning model for automated ischemic stroke lesion segmentation. *Biomedical Signal Processing and Control*, *118*, 109658. DOI: 10.1016/j.bspc.2026.109658

Pailoplee, S. (2014). Mapping b-Value Anomalies Along the Indonesian Island Chain: Implications for Upcoming Earthquakes. *Journal of Earthquake and Tsunami*, *8*(4), 1450010. DOI: 10.1142/S1793431114500109

Pailoplee, S. (2017). Mapping of b-Value Anomalies Along the Strike-Slip Fault System on the Thailand-Myanmar Border: Implications for Upcoming Earthquakes. *Journal of Earthquake and Tsunami*, *11*(2), 1671001. DOI: 10.1142/S1793431116710019

Pandeya, B., Buytaert, W., Zulkafli, Z., Karpouzoglou, T., Mao, F., & Hannah, D. M. (2016). A comparative analysis of ecosystem services valuation approaches for application at the local scale and in data scarce regions. *Ecosystem Services*, *22*, 250–259. DOI: 10.1016/j.ecoser.2016.10.015

Pang, B., Cheng, S., Huang, Y., Jin, Y., Guo, Y., Prentice, I. C., Harrison, S. P., & Arcucci, R. (2025). Fire-Image-DenseNet (FIDN) for predicting wildfire burnt area using remote sensing data. *Computers & Geosciences*, *195*, 105783. DOI: 10.1016/j.cageo.2024.105783

Parrot, M. (2022). Ionospheric variations at the time of the M8.8 Chile earthquake and statistical analysis of plasma parameters recorded by DEMETER. In American Geophysical Union, 2010, pp. 1542.

Parrot, M., Tramutoli, V., Liu, T. J. Y., Pulinets, S., Ouzounov, D., Genzano, N., Lisi, M., Hattori, K., & Namgaladze, A. (2021). Atmospheric and ionospheric coupling phenomena associated with large earthquakes. *The European Physical Journal. Special Topics*, *230*(1), 197–225. DOI: 10.1140/epjst/e2020-000251-3

Pathak, P., & Srivastava, S. (2026). Smart material design: Integrating data-driven optimization and complexity analysis for next-generation materials. In *Smart Materials Engineering* (pp. 41–63). Data-Driven Approaches and Multiscale Modelling. DOI: 10.1007/978-3-032-09540-4_3

Peng, C., Song, Y., Huang, Q., Niu, Z., Zeng, G., Kang, H.-J., & Ying, M. (2025). Estimating coastal mangrove aboveground biomass of Zhejiang province based on sentinel-2 images and multiple machine learning algorithms. *Frontiers in Forests and Global Change*, *8*, 1662546. Advance online publication. DOI: 10.3389/ffgc.2025.1662546

Pham, T. D., Ha, N. T., Saintilan, N., Skidmore, A. K., Phan, D. C., Le, N. N., Viet, H. L., Takeuchi, W., & Friess, D. A. (2023). Advances in Earth observation and machine learning for quantifying blue carbon. *Earth-Science Reviews*, *243*, 104501. DOI: 10.1016/j.earscirev.2023.104501

Pham, T. D., Xia, J., Ha, N. T., Bui, D. T., Le, N. N., & Tekeuchi, W. (2019). A Review of Remote Sensing Approaches for Monitoring Blue Carbon Ecosystems: Mangroves, Seagrassesand Salt Marshes during 2010–2018. *Sensors (Basel)*, *19*(8), 1933. Advance online publication. DOI: 10.3390/s19081933

Pham, T. D., Yokoya, N., Bui, D. T., Yoshino, K., & Friess, D. A. (2019). Remote Sensing Approaches for Monitoring Mangrove Species, Structure, and Biomass: Opportunities and Challenges. *Remote Sensing (Basel)*, *11*(3), 230. DOI: 10.3390/rs11030230

Phinn, S. R., Roelfsema, C. M., & Mumby, P. J. (2012). Multi-scale, object-based image analysis for mapping geomorphic and ecological zones on coral reefs. *International Journal of Remote Sensing*, *33*(12), 3768–3797. DOI: 10.1080/01431161.2011.633122

Phinzi, K., Ngetar, N. S., Pham, Q. B., Chakilu, G. G., & Szabó, S. (2023). Understanding the role of training sample size in the uncertainty of high-resolution LULC mapping using random forest. *Earth Science Informatics*, *16*(4), 3667–3677. DOI: 10.1007/s12145-023-01117-1

Pimple, U., Simonetti, D., Peters, R., Berger, U., Podest, E., & Gond, V. (2022). Enhancing monitoring of mangrove spatiotemporal tree diversity and distribution patterns. *Land Degradation & Development*, *34*(5), 1265–1282. DOI: 10.1002/ldr.4537

Piyush, S., Rawat, V., & Singh, K. (2024). Temporal reasoning networks for flood-induced geomorphic changes. *ISPRS Journal of Photogrammetry and Remote Sensing*. Kemarau, R. A.,

Polidoro, B. A., Carpenter, K. E., Collins, L., Duke, N. C., Ellison, A. M., Ellison, J. C., Farnsworth, E. J., Fernando, E. S., Kathiresan, K., Koedam, N. E., Livingstone, S. R., Miyagi, T., Moore, G. E., Nam, V. N., Ong, J. E., Primavera, J. H., Salmo, S. G., Sanciangco, J. C., Sukardjo, S., & Yong, J. W. H. (2010). The Loss of Species: Mangrove Extinction Risk and Geographic Areas of Global Concern. *PLoS One*, *5*(4), e10095–e10095. DOI: 10.1371/journal.pone.0010095

Porcar-Castell, A., Arthur, A. M., Rossini, M., Eklundh, L., Pacheco-Labrador, J., Anderson, K., Balzarolo, M., Martín, M. P., Jin, H., Tomelleri, E., Cerasoli, S., Sakowska, K., Hueni, A., Julitta, T., Nichol, C., & Vescovo, L. (2015). EUROSPEC: At the interface between remote-sensing and ecosystem CO₂ flux measurements in Europe. *Biogeosciences*, *12*(20), 6103–6124. DOI: 10.5194/bg-12-6103-2015

Pradeep, J., Shaji, E., & Chandran, C. S. (2022). Assessment of coastal variations due to climate change using remote sensing and machine learning techniques: A case study from west coast of India. *Estuarine, Coastal and Shelf Science*, *275*, 107968. DOI: 10.1016/j.ecss.2022.107968

Pricope, N. G., Halls, J. N., Dalton, E. G., Minei, A., Chen, C., & Wang, Y. (2024). Precision Mapping of Coastal Wetlands: An Integrated Remote Sensing Approach Using UAS LiDAR and Multispectral Data. *Yaogan Xuebao*, *4*. Advance online publication. DOI: 10.34133/remotesensing.0169

Prodhan, F. A., Hoque, M. Z., Nafee, K. M., Al Fahad, M. S., & Rahman Sakib, M. N. (2025). Machine Learning-Based Analysis of Community Perceptions on Coastal Forest Ecosystem Services, Restoration Willingness and Their Determinants in Bangladesh. *Wild*, *2*(3), 26. DOI: 10.3390/wild2030026

Psaros, A. F., Meng, X., Zou, Z., Guo, L., & Karniadakis, G. E. (2023). Uncertainty quantification in scientific machine learning: Methods, metrics, and comparisons. *Journal of Computational Physics*, *477*, 111902. DOI: 10.1016/j.jcp.2022.111902

Pulinets, S., & Ouzounov, D. (2011). Lithosphere-Atmosphere-Ionosphere Coupling (LAIC) model - An unified concept for earthquake precursors validation. *Journal of Asian Earth Sciences*, *41*(4-5), 371–382. DOI: 10.1016/j.jseaes.2010.03.005

Rabiei, S., Babaeian, E., & Grunwald, S. (2025). Surface and Subsurface Soil Moisture Estimation Using Fusion of SMAP, NLDAS-2, and SOLUS100 Data with Deep Learning. *Remote Sensing (Basel)*, *17*(4), 659. DOI: 10.3390/rs17040659

Rahimi, I., Ahmad, I., & Azeez, S. (2019). Urban growth mapping using remote sensing and GIS techniques: Case study Darbandikhan, Kurdistan Region, Iraq. *Magallat Tikrit Li-l-'Ulum Al-Zira'Iyyat*, *19*(4), 50–59. DOI: 10.25130/tjas.19.4.8

Rahim, N. H. A., Satyanarayana, B., Hossain, M. S., Idris, I., Jani, J. M., Wolswijk, G., Aminah, I. S., Zanzuri, N. A., A'in, M. N. N., Cannicci, S., & Dahdouh-Guebas, F. (2026). Mangroves affected by shifted river mouth underscore the importance of sustainable conservation of Setiu Wetlands, Malaysia. *Marine Pollution Bulletin*, *224*, 119183. DOI: 10.1016/j.marpolbul.2025.119183

Raj, N., & Pasfield-Neofitou, S. (2024). Assessment and Prediction of Sea Level and Coastal Wetland Changes in Small Islands Using Remote Sensing and Artificial Intelligence. *Remote Sensing (Basel)*, *16*(3), 551. DOI: 10.3390/rs16030551

Raj, N., Singh, N., Downs, N., & Singh-Peterson, L. (2025). Predicting Sea-Level Extremes and Wetland Change in the Maroochy River Floodplain Using Remote Sensing and Deep Learning Approach. *Remote Sensing (Basel)*, *17*(17), 2988. DOI: 10.3390/rs17172988

Ramachandra, T. V., Mondal, T., & Setturu, B. (2023). Relative performance evaluation of machine learning algorithms for land use classification using multispectral moderate resolution data. *SN Applied Sciences*, *5*(10), 274. DOI: 10.1007/s42452-023-05496-4

Ran, B., Xing, Z., Fang, J., Yang, Y., He, Y., Li, X., Liu, X., & Liu, G. (2025). Analysis of soil moisture dynamics and its response to rainfall in the MU US sandy land. *Water (Basel)*, *17*(1), 105. DOI: 10.3390/w17010105

Ranith, R., Menon, N., Nobi, E. P., Haris, F., Raj, R. P., & Pettersson, L. H. (2025). Decadal-scale urban footprints on seagrass blue carbon dynamics in Palk Bay, India: A remote sensing based framework. *Frontiers in Marine Science*, *12*, 1690700. Advance online publication. DOI: 10.3389/fmars.2025.1690700

Rasheed, M. W., Tang, J., Sarwar, A., Shah, S., Saddique, N., Khan, M. U., Khan, M. I., Nawaz, S., Shamshiri, R. R., Aziz, M., & Sultan, M. (2022). Soil Moisture Measuring Techniques and factors Affecting the moisture dynamics: A Comprehensive review. *Sustainability (Basel)*, *14*(18), 11538. DOI: 10.3390/su141811538

Razzaq, I., Ullah, S., Akram, M., Ashraf, M. W., Shaaban, I. A., Gurbanova, L., Mohammad, A., Shah, S. S. A., & Nazir, M. A. (2025). Graphene-based polymer composites for high-performance chemical sensing and detection: A critical review. *Advanced Materials Technologies*, *10*(18), e00578. DOI: 10.1002/admt.202500578

Reguero, B. G., Beck, M. W., Schmid, D., Stadtmüller, D., Raepple, J., Schüssele, S., & Pfliegner, K. (2020). Financing coastal resilience by combining nature-based risk reduction with insurance. *Ecological Economics*, *169*, 106487. DOI: 10.1016/j.ecolecon.2019.106487

Reichstein, M., Camps-Valls, G., Stevens, B., Jung, M., Denzler, J., Carvalhais, N., & Prabhat, . (2019). Deep learning and process understanding for data-driven Earth system science. *Nature*, *566*(7743), 195–204. DOI: 10.1038/s41586-019-0912-1 PMID: 30760912

Reithmaier, G. M. S., Cabral, A., Akhand, A., Bogard, M. J., Borges, A. V., Bouillon, S., Burdige, D. J., Call, M., Chen, N., Chen, X., Cotovicz, L. C.Jr, Eagle, M. J., Kristensen, E., Kroeger, K D., Lu, Z., Maher, D. T., Pérez-Lloréns, J. L., Ray, R., Taillardat, P., & Santos, I. R. (2023). Carbonate chemistry and carbon sequestration driven by inorganic carbon outwelling from mangroves and saltmarshes. *Nature Communications*, *14*(1), 8196. DOI: 10.1038/s41467-023-44037-w PMID: 38081846

Ribeiro, R. J. da C., Goulart, F. F., Jacobson, T. K. B., da Rocha, D. M. S., Jaloretto, D., Gonçalves, B. E. F., & Gerber, D. (2024). Beyond climate villains: Estimating carbon storage in urban parks using high-resolution imagery combined with field sampling. *Research Square (Research Square)*. DOI: 10.21203/rs.3.rs-3817014/v1

Richard, M. O. (2024). Blue Carbon and the Role of Mangroves in Carbon Sequestration: Its Mechanisms, Estimation, Human Impacts and Conservation Strategies for Economic Incentives Among African Countries Along the Indian Ocean Belt. SSRN *Electronic Journal*. DOI: 10.2139/ssrn.4912339

Rogers, K., Macreadie, P. I., Kelleway, J. J., & Saintilan, N. (2019). Blue carbon in coastal landscapes: A spatial framework for assessment of stocks and additionality. *Sustainability Science*, *14*(2), 453–467. DOI: 10.1007/s11625-018-0575-0

Rolland, L., ... (2021). Worldwide GNSS ionospheric response of the magnitude 8.8 2010 Chilean earthquake and tsunami: a revisit. In EGU General Assembly Conference Abstracts, 2021, EGU21-10958.

Rondón, M., Ewane, E. B., Abdullah, M. M., Watt, M. S., Blanton, A., Abulibdeh, A., Burt, J. A., Rogers, K., Ali, T., Reef, R., Mohtar, R. H., Sidik, F., Fahrenberg, M., de-Miguel, S., Galgamuwa, G. A. P., Charabi, Y., Arachchige, P. S. P., Velasquez-Camacho, L., Al-Awadhi, T., … Mohan, M. (2023). Remote sensing-based assessment of mangrove ecosystems in the Gulf Cooperation Council countries: a systematic review [Review of *Remote sensing-based assessment of mangrove ecosystems in the Gulf Cooperation Council countries: a systematic review*]. *Frontiers in Marine Science, 10*. Frontiers Media. DOI: 10.3389/fmars.2023.1241928

Rosas-Muñoz, J., & Baquedano-Rodríguez, M. (2022). What can we learn about the natural disasters and multidimensional poverty link for artisanal fishery in Chile? *Marine Policy*, *144*, 105225. DOI: 10.1016/j.marpol.2022.105225

Rowley, A., & Karakuş, O. (2023). Predicting air quality via multimodal AI and satellite imagery. *Remote Sensing of Environment*, *293*, 113609. DOI: 10.1016/j.rse.2023.113609

Russell, S. K., Gillanders, B. M., Detmar, S., Fotheringham, D., & Jones, A. R. (2023). Determining Environmental Drivers of Fine-Scale Variability in Blue Carbon Soil Stocks. *Estuaries and Coasts*, *47*(1), 48–59. DOI: 10.1007/s12237-023-01260-4

Saeed, W., & Omlin, C. (2023). Explainable AI (XAI): A systematic meta-survey of current challenges and future opportunities. *Knowledge-Based Systems*, *263*, 110273. DOI: 10.1016/j.knosys.2023.110273

Saha, L., & Dubey, A. (2025). Innovative AI solutions for Water-Efficient smart irrigation in agriculture. *SHS Web of Conferences, 216*, 01052. DOI: 10.1051/shsconf/202521601052

Salem, A., & Hashemi-Beni, L. (2022). Inundated Vegetation Mapping Using SAR Data: A Comparison of Polarization Configurations of UAVSAR L-Band and Sentinel C-Band. *Remote Sensing (Basel)*, *14*(24), 6374. DOI: 10.3390/rs14246374

Salh, H., Muhammad, A., Ghafar, M. M., & Külahcı, F. (2022). Potential utilization of air temperature, total electron content, and air relative humidity as possible earthquake precursors: A case study of Mexico M7.4 earthquake. *Journal of Atmospheric and Solar-Terrestrial Physics*, *237*, 105927. DOI: 10.1016/j.jastp.2022.105927

Salman, H. A., Kalakech, A., & Steiti, A. (2024). Random forest algorithm overview. *Babylonian Journal of Machine Learning*, 69–79.

Sangeetha, J., & Govindarajan, P. (2023). Prediction of agricultural waste compost maturity using fast regions with convolutional neural network (R-CNN). *Materials Today: Proceedings*, *10*, 100523. DOI: 10.1016/j.matpr.2023.01.112

Sarang, P. (2023). Support vector machines: A supervised learning algorithm for classification and regression. In *Thinking data science: A data science practitioner's guide* (pp. 153–165). Springer. DOI: 10.1007/978-3-031-02363-7_8

Sartono, A., Darmawan, M., Ramadhani, F., Ramdhan, M., Safitri, S., & Sutejo, B. (2025). Environmental Monitoring of Shallow Marine Waters Using AI and Remote Sensing: A Systematic Literature Review. *Environmental Quality Management*, *35*(1), e70148. Advance online publication. DOI: 10.1002/tqem.70148

Saxena, P., & Shukla, P. (2023). A review on recent developments and advances in environmental gas sensors to monitor toxic gas pollutants. *Environmental Progress & Sustainable Energy*, *42*(5), e14126. DOI: 10.1002/ep.14126

Senaras, C., Grady, M., Rana, A. S., Nieto, L., Ciampitti, I., Holden, P., & Wania, A. (2024). Detection of maize crop phenology using Planet Fusion. *Remote Sensing (Basel)*, *16*(15), 2730. DOI: 10.3390/rs16152730

Sensale-Rodriguez, B., Yan, R., Kelly, M. M., Fang, T., Tahy, K., Hwang, W. S., Jena, D., Liu, L., & Xing, H. G. (2012). Broadband graphene terahertz modulators enabled by intraband transitions. *Nature Communications*, *3*(1), 780. DOI: 10.1038/ncomms1787 PMID: 22510685

Sharma, S., & Khanal, P. (2024). Forest fire prediction: A spatial machine learning and neural network approach. *Fire (Basel, Switzerland)*, *7*(6), 205. DOI: 10.3390/fire7060205

Sheng, L., Lv, Y., Ren, Z., Zhou, H., & Deng, X. (2024). Detection of the optimal temporal windows for mapping paddy rice under a double-cropping system using Sentinel-2 imagery. *Remote Sensing (Basel)*, *17*(1), 57. DOI: 10.3390/rs17010057

Shetty, S., Schneider, P., Stebel, K., Hamer, P. D., Kylling, A., & Berntsen, T. K. (2024). Estimating surface NO_2 concentrations over Europe using Sentinel-5P TROPOMI observations and machine learning. *Remote Sensing of Environment*, *312*, 114321. DOI: 10.1016/j.rse.2024.114321

Shirmohammadi, S., Wang, F., & Hsu, C.-H. (2025). Review and Performance Evaluation of Uncertainty Quantification in Data-Driven AI-Assisted Measurements. *IEEE Open Journal of Instrumentation and Measurement*, *4*, 1–22. DOI: 10.1109/OJIM.2025.3621742

Shojaeezadeh, S. A., Elnashar, A., & Weber, T. K. D. (2024). A novel fusion of optical and radar satellite data for crop phenology estimation using machine learning. *Science of Remote Sensing*, *11*, 100227. DOI: 10.1016/j.srs.2025.100227

Shokati, H., Mashal, M., Noroozi, A., Mirzaei, S., Mohammadi-Doqozloo, Z., Nabiollahi, K., Taghizadeh-Mehrjardi, R., Khosravani, P., Adhikari, R., Hu, L., & Scholten, T. (2025). Comparing UAV-Based hyperspectral and Satellite-Based multispectral data for soil moisture estimation using machine learning. *Water (Basel), 17*(11), 1715. DOI: 10.3390/w17111715

Silberstein, J., & Maser, C. (2013). *Land-use planning for sustainable development.* CRC Press. DOI: 10.1201/b15958

Simarmata, N., Wikantika, K., Tarigan, T. A., Aldyansyah, M., Tohir, R. K., Fauzi, A. I., & Fauzia, A. R. (2025). Comparison of random forest, gradient tree boosting, and classification and regression trees for mangrove cover change monitoring using Landsat imagery. *The Egyptian Journal of Remote Sensing and Space Sciences, 28*(1), 138–150. DOI: 10.1016/j.ejrs.2025.02.002

Simpson, J., Bruce, E., Davies, K., & Barber, P. A. (2022). A Blueprint for the Estimation of Seagrass Carbon Stock Using Remote Sensing-Enabled Proxies. *Remote Sensing (Basel), 14*(15), 3572. DOI: 10.3390/rs14153572

Singh, E., Meyyappan, M., & Nalwa, H. S. (2017). Flexible graphene-based wearable gas and chemical sensors. *ACS Applied Materials & Interfaces, 9*(40), 34544–34586. DOI: 10.1021/acsami.7b07063 PMID: 28876901

Sit, M., Demiray, B., Xiang, Z., Ewing, G., Sermet, Y., & Demir, I. (2020). A Comprehensive Review of Deep Learning Applications in Hydrology and Water Resources. *Water Science and Technology: A Journal of the International Association on Water Pollution Research.* https://doi.org/DOI: 10.31223/osf.io/xs36g

Sobrino, J. A., Jiménez-Muñoz, J. C., & Paolini, L. (2004). Land surface temperature retrieval from Landsat TM 5. *Remote Sensing of Environment, 90*(4), 434–440. DOI: 10.1016/j.rse.2004.02.003

Song, L., Zhao, J., Li, Y., Liu, L., Duan, J., He, L., Wang, Y., Guo, T., & Feng, W. (2026). Early detection of wheat powdery mildew: A multi-source in situ remote sensing approach enabled by stacked ensemble learning. *Artificial Intelligence in Agriculture, 16*(1), 124–138. DOI: 10.1016/j.aiia.2025.10.004

Sridharan, A., Ferrario, M. F., & Gopalan, S. (2023). Earthquake-triggered landslides and Environmental Seismic Intensity: Insights from the 2018 Papua New Guinea earthquake (Mw 7.5). *All Earth, 35*(1), 195–209. DOI: 10.1080/27669645.2023.2233140

Stankovic, M., Ambo-Rappe, R., Carly, F., Dangan-Galon, F., Fortes, M. D., Hossain, M. S., Kiswara, W., Van Luong, C., Minh-Thu, P., Mishra, A. K., Noiraksar, T., Nurdin, N., Panyawai, J., Rattanachot, E., Rozaimi, M., Htun, U. S., & Prathep, A. (2021). Quantification of blue carbon in seagrass ecosystems of Southeast Asia and their potential for climate change mitigation. *The Science of the Total Environment, 783*, 146858. https://doi.org/https://doi.org/10.1016/j.scitotenv.2021.146858. DOI: 10.1016/j.scitotenv.2021.146858

Stojković, S. (2017). GIS analysis of land use changes: Case study: The Stara Pazova municipality, Serbia. *Zbornik Radova – Geografski Fakultet Univerziteta u Beogradu*, 295–306.

Suárez-Fernández, G. E., Martínez-Sánchez, J., & Arias, P. (2025). Enhancing carbon stock estimation in forests: Integrating multi-data predictors with random forest method. *Ecological Informatics, 86*, 102997. DOI: 10.1016/j.ecoinf.2025.102997

Suchitra, M. R., Balachandar, S., Govindarajan, P., & Parthasarathy, S. (2023). Introduction and Validation of a New Suchitra Scoring System and Determining the Cut-off Value for Healthy Lifestyle Among College Students – Kumbakonam Urban Rural Epidemiological Study-KURES-8. *Current Research in Nutrition and Food Science Journal, 11*(3), 1039–1049.

Sun, B., Lei, Y., Cui, L., Li, W., Kang, X., & Zhang, M. (2018). Addressing the Modelling Precision in Evaluating the Ecosystem Services of Coastal Wetlands. *Sustainability (Basel), 10*(4), 1136. DOI: 10.3390/su10041136

Sunkur, R., Kantamaneni, K., Bokhoree, C., Rathnayake, U., & Fernando, M. (2024). Mangrove mapping and monitoring using remote sensing techniques towards climate change resilience. *Scientific Reports, 14*(1), 6949. DOI: 10.1038/s41598-024-57563-4 PMID: 38521843

Suo, L., Huang, M., Zhang, Y., Duan, L., & Shan, Y. (2018). Soil moisture dynamics and dominant controls at different spatial scales over semiarid and semi-humid areas. *Journal of Hydrology (Amsterdam), 562*, 635–647. DOI: 10.1016/j.jhydrol.2018.05.036

Tahara, S., Sudo, K., Yamakita, T., & Nakaoka, M. (2022). Species level mapping of a seagrass bed using an unmanned aerial vehicle and deep learning technique. *PeerJ, 10*, e14017. DOI: 10.7717/peerj.14017

Tâm, N. T. T., & Ngân, N. V. C. (2025). Multi-sensor remote sensing and machine learning for aboveground biomass mapping in Vietnam's Melaleuca wetlands: A review. *International Journal of Research and Innovation in Applied Science, 10*(10), 473–487. DOI: 10.51584/IJRIAS.2025.1010000035

Tao, L., Chen, Z., Li, Z., Wang, J., Xu, X., & Xu, J.-B. (2021). Enhancing light–matter interaction in 2D materials by optical micro/nano architectures for high-performance optoelectronic devices. *InfoMat*, *3*(1), 36–60. DOI: 10.1002/inf2.12148

Tessler, Z. D., Vörösmarty, C. J., Grossberg, M., Gladkova, I., Aizenman, H., Syvitski, J. P. M., & Foufoula-Georgiou, E. (2015). Profiling risk and sustainability in coastal deltas of the world. *Science*, *349*(6248), 638–643. DOI: 10.1126/science.aab3574

Tian, Z., Wei, J., & Li, Z. (2023). How important is satellite-retrieved aerosol optical depth in deriving surface PM2.5 using machine learning? *Remote Sensing (Basel)*, *15*(15), 3780. DOI: 10.3390/rs15153780

Tokluoğlu, E., Kurtuluş, B., Sağır, Ç., Erdem Altın, G., Yurdakul, E., Ateş, E., Canoğlu, M. C., & Yamaç, S. S. (2026). Detection of potato fields using Sentinel-2 and Landsat 8 data-based machine learning models in semi-arid region of Central Anatolia, Türkiye. *Potato Research*, *69*(1), 14. DOI: 10.1007/s11540-025-09985-4

Traganos, D., Lee, C. B., Blume, A., Poursanidis, D., Čižmek, H., Déter, J., Mačić, V., Montefalcone, M., Pergent, G., Pergent-Martinì, C., Ricart, A. M., & Reinartz, P. (2022). Spatially Explicit Seagrass Extent Mapping Across the Entire Mediterranean. *Frontiers in Marine Science*, *9*, 871799. Advance online publication. DOI: 10.3389/fmars.2022.871799

Traganos, D., & Reinartz, P. (2018). Machine learning-based retrieval of benthic reflectance and Posidonia oceanica seagrass extent using a semi-analytical inversion of Sentinel-2 satellite data. *International Journal of Remote Sensing*, *39*(24), 9428–9452. DOI: 10.1080/01431161.2018.1519289

Tripathy, K. P., & Mishra, A. K. (2023). Deep learning in hydrology and water resources disciplines: Concepts, methods, applications, and research directions. *Journal of Hydrology (Amsterdam)*, *628*, 130458. DOI: 10.1016/j.jhydrol.2023.130458

Tseng, K.-Y., Hsieh, Y.-T., & Lin, H.-C. (2024). Machine learning prediction on wetland succession and the impact of artificial structures from a decade of field data. *The Science of the Total Environment*, *937*, 173426. DOI: 10.1016/j.scitotenv.2024.173426

Tuğaç, M. G., Şimşek, F. F., & Torunlar, H. (2024). Classification of agricultural crops with random forest and support vector machine algorithms using Sentinel-2 and Landsat 8 images. *International Journal of Environment and Geoinformatics*, *11*(3), 106–118. DOI: 10.30897/ijegeo.1479116

Ualiyeva, R. M., Kaverina, M. M., Osipova, A. V., Kairbayev, Y. B., Zhangazin, S B., Iksat, N. N., & Mapitov, N. B. (2025). VNIR hyperspectral signatures for early detection and machine learning classification of wheat diseases. *Plants*, *14*(23), 3644. DOI: 10.3390/plants14233644 PMID: 41375355

Ullah, W., Ahmad, K., Ullah, S., Tahir, A. A., Javed, M. F., Nazir, A., & Mohamed, A. (2023). Analysis of the relationship among land surface temperature, land use land cover, and normalized difference vegetation index with topographic elements in the lower Himalayan region. *Heliyon*, *9*(2), e13322. DOI: 10.1016/j.heliyon.2023.e13322 PMID: 36825192

Unnikrishnan, V. J., & Jothiprakash, V. (2018). Daily rainfall forecasting for one year in a single run using Singular Spectrum Analysis. *Journal of Hydrology (Amsterdam)*, *609*, 609–621. DOI: 10.1016/j.jhydrol.2018.04.032

Utku, A. (2025). Data-driven prediction of climate variables in agricultural cities of India with hybrid GA-TCN-LSTM model. *Journal of Forecasting*. Advance online publication. DOI: 10.1002/for.70088

Valderrama-Landeros, L., Flores-de-Santiago, F., Álvarez-Sánchez, L. F., Flores-Verdugo, F., & Rodríguez-Sobreyra, R. (2024). The influence of spatial resolution on coastline detection by means of multisource remote sensing data. *Remote Sensing Applications: Society and Environment*, *35*, 101258. DOI: 10.1016/j.rsase.2024.101258

van Klompenburg, T., Kassahun, A., & Catal, C. (2020). Crop yield prediction using machine learning: A systematic literature review. *Computers and Electronics in Agriculture*, *177*, 105709. DOI: 10.1016/j.compag.2020.105709

Venkatanathan, N., Senthil Kumar, M., & Jeyaraman, R. (2021). OLR Anomaly and Tidal Amplitude Irregularities (Tai) as Short-Term Earthquake Precursors—A Study of Doublet Earthquakes (1990-2010). In Seismic Hazards and Risk (pp. 361-375). Springer.

Vergopolan, N., Chaney, N. W., Pan, M., Sheffield, J., Beck, H. E., Ferguson, C. R., Torres-Rojas, L., Sadri, S., & Wood, E. F. (2021). SMAP-HydroBlocks, a 30-m satellite-based soil moisture dataset for the conterminous US. *Scientific Data*, *8*(1), 264. DOI: 10.1038/s41597-021-01050-2 PMID: 34635675

Viana, C. M., Oliveira, S., Oliveira, S. C., & Rocha, J. (2019). Land use/land cover change detection and urban sprawl analysis. In *Spatial modeling in GIS and R for earth and environmental sciences* (pp. 621–651). Elsevier. DOI: 10.1016/B978-0-12-815226-3.00029-6

Vigny, C., & Klein, E. (2022). The 1877 megathrust earthquake of North Chile two times smaller than thought? A review of ancient articles. *Journal of South American Earth Sciences*, *117*, 103878. DOI: 10.1016/j.jsames.2022.103878

Vij, P., & Tiwari, A. (2025). AI-Driven Drought Monitoring: Advanced machine learning techniques for early prediction. *SHS Web of Conferences, 216*, 01024. DOI: 10.1051/shsconf/202521601024

Vijayasekaran, D. (2019). SEN2 AGRI – Crop type mapping pilot study using Sentinel-2 satellite imagery in India. *The International Archives of the Photogrammetry, Remote Sensing and Spatial Information Sciences*, *XLII-3*(W6), 175–180. DOI: 10.5194/isprs-archives-XLII-3-W6-175-2019

Volvach, A. E., Kogan, L. P., Kanonidi, K. H., Bubukin, I. T., Shtenberg, V. B., Volvach, L. N., & Biazitov, D. T. (2022). Statistical precursors of a strong earthquake on April 6, 2009 on the Apennine Peninsula. *Heliyon*, *8*(8), e10200. DOI: 10.1016/j.heliyon.2022.e10200 PMID: 36042722

Vulova, S., Meier, F., Rocha, A. D., Quanz, J., Nouri, H., & Kleinschmit, B. (2021). Modeling urban evapotranspiration using remote sensing, flux footprints, and artificial intelligence. *The Science of the Total Environment*, *786*, 147293. DOI: 10.1016/j.scitotenv.2021.147293 PMID: 33975115

Wallapure Manikrao, K., Allagi, S., Leong, W. Y., & Laddi, M. (2025). An Explainable AI for Stock Market Prediction: A Machine Learning Approach with XAI and Deep Neural Networks. *Journal of Computational and Cognitive Engineering*. Advance online publication. DOI: 10.47852/bonviewJCCE52026428

Wang, C., Jiang, S., Zheng, Y., Han, F., Kumar, R., Rakovec, O., & Li, S. (2024). Distributed Hydrological Modeling with Physics-Encoded Deep Learning: A General Framework and its application in the Amazon. *Water Resources Research*, *60*(4), e2023WR036170. Advance online publication. DOI: 10.1029/2023WR036170

Wang, F., Qin, Z., Song, C., Tu, L., Karnieli, A., & Zhao, S. (2015). An improved mono-window algorithm for land surface temperature retrieval from Landsat 8 thermal infrared sensor data. *Remote Sensing (Basel)*, *7*(4), 4268–4289. DOI: 10.3390/rs70404268

Wang, J., Li, X., Jin, L., Li, J., Sun, Q., & Wang, H. (2022). An air quality index prediction model based on CNN-ILSTM. *Scientific Reports*, *12*(1), 8373. DOI: 10.1038/s41598-022-12355-6 PMID: 35589914

Wang, J., & Wang, R. (2025). Development of gas sensors and their applications in health safety, medical detection, and diagnosis. *Chemosensors (Basel, Switzerland)*, *13*(5), 190. DOI: 10.3390/chemosensors13050190

Wang, J., Yin, H., Feng, Z., Ma, P., & Wang, L. (2021). Fast Inversion of the Earthquake Rupture Processes with Complicated Velocity Structure: An Application to the Earthquake of 2017 Mw 6.5 Jiuzhaigou, China. *Journal of Earthquake and Tsunami*, *15*(6), 2150030. DOI: 10.1142/S1793431121500305

Wang, K., Jia, M., Zhang, X., Zhao, C., Zhang, R., & Wang, Z. (2024). Evaluating Ecosystem Service Value Changes in Mangrove Forests in Guangxi, China, from 2016 to 2020. *Remote Sensing (Basel)*, *16*(3), 494. DOI: 10.3390/rs16030494

Wang, L., Cheng, Y., Gopalan, S., Luo, F., Amreen, K., Singh, R. K., Goel, S., Lin, Z., & Naidu, R. (2023). Review and perspective: Gas separation and discrimination technologies for current gas sensors in environmental applications. *ACS Sensors*, *8*(4), 1373–1390. DOI: 10.1021/acssensors.2c02810 PMID: 37074840

Wang, N., Chen, Q., Wang, H., Capurso, W. D., Niemoczynski, L. M., Zhu, L., & Snedden, G. A. (2023). Field observations and long short-term memory modeling of spectral wave evolution at living shorelines in Chesapeake Bay, USA. *Applied Ocean Research*, *141*, 103782. DOI: 10.1016/j.apor.2023.103782

Wang, Q., Kavhiza, N. J., Islam, F., Huqqani, I. A., Abbas, M., & Barman, S. (2025). Multisensor Data Fusion for Coastal Boundary Detection by Res-U-Net Implementation Using High-Resolution UAV Imagery. *IEEE Journal of Selected Topics in Applied Earth Observations and Remote Sensing*, *18*, 16722–16732. DOI: 10.1109/JSTARS.2025.3584853

Wang, R., Ma, L., He, G., Johnson, B., Yan, Z., Chang, M., & Liang, Y. (2024). Transformers for Remote Sensing: A Systematic Review and Analysis. *Sensors (Basel)*, *24*(11), 3495. DOI: 10.3390/s24113495

Wang, S., Fu, Z., Chen, H., Nie, Y., & Xu, Q. (2019). Mechanisms of surface and subsurface runoff generation in subtropical soil-epikarst systems: Implications of rainfall simulation experiments on karst slope. *Journal of Hydrology (Amsterdam)*, *580*, 124370. DOI: 10.1016/j.jhydrol.2019.124370

Wang, S., Yan, D., Wang, C., Wu, L., & Huang, Y. (2024). A bibliometric analysis of blue carbon (1993–2023): Evolution of research hot topics and trends. *Frontiers in Marine Science*, *11*, 1430545. Advance online publication. DOI: 10.3389/fmars.2024.1430545

Wang, W., Liu, X., Bi, J., & Liu, Y. (2022). A machine learning model to estimate ground-level ozone concentrations in California using TROPOMI data and high-resolution meteorology. *Environment International, 158*, 106917. DOI: 10.1016/j.envint.2021.106917 PMID: 34624589

Wang, W., Mao, W., Tong, X., & Xu, G. (2021). A novel recursive model based on a convolutional long short-term memory neural network for air pollution prediction. *Remote Sensing (Basel), 13*(7), 1284. DOI: 10.3390/rs13071284

Wang, Y., Fang, S., Zhao, L., Huang, X., & Jiang, X. (2022). Parcel-based summer maize mapping and phenology estimation combined using Sentinel-2 and time series Sentinel-1 data. *International Journal of Applied Earth Observation and Geoinformation, 108*, 102720. DOI: 10.1016/j.jag.2022.102720

Wei, J., Liu, S., Li, Z., Liu, C., Qin, K., Liu, X., Pinker, R. T., Dickerson, R. R., Lin, J., Boersma, K. F., Sun, L., Li, R., Xue, W., Cui, Y., Zhang, C., & Wang, J. (2022). Ground-level NO_2 surveillance from space across China for high resolution using interpretable spatiotemporally weighted artificial intelligence. *Environmental Science & Technology, 56*(14), 9988–9998. DOI: 10.1021/acs.est.2c03834 PMID: 35767687

Wei, J., Li, Z., Lyapustin, A., Sun, L., Peng, Y., Xue, W., Su, T., & Cribb, M. (2021). Reconstructing 1-km-resolution high-quality PM2.5 data records from 2000 to 2018 in China: Spatiotemporal variations and policy implications. *Remote Sensing of Environment, 252*, 112136. DOI: 10.1016/j.rse.2020.112136

Wei, Y., Wang, R., & Feng, P. (2024). Improving Hydrological Modeling with Hybrid Models: A Comparative Study of Different Mechanisms for Coupling Deep Learning Models with Process-based Models. *Water Resources Management, 38*(7), 2471–2488. DOI: 10.1007/s11269-024-03780-5

Windle, A. E., Staver, L. W., Elmore, A. J., Scherer, S., Keller, S. M., Malmgren, B., & Silsbe, G. M. (2023). Multi-temporal high-resolution marsh vegetation mapping using unoccupied aircraft system remote sensing and machine learning. *Frontiers in Remote Sensing, 4*, 1140999. Advance online publication. DOI: 10.3389/frsen.2023.1140999

Workneh, H., & Jha, M. (2025). Utilizing hybrid deep learning models for streamflow prediction. *Water (Basel), 17*(13), 1913. DOI: 10.3390/w17131913

Xiang, X., Li, Q., Khan, S., & Khalaf, O. I. (2020). Urban water resource management for sustainable environment planning using artificial intelligence techniques. *Environmental Impact Assessment Review, 86*, 106515. DOI: 10.1016/j.eiar.2020.106515

Xia, X., Wang, Y., Zhong, Z., & Weng, G. J. (2017). A frequency-dependent theory of electrical conductivity and dielectric permittivity for graphene-polymer nanocomposites. *Carbon, 111*, 221–230. DOI: 10.1016/j.carbon.2016.09.078

Xie, Y., Tian, Y., Zhang, Q., Zhang, Y., Jin, T., & lin, J. (2025). *A New Method for Mangrove Biomass Prediction Based on Automatic Machine Learning and Multi-Source Uav Data Fusion*. DOI: 10.2139/ssrn.5168622

Xie, K., Liu, P., Zhang, J., Han, D., Wang, G., & Shen, C. (2021). Physics-guided deep learning for rainfall-runoff modeling by considering extreme events and monotonic relationships. *Journal of Hydrology (Amsterdam), 603*, 127043. DOI: 10.1016/j.jhydrol.2021.127043

Xiong, P., Bi, Y., & Shen, X. (2009). Study of outgoing longwave radiation anomalies associated with two earthquakes in China using wavelet maxima. Lecture Notes in Computer Science, 5572 LNAI, 77-87. DOI: 10.1007/978-3-642-02319-4_10

Xiong, P., Shen, X. H., Bi, Y. X., Kang, C. L., Chen, L. Z., Jing, F., & Chen, Y. (2010). Study of outgoing longwave radiation anomalies associated with Haiti earthquake. *Natural Hazards and Earth System Sciences, 10*(10), 2169–2178. DOI: 10.5194/nhess-10-2169-2010

Xu, H., Zlatanova, S., Liang, R., & Canbulat, I. (2025). Generative AI as a Pillar for Predicting 2D and 3D Wildfire Spread: Beyond Physics-Based Models and Traditional Deep Learning. *Fire (Basel, Switzerland), 8*(8), 293. DOI: 10.3390/fire8080293

Xu, L., Ye, Z., Dai, J., Li, Q., Hong, Y., Tao, Y., Yu, H., Zhang, C., Chen, Z., & Chen, N. (2025). Spatiotemporal seamless estimation of global surface soil moisture using triple collocation, machine learning, and data assimilation. *IEEE Transactions on Geoscience and Remote Sensing, 63*, 1–16. DOI: 10.1109/TGRS.2025.3568034

Xu, T., & Liang, F. (2021). Machine learning for hydrologic sciences: An introductory overview. *WIREs. Water, 8*(5), e1533. Advance online publication. DOI: 10.1002/wat2.1533

Xu, W., Chen, J., Corzo, G., Xu, C., Zhang, X. J., Xiong, L., Liu, D., & Xia, J. (2024). Coupling deep learning and physically based hydrological models for monthly streamflow predictions. *Water Resources Research, 60*(2), e2023WR035518. Advance online publication. DOI: 10.1029/2023WR035618

Xu, Z., Li, J., Cheng, S., Rui, X., Zhao, Y., He, H., Guan, H., Sharma, A., Erxleben, M., Chang, R., & Xu, L. L. (2025). Deep learning for wildfire risk prediction: Integrating remote sensing and environmental data. *ISPRS Journal of Photogrammetry and Remote Sensing, 227*, 632–677. DOI: 10.1016/j.isprsjprs.2025.06.002

Yadav, B., Sharma, L. K., & Bijarniya, B. (2025). Predicting Forest Evapotranspiration using Remote Sensing and Machine Learning. *The International Archives of the Photogrammetry, Remote Sensing and Spatial Information Sciences/International Archives of the Photogrammetry. The International Archives of the Photogrammetry, Remote Sensing and Spatial Information Sciences, 2025*, 1611–1619. DOI: 10.5194/isprs-archives-XLVIII-G-2025-1611-2025

Yang, L., Xu, H., & Yu, S. (2021). Estimating PM2.5 concentrations in contiguous eastern coastal zone of China using MODIS AOD and a two-stage random forest model. *Journal of Atmospheric and Oceanic Technology*, *38*(12), 2071–2080. DOI: 10.1175/JTECH-D-20-0214.1

Ye, Z., Yin, S., Cao, Y., & Wang, Y. (2024). AI-driven optimization of agricultural water management for enhanced sustainability. *Scientific Reports*, *14*(1), 25721. DOI: 10.1038/s41598-024-76915-8 PMID: 39468244

Yıkmış, S., Demirok, N. T., Altan, A. D., Pacal, I., Türkol, M., Tokatlı, N., & Aadil, R. M. (2025). ThermoMicrowave-sonication improves the stability and digestive bioaccessibility of phenolic compounds in parsley juice. *Food Chemistry: X*, •••, 103406. PMID: 41510132

Yu, Z., Li, T., Zhu, Y., & Pan, R. (2024). Exploring foundation models in remote sensing image change detection: comprehensive survey. *arXiv preprint*.

Yuan, H., Liang, S., Gao, Y., Gao, Y., & Lian, X. (2025). Experimental study on estimating bare soil moisture content based on UAV multi-source remote sensing. *Geocarto International*, *40*(1), 2448985. Advance online publication. DOI: 10.1080/10106049.2024.2448985

Yulianto, F., Basyuni, M., Mardiatno, D., Ardiati, A. S., Kusuma, I. K. T. W., Purnomo, M. B., Kongko, W., Wardani, K. S., Fachrudin, I., Baba, S., & Miyagi, T. (2025). Long-term monitoring of mangrove restoration and carbon dynamics in Tanjung Benoa, Bali, Indonesia: An assessment approach using combination remote sensing and field surveys. *Research Square (Research Square)*. DOI: 10.21203/rs.3.rs-6779800/v1

Yun, S.-M., Hamm, S.-Y., Jeon, H.-T., & Cheong, J.-Y. (2021). Trend analysis of earthquake researches in the world. *Journal of the Korean Earth Science Society*, *42*(1), 76–87. DOI: 10.5467/JKESS.2021.42.1.76

Yu, S., & Singh, M. (2025). Deep Learning-Based Remote Sensing Image Analysis for Wildfire Risk Evaluation and Monitoring. *Fire (Basel, Switzerland)*, *8*(1), 19. DOI: 10.3390/fire8010019

Zafar, Z., Zubair, M., Zha, Y., Fahd, S., & Nadeem, A. A. (2024). Performance assessment of machine learning algorithms for mapping of land use/land cover using remote sensing data. *The Egyptian Journal of Remote Sensing and Space Sciences*, *27*(2), 216–226. DOI: 10.1016/j.ejrs.2024.03.003

Zaki, N. H. M., Chong, W. S., Muslim, A. M., Reba, M. N. M., & Hossain, M. S. (2022). Assessing optimal UAV-data pre-processing workflows for quality ortho-image generation to support coral reef mapping. *Geocarto International*, *37*(25), 1–25. DOI: 10.1080/10106049.2022.2037732

Zaki, N. H. M., & Hossain, M. S. (2024). Optimum image alignment setting selection for structure-from-motion photogrammetry using Remotely Piloted Aircraft Systems (RPAS) to support coral habitat classification. *Remote Sensing Applications: Society and Environment*, *35*, 101233. https://doi.org/https://doi.org/10.1016/j.rsase.2024.101233. DOI: 10.1016/j.rsase.2024.101233

Zaki, N. H. M., Khalil, I., & Hossain, M. S. (2025). Assessing the effects of image alignment on bathymetry and zonation mapping of coral reefs using UAV. *Remote Sensing Applications: Society and Environment*, *37*, 101515. DOI: 10.1016/j.rsase.2025.101515

Zaman, T. A., Haque, T., Gupta, H. S., Sameer, M., & Nofal, O. (2025). Computational review of disaster resilience research in Bangladesh. *Progress in Disaster Science*, *28*, 100486. DOI: 10.1016/j.pdisas.2025.100486

Zhang, T., & Yang, Y. (2025). Mapping Very-High Resolution Evapotranspiration Data Using a Dual GAN Model. *IEEE 15th Annual Computing and Communication Workshop and Conference (CCWC)*, 00117–00125. DOI: 10.1109/CCWC62904.2025.10903799

Zhang, Y., Hou, Y., & Yang, X. (2018). Sea Level Rise and Coastal Infrastructure. In *InTech eBooks*. DOI: 10.5772/intechopen.68572

Zhang, B., Ding, X., Amelung, F., Wang, C., Xu, W., Zhu, W., Shimada, M., Zhang, Q., & Baba, T. (2021). Impact of ionosphere on InSAR observation and coseismic slip inversion: Improved slip model for the 2010 Maule, Chile, earthquake. *Remote Sensing of Environment*, *267*, 112733. DOI: 10.1016/j.rse.2021.112733

Zhang, H., Bauters, M., Boeckx, P., & Van Oost, K. (2021). Mapping Canopy Heights in Dense Tropical Forests Using Low-Cost UAV-Derived Photogrammetric Point Clouds and Machine Learning Approaches. *Remote Sensing (Basel)*, *13*(18), 3777. DOI: 10.3390/rs13183777

Zhang, L. (2024). Environmental hotspots, frontiers and analytical framework of Blue Carbon research: A quantitative analysis of knowledge map based on CiteSpace. *Frontiers in Environmental Science*, *12*, 1387655. Advance online publication. DOI: 10.3389/fenvs.2024.1387655

Zhang, Q., Li, P., Ren, X., Ning, J., Li, J., Liu, C., Wang, Y., & Wang, G. (2023). A new real-time groundwater level forecasting strategy: Coupling hybrid data-driven models with remote sensing data. *Journal of Hydrology (Amsterdam)*, *625*, 129962. DOI: 10.1016/j.jhydrol.2023.129962

Zhang, X., Treitz, P. M., Chen, D., Quan, C., Shi, L., & Li, X. (2017). Mapping mangrove forests using multi-tidal remotely-sensed data and a decision-tree-based procedure. *International Journal of Applied Earth Observation and Geoinformation*, *62*, 201–214. https://doi.org/https://doi.org/10.1016/j.jag.2017.06.010. DOI: 10.1016/j.jag.2017.06.010

Zhang, Y., Hou, J., & Huang, C. (2023). Basin Scale Soil Moisture Estimation with Grid SWAT and LESTKF Based on WSN. *Sensors (Basel)*, *24*(1), 35. DOI: 10.3390/s24010035 PMID: 38202901

Zhang, Y., Peña-Arancibia, J. L., McVicar, T. R., Chiew, F. H. S., Vaze, J., Liu, C., Lu, X., Zheng, H., Wang, Y., Liu, Y. Y., Miralles, D. G., & Pan, M. (2016). Multi-decadal trends in global terrestrial evapotranspiration and its components. *Scientific Reports*, *6*(1), 19124. DOI: 10.1038/srep19124 PMID: 26750505

Zhao, Q., Yu, L., Du, Z., Peng, D., Hao, P., Zhang, Y., & Gong, P. (2022). An Overview of the Applications of Earth Observation Satellite Data: Impacts and Future Trends. *Remote Sensing (Basel)*, *14*(8), 1863. DOI: 10.3390/rs14081863

Zhao, T., Song, C., Yu, J., Xing, L., Xu, F., Li, W., & Wang, Z. (2025). Leveraging immersive digital twins and AI-Driven decision support Systems for sustainable water reserves Management: A Conceptual framework. *Sustainability (Basel)*, *17*(8), 3754. DOI: 10.3390/su17083754

Zhao, Z., Li, Y., Liu, K., Wu, C., Yu, B., Liu, G., & Wang, Y. (2025). Layered soil moisture retrieval and agricultural application based on Multi-Source Remote Sensing and vegetation suppression technology: A case study of Youyi Farm, China. *Remote Sensing (Basel)*, *17*(13), 2130. DOI: 10.3390/rs17132130

Zhong, L., Hu, L., & Zhou, H. (2020). Deep learning-based multi-temporal crop classification with Sentinel-2 imagery. *Remote Sensing*, *12*(1), 149. DOI: 10.3390/rs12010149

Zhong, M., Zhang, H., Jiang, T., Guo, J., Zhu, J., Wang, D., & Chen, X. (2023). A hybrid model combining the Cama-Flood model and deep learning methods for streamflow prediction. *Water Resources Management*, *37*(12), 4841–4859. DOI: 10.1007/s11269-023-03583-0

Zhou, S., Xu, L., & Chen, N. (2023). Rice yield prediction in Hubei Province based on deep learning and the effect of spatial heterogeneity. *Remote Sensing (Basel)*, *15*(5), 1361. DOI: 10.3390/rs15051361

Zhu, C., Lu, D., Victoria, D., & Dutra, L. (2015). Mapping fractional cropland distribution in Mato Grosso, Brazil using time series MODIS enhanced vegetation index and Landsat Thematic Mapper data. *Remote Sensing (Basel)*, *8*(1), 22–34. DOI: 10.3390/rs8010022

Zhu, M., Jiao, Y., Wu, C., Shi, W., Huang, H., Zhang, Y., & Yue, T. (2025). Simulation and optimization of double-season rice yield in Jiangxi Province based on high-accuracy surface modeling–APSIM model. *Agriculture*, *15*(10), 1034. DOI: 10.3390/agriculture15101034

Zhuo, L., Dai, Q., Zhao, B., & Han, D. (2020). Soil Moisture Sensor Network Design for Hydrological Applications. *Hydrology and Earth System Sciences*, *24*(5), 2577–2591. Advance online publication. DOI: 10.5194/hess-24-2577-2020

Zhu, Z., & Woodcock, C. E. (2014). Continuous change detection and classification of land cover using all available Landsat data. *Remote Sensing of Environment*, *144*, 152–171. DOI: 10.1016/j.rse.2014.01.011

Ziv, L., & Nakash, M. (2025). Behind the algorithm: International insights into data-driven AI model development. *Machine Learning and Knowledge Extraction*, *7*(4), 122. DOI: 10.3390/make7040122

Zong, B., Wu, S., Yang, Y., Li, Q., Tao, T., & Mao, S. (2025). Smart gas sensors: Recent developments and future prospective. *Nano-Micro Letters*, *17*(1), 54. DOI: 10.1007/s40820-024-01543-w PMID: 39489808

Zoran, M., Savastru, R., & Savastru, D. (2016). Earthquake anomalies recognition through satellite and in-situ monitoring data. *European Journal of Remote Sensing*, *49*(1), 1011–1032. DOI: 10.5721/EuJRS20164952

Zou, S., Ju, H., & Zhang, J. (2025). Water quality Management in the age of AI: Applications, challenges, and Prospects. *Water (Basel)*, *17*(11), 1641. DOI: 10.3390/w17111641

About the Contributors

Mehmet Ali Çelik is an Assoc. Prof. of Physical Geography. He focuses on the interaction between climate variables, nature and the environment. His multidisciplinary research interests include remote sensing and GIS studies involving climate, natural resource management, and disasters. Dr. Çelik uses multispectral and hyperspectral remote sensing and GIS to monitor changes to ecosystems and analyse the contribution of natural and anthropogenic factors to these changes.

***

Sharif Eime Akhter is a CSE student at American International University-Bangladesh (AIUB), focused on learning and exploring Machine Learning and Deep Learning for academic research and practical applications.

Anorgul Ashirova is working as an Associate Professor at the Department of General Professional Sciences, Mamun University, Khiva, Uzbekistan

Rubidha Devi D works as an Assistant Professor in the Computer Science and Engineering Department at SASTRA Deemed University, Kumbakonam, India. She has published several research works in peer-reviewed journals. Her research area includes Deep Learning, Machine learning, and Temporal and Spatial Time-series Analysis.

Selvaraj Damodaran is working as a professor in the Department of Electronics and Communication Engineering at Panimalar Engineering College. He obtained his B.E Degree from Madras University, ME degree from Anna University and Ph.D. from Sathyabama University. He has a vast teaching experience of 21 years. He has published more than 75 papers at various refereed journals and conferences. His area of interest is in the field of signal processing, image processing, Biomedical

Instrumentation, wireless communication, Artificial Intelligence, Machine learning, Data Mining and Antenna Design. He is a member of IEEE and life member of IETE.

R. Deepa is an Assistant Professor in the Department of Computing Technologies at SRM Institute of Science and Technology. She holds a Ph.D. from Sathyabama Institute of Science and Technology and has a strong background in computing and information systems. Her research interests include Natural Language Processing (NLP) and Computer Vision, and she has authored multiple publications in journals, book chapters, and conference proceedings. Dr. Deepa has been contributing to teaching and research in computing disciplines since 2011, and she currently continues her academic work at SRMIST.

Pradeepa G is an Assistant Professor of Department of Information Technology in Vivekanandha College of Technology for Women, Tiruchengode and has 11 years of teaching experience. I have obtained her B.E in Computer Science and Engineering from Anna University, Chennai in 2007, and M.E in Computer Science Engineering from Anna University, Chennai during 2013.I have published many papers in National and Interna

Yonis Gulzar is an Associate Professor at King Faisal University, Saudi Arabia. He received his Ph.D. in Computer Science from the International Islamic University Malaysia. His research focuses on artificial intelligence, deep learning, database systems, and AI applications in healthcare and agriculture. He has published extensively in journals, conferences, and books and has led multiple funded research projects.

Deepak Gupta is a distinguished academician with over 20 years of teaching experience and expertise in Computer Science and Engineering. He holds a Ph.D. in Computer Engineering and has an extensive research portfolio, including many international journal papers, multiple conference proceedings, and several books. His research interests span machine learning, artificial intelligence, and data mining. Currently, Dr. Gupta serves as an Associate Professor and Head of Department of CSE (Emerging Branches) at ITM Gwalior, MP, India. He has also contributed significantly to the field through patents and copyrights and actively engages as a reviewer, conference chair and editor.

Mohammad Shawkat Hossain is a distinguished environmental scientist and academic with over two decades of expertise in coastal and marine ecosystem management, remote sensing, and sustainable forestry. His career is dedicated to applying geospatial technologies for biodiversity conservation and climate resilience across South and Southeast Asia. Dr. Hossain's academic foundation is in forestry, earning his B.Sc. (Hons) and M.Sc. from the Institute of Forestry and Environmental Sciences at Chittagong University, Bangladesh. He obtained his Ph.D. in Marine Ecology and Biology from Universiti Putra Malaysia (UPM) in 2015, where his research focused on pioneering remote sensing techniques for seagrass and coral reef mapping.

M. Robinson Joel received the Doctorate Ph.D degree in Computer and Information Technology from the Manonmaniam Sundaranar University, Tirunelveli in March 2019. He completed his M.Tech and MCA from the same university year 2008 and 2002 respectively. He is currently working as an Associate Professor in the Department of Computer Science and Engineering at KCG College of Technology, Chennai. He has 16 years of teaching experience.

Daniel Raj K is a faculty member in the Department of Computer Science and Engineering (AI & ML) at Kalaignar Karunanidhi Institute of Technology, Coimbatore. He holds a B.Tech in Information Technology and an M.E in Computer Science Engineering. His work includes journal publications on neuromorphic computing and intelligent systems, as well as reviewing contributions for IEEE, IGI Global, and other SCOPUS-indexed venues. He has authored book chapters and continues to work on applied projects in machine learning and resilient communication systems.

Atin Kumar is an Assistant Professor and Soil Scientist at Uttaranchal University, Dehradun, with expertise in soil fertility, soil remediation, water quality, and heavy metal contamination. He is the author and editor of 3 academic books, a patent holder, and an active reviewer for several recognized international journals. With over 50 publications and multiple book chapters, he has been honored with the Young Scientist Awards. His research integrates innovative approaches such as nanotechnology to improve soil–water–environment interactions, with a focus on sustainable agriculture and environmental resilience.

Sanjeev Kumar is a Professor in Lovely Professional University Punjab India. He received Undergraduate in Hotel Management from Osmania University, Postgraduate in Tourism Management from Madurai Kamaraj University. He completed Doctor of Philosophy in Hospitality Management in 2019 from Amity University Jaipur India. His research areas lies in food Service Industry, his research focuses on Alcoholic beverages, Event management and Sustainable Management Practices, Metaverse, AI Machine Learning and Artificial Intelligence. He published more than 80 research papers, articles and chapters in Scopus Indexed, UGC Approved and peer reviewed Journals and books. Dr. Sanjeev Kumar participated and acted as resource person in various National and International conferences, seminars, session chair, research workshops and industry talks and his work has been widely cited. Dr. Sanjeev also edited more than 20 books with various International Publishers such as IGI Global, Cambridge Scholars, Emerald, CABI, Bentham Science, Taylor & Francis. Dr Sanjeev is an active researcher and has published numerous research articles in prestigious National and International journals and books.

Anita M is an Assistant Professor in the Department of Computing Technologies at SRM Institute of Science and Technology, Kattankulathur, Tamil Nadu, India. She earned her Ph.D. in Natural Language Processing from SRMIST and has over 18 years of teaching experience. Her research interests include Natural Language Processing, Computational Linguistics, Machine Learning, and Deep Learning, and she has published multiple research papers in international journals and conferences

Thiyagarajan Manoharan, B.E., M.Tech., (Ph.D.), is currently working as an Assistant Professor in the Department of Computer Science and Engineering at R.M.K. Engineering College, Kavaraipettai, Chennai. He is pursuing his Ph.D. at Anna University, Chennai, with a research focus on Precision Agriculture and Information Security.He completed his M.Tech. in Computer Science and Engineering from SRM University and his B.E. in Computer Science and Engineering from Anna University, Chennai. His research interests include Machine Learning, AIoT, Cyber Security, and Cloud Computing. He has 17 years of teaching and research experience across various engineering colleges and universities. He has published several research papers in international journals and has also presented his work at international conferences. He can be contacted at mtn.cse@rmkec.ac.in.

Parimala Devi Muthusamy received B.E. (Electronics and Communication) from Bharathiyar University, Coimbatore, M.E. (VLSI Design) from Anna University, Chennai, M.B.A. (Human Resource Management) from Anna University and Ph.D (Information and Communication Engineering) from Anna University, Chennai. Currently, she is Associate Professor in the Velalar College of Engineering and Technology, Erode. She has published 15 papers in national and international journals (SCI, SCOPUS) and conferences. She has published 15 book chapters with IEEE, Springer, IET, Taylor & Francis. She has been the reviewer for many SCI, SCIE, SCOPUS indexed journals and received best paper awards at international conferences. She has two Grant Patents and 5 Copyrights for software (Government of India) She has published a book named "Advanced Microprocessors and Microcontrollers" with University Science Press, New Delhi. She is a lifetime member of IETE. Her research interests include VLSI &SoC, IoT, Block chain and Machine & Deep learning.

Giridharan N. is a researcher in remote sensing and geospatial analysis whose work includes studies on forest canopy density and FPAR changes in Sathyamangalam Reserve Forest, contributing to environmental monitoring and forest change assessment.

G. Indra Navaroj is currently working as an Assistant Professor in the Department of Information Technology, Vel Tech Rangarajan Dr. Sagunthala R&D Institute of Science and Technology, Chennai, Tamil Nadu, India. She received a Ph.D degree in Information and Communication Engineering from Anna University, Chennai, in the year 2022. She received her B.Tech. degree in Information Technology in 2007 from Anna University, Chennai, and her M.Tech., degree in Computer and Information Technology in 2010 from Manonmaniam Sundaranar University, Tirunelveli. She has 17.10 years of teaching experience. She completed more than ten NPTEL courses and an AICTE FDP. "Blockchain Architecture Design and Use Case", "Introduction to Internet of Things ", "Industrial 4.0 and Industrial Internet of Things", "Object Oriented Analysis and Design", "Cloud Computing", "Object Oriented System Development using UML, Java and Pattern". She received a certificate of Appreciation from NPTEL to be recognized as an "NPTEL Star" –NPTEL_DISPLINE, Dec 2020. She received the 'Certificate of Recognition' from the Research Center, Anna University, Chennai, for publishing her research in a Q1 journal. She has participated in more than 45 Faculty Development Programs (FDPs), seminars, and webinars hosted by various colleges. She has written two book chapters and published many research papers on IoT and Wireless Sensor Networks. Her research area includes Wireless Sensor Networks, Blockchain, and IoT.

Brindha P. received B.E. (Electronics and Communication Engineering) from Anna University of Technology, Chennai, M.E. (Applied Electronics) from Anna University, Chennai and Ph.D (Information and Communication Engineering) from Anna University, Chennai. Currently, she is Associate Professor in the Erode Sengunthar Engineering College, Erode. She has published 10 papers in national and international journals and conferences. She has one Grant Patent and 1 Copyrights for software (Government of India). She has published a book named "Advanced Microprocessors and Microcontrollers" with University Science Press, New Delhi. He is a lifetime member of IETE. Her research interests include VLSI, IoT, AI, Machine Learning & Deep learning.

Md Sadiqur Rahman is an undergraduate Student in Computer Science and Engineering at American International University - Bangladesh. His interests include artificial intelligence and data science. He is actively learning foundational skills in computer science.

Khurramov Ruslan is affiliated with the Department of Information Technology and Exact Sciences, Termez University of Economics and Service, Termez, Uzbekistan. His academic interests span information technology, applied computing, and exact sciences, with a focus on developing analytical and technology-driven solutions for contemporary educational and socio-economic challenges. He is actively involved in teaching, research, and academic development initiatives within the university.

Edwin Raja S has completed his Ph.D. in Information and Communication Engineering from Anna University-chennai in the year 2021. He obtained his M.Tech. (Network Engineering) in the year 2010 from Kalasalingam University, Krishnan kovil and B.E. (Computer Science and Engineering) in the year 2006 from C.S.I Institute of Technology, Thovalai. He has more than thirteen plus years of teaching experience and his area of interests are Cyber Security, Big Data Analytics, Machine Learning & Computer Networks. He has published more than 25 research articles in various journals and conferences. Also, he has published 3 patents, acted as reviewer in reputed journals.

Brindha Samarasam received her Undergraduate degree in Biomedical Instrumentation Engineering and Postgraduate degree in Applied Electronics. She is currently working as an Assistant Professor at Nandha Engineering College, Erode, with over 13 years of teaching experience, and is a Research Scholar at Avinashilingam Institute for Home Science and Higher Education for Women, School of Engineering. She has published 15 research papers in reputed national and international journals and conferences, one book chapter, and has served as a reviewer for six journals. She has published three patents and has actively participated in more than 20 national and international conferences. Her research interests include medical image analysis, machine learning, deep learning, Explainable AI (XAI), and computer-aided diagnosis, with a focus on uterine carcinoma.

Turobova Sevara is a faculty member in the Department of Social Sciences Education at Termez University of Economics and Service, Termez, Uzbekistan. She is engaged in teaching and academic activities related to social sciences education, with a focus on strengthening students' conceptual understanding, critical thinking, and civic awareness. Her academic interests include social science pedagogy, educational development, and the application of innovative teaching methodologies to improve learning outcomes in higher education. She actively contributes to curriculum implementation and scholarly initiatives that support quality education and interdisciplinary learning.

Maheswari Subramanian completed Under Graduation in Biomedical Engineering at Sri Ramakrishna Engineering College, Coimbatore. She has completed her Post Graduation in Applied Electronics at Anna University Regional Centre, Coimbatore and Pursuing Ph.D in the gynecological disorders using imaging modality. She is having 14 years of Teaching Experience. At present she is working as an Assistant Professor in the department of Biomedical Engineering at Velalar College of Engineering and Technology, Thindal, Erode, Tamilnadu. She published her research works in 8 International Journals and presented papers in many National and International IEEE Conferences. She has been the reviewer for many indexed journals and reviewed papers related to gynecological disorders. She organized FDP's, National Conferences and STTP's funded by CSIR, INSA and AICTE respectively. She is an active member of BMESI, ISTE, IEEE – EMB Society and IEEE – WIE. Also the Branch coordinator of IEEE-EMBS Student branch & WIE affinity group of VCET. She had conducted many events for the benefit of IEEE students firm. She received Fund for the project development under Niral Thiruvizha 2.0 and exhibited in the zonal presentation. She received three patents and three copyrights for lab manuals approved by Government of India. Her area of interests are Biomaterials, Biomechanics, Diagnostic and Therapeutic Equipments, Rehabilitation Engineering, Radiological Equipments, medical image processing and analysis using Machine Learning and Deep Learning Techniques. She gave lectures on artificial intelligence, machine Learning and Big Data Analytics in AICTE sponsored STTPs.

Akbar Toyirov is working in the Department of Information Technology and Exact Sciences at Termez University of Economics and Service, Termez, Uzbekistan.

Index

M

N

P

R

S

T

U

W

www.ingramcontent.com/pod-product-compliance
Lightning Source LLC
LaVergne TN
LVHW080905300926
849162LV00003B/558